Lecture Notes in Civil Engineering

Volume 155

Lecture Notes in Civil Engineering (LNCE) publishes the latest developments in Civil Engineering - quickly, informally and in top quality. Though original research reported in proceedings and post-proceedings represents the core of LNCE, edited volumes of exceptionally high quality and interest may also be considered for publication. Volumes published in LNCE embrace all aspects and subfields of, as well as new challenges in, Civil Engineering. Topics in the series include:

- Construction and Structural Mechanics
- Building Materials
- Concrete, Steel and Timber Structures
- Geotechnical Engineering
- Earthquake Engineering
- Coastal Engineering
- Ocean and Offshore Engineering; Ships and Floating Structures
- Hydraulics, Hydrology and Water Resources Engineering
- Environmental Engineering and Sustainability
- Structural Health and Monitoring
- Surveying and Geographical Information Systems
- Indoor Environments
- Transportation and Traffic
- Risk Analysis
- Safety and Security

To submit a proposal or request further information, please contact the appropriate Springer Editor:

– Pierpaolo Riva at pierpaolo.riva@springer.com (Europe and Americas);
– Swati Meherishi at swati.meherishi@springer.com (Asia - except China, and Australia, New Zealand);
– Wayne Hu at wayne.hu@springer.com (China).

All books in the series now indexed by Scopus and EI Compendex database!

More information about this series at http://www.springer.com/series/15087

Amadeo Benavent-Climent · Fabrizio Mollaioli
Editors

Energy-Based Seismic Engineering

Proceedings of IWEBSE 2021

Springer

Editors
Amadeo Benavent-Climent
Escuela Técnica Superior de Ingenieros Industriales
Technical University of Madrid
Madrid, Spain

Fabrizio Mollaioli
Dipartimento di Ingegneria Strutturale e Geotecnica (DISG)
Sapienza University of Rome
Rome, Italy

ISSN 2366-2557 ISSN 2366-2565 (electronic)
Lecture Notes in Civil Engineering
ISBN 978-3-030-73931-7 ISBN 978-3-030-73932-4 (eBook)
https://doi.org/10.1007/978-3-030-73932-4

This Springer imprint is published by the registered company Springer Nature Switzerland AG
The registered company address is: Gewerbestrasse 11, 6330 Cham, Switzerland

Contents

Assessment of Plastic Energy Demand Spectra on Frame Systems

Ahmet Anıl Dindar, Gökhan Polat, Cem Yalçın, Ercan Yüksel, Hasan Özkaynak, and Oral Büyüköztürk

Abstract Determination of structural resistance to seismic loads is a complex problem. To overcome the complexity, simple but efficient methods have been developed for engineers. In that process for the sake of simplicity and practicality, certain assumptions are made in defining seismic demand. One of the them is energy dissipation by structural members during seismic actions. However, energy dissipation is directly related to damage occurrence and propagation in the member. Calculation of plastic energy to be dissipated by structural member requires definition of energy demand on the structural system. This study aims to assess plastic energy spectrum approach on frame type reinforced concrete structures. Plastic energy demand values on three frame systems representing low- to mid-rise buildings are obtained from plastic energy spectrum and also nonlinear time history analysis (NLTH). Plastic energy spectrum is taken from the previous study of the authors. Comprehensive NLTH analyses on selected frame systems are conducted on the pre-designed systems which consists of concentrated plastic hinges. Moment and rotation response time histories of the structural members are used in calculation of energy dissipation.

A. A. Dindar (✉) · G. Polat
Gebze Technical University, Kocaeli 41400, Turkey
e-mail: adindar@gtu.edu.tr

G. Polat
e-mail: g.polat@gtu.edu.tr

C. Yalçın
Bogazici University, Istanbul 34342, Turkey
e-mail: yalcince@boun.edu.tr

E. Yüksel
Istanbul Technical University, Istanbul 34469, Turkey
e-mail: yukselerc@itu.edu.tr

H. Özkaynak
Beykent University, Istanbul 34387, Turkey
e-mail: hasanozkaynak@beykent.edu.tr

O. Büyüköztürk
Massachusetts Institute of Technology, Cambridge, MA, USA
e-mail: obuyuk@mit.edu

A. Benavent-Climent and F. Mollaioli (eds.), *Energy-Based Seismic Engineering*,
Lecture Notes in Civil Engineering 155,
https://doi.org/10.1007/978-3-030-73932-4_1

Comparisons between spectra- and NLTH-based results are made on three systems. For low-rise system, both plastic energy dissipation values are found approximate whereas for mid-rise systems, plastic energy spectrum is found conservative. The numerical examples reveal that plastic energy spectrum is a robust concept for energy-based design methodologies.

Keywords Plastic energy demand spectrum · Energy dissipation · Damage · Nonlinear time-history analysis · RC frame

1 Introduction

Seismic analysis and design of engineering structures include several uncertainties such as definition of the seismic intensity and the resistance capacity of the structural member. To overcome the complexity of the seismic design, conventional methods lead the engineers to the use of elastic response analysis in determining the seismic demand.

Forced-based methodology (FBD) aims to equate the resistance capacity of the structure to pre-determined seismic forces up to an upper bound of lateral drift limits. Once the lateral drift response is sufficient, the load resistance capacities of the members are calculated with respect to the internal forces. Occurrence and propagation of the plastic deformations are implicitly included in the methodology. On the other hand, displacement-based (DBD) design is a variation of the performance-based design (PBD) where the member plastic deformations are considered as a design parameter in the analysis. This approach requires nonlinear analysis of the structure using either a selected static loading pattern or a set of ground motions. Through the analysis steps, the plastic deformation are controlled. At the end of the analysis, performance of the structure is concluded with respect to the level of plastic deformation at each member and cumulatively in the system. Plastic deformation is a measure of the damage occurrence at the section where the flexural forces since modern design codes prevails brittle failure of structural members.

Damage occurrence on the member is a result of energy dissipation. Thus, a next generation design approach should consider the energy terms in both of the demand and resistance. Energy balance equation has a potential of providing a basis for such purpose. Several studies proposed definitions of seismic demand [1] and capacity in energy terms [2]. However, a complete energy based design of engineering structures is not yet completed except for few novel studies [3, 4].

Formulation of input and plastic energy spectrum was proposed by the authors [5]. The formula was validated for single-degree-of-freedom (SDOF) systems. In this paper we describe the numerical assessment of plastic energy demand spectrum for a multi-degree-of-freedom (MODF) system. Comparison is made of plastic energy demand values found from proposed formula and those from the nonlinear time history analyses were found consistent.

2 Plastic Energy Based Demand

Conventional design methodologies (FBD and DBD) rely on elastic response spectrum which considers the seismic intensity, site conditions, damping ratio and focal distance. However, energy demand spectrum includes not only the parameters given above but also duration and frequency content of selected ground motions. The assessment of comparison between response and energy spectrum was accomplished in Fajfar et al. [6, 7]. Use of plastic energy spectrum derived directly from the energy balance equation could be a rational in designing structural members.

Structural properties including constitutive models, damping, ductility and ground motion characteristics namely soil conditions and focal distance considered in construction of input and plastic energy spectrum were extensively studied in Dindar [8]. Mass normalized spectral formulations of both input and plastic energy with specific scaling procedure, given in Eq. (1), were proposed in Dindar et al. [5].

$$E_P^{PGA} = \left(\frac{PGA}{0.1g}\right)^2 \times E_P^{0.1g} \times m \tag{1}$$

where PGA is design seismic intensity, $E_P^{0.1g}$ is basis value for PGA $= 0.1\ g$ and m is modal mass. Structural and ground motion characteristics were incorporated into the basis value and form of the spectra is given in Fig. 1.

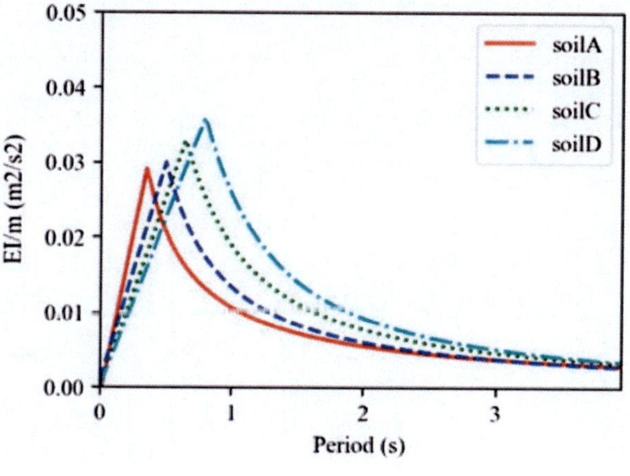

Fig. 1 Proposed plastic energy spectrum for $\mu = 6$

3 Explanation of Frame System and Ground Motions

3.1 *Frame System*

A set of reinforced concrete (RC) frame systems designed according to recent Turkish Building Seismic Code (TBSC-2018) [9] are modeled in SAP2000 program [10]. The set which includes three frame type systems with 3-, 5- and 5-stories, given in Fig. 2, is representative of the low- to mid-rise buildings.

Frame systems are considered as the interim part of three dimensional regular residential building having same span length between adjacent frames. Material quality of concrete and reinforcing bars are taken as 25 MPa and 420 MPa, respectively. Cross-section dimensions of the columns of the systems were arranged that axial load ratio is set to approximately 10%. However, cross-section of beams are kept constant as 250 mm × 500 mm. Volumetric ratio of the members were sufficient to prevent shear failure at critical sections. Total mass and first three vibrational periods of all buildings are given in Table 1.

The location of the buildings is selected in İzmit, Turkey (Lat: 40.805936, Lon: 29.358273) where the expected peak ground acceleration of design earthquake having a return period of 475 years (probability of exceedance is 10% in 50 years) is 0.39 g. Soil condition of the location is chosen as hard soil "B" classified according to soil shear velocity (V_{s30} = 800 m/s).

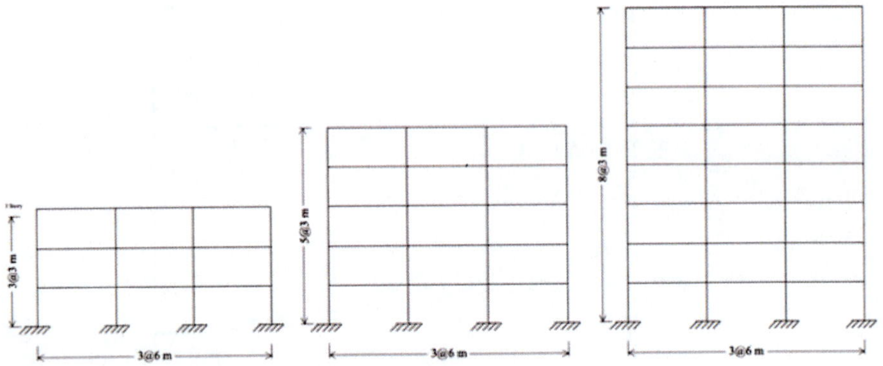

Fig. 2 Schematic drawings of the frame systems

Table 1 Mass and modal properties of the frame systems

Story	Mass (metric ton)	T_1 (s)	M_1/mass (%)	T_2 (s)	M_2/mass (%)	T_3 (s)	M_3/mass (%)
3	123	0.32	87	0.10	11	0.05	5
5	206	0.42	79	0.14	12	0.07	5
8	330	0.56	75	0.17	11	0.09	5

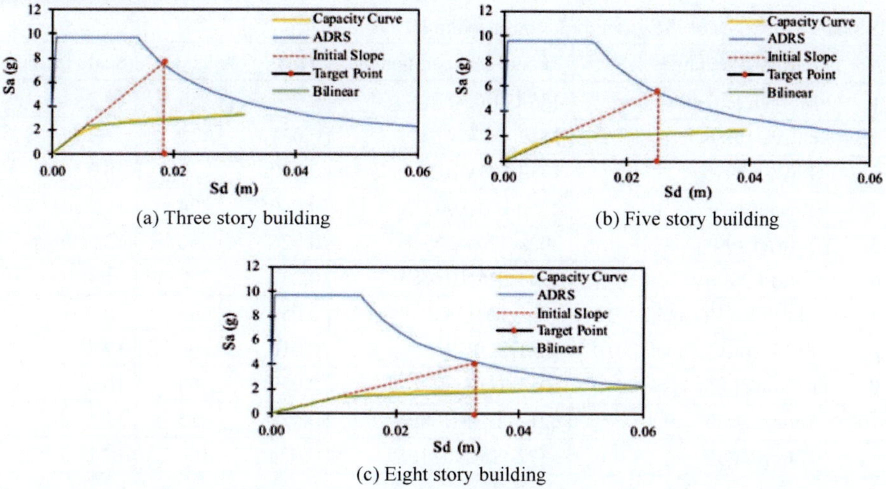

(a) Three story building

(b) Five story building

(c) Eight story building

Fig. 3 Determination of target displacement of roof for all buildings

Table 2 Target, yield displacement and resulting ductility of frame systems

Story Number	d_{roof} (m)	d_y (m)	μ
3	0.211	0.066	3.21
5	0.325	0.102	3.18
8	0.441	0.152	2.91

Strucutral members in all buildings are modeled with concentrated plastic hinges with moment and rotation relation. Constant mode incerermental nonlinear push-over analyses are conducted on all buildings in order to determine displacement ductility demand under design earthquake, as seen in Fig. 3. Procedure given in TBSC-2018 is followed in push-over analyses.

Calculated lateral displacement ductility of the systems are found approximately $\mu = 3$ as seen in Table 2.

3.2 *Ground Motion Selection and Scaling*

A set of ground motions used in nonlinear time history (NLTH) analyses is selected and scaled as per TBSC-2018. Properties of 11 single component ground motions recorded in historical earthquakes are presented in Table 3.

To provide the consistency among the ground motions, all records' amplitudes are linearly scaled to given design spectrum as mentioned in TBSC-2018. Amplitude scaling is preferred to spectra matching for the sake of not losing the frequency content of the original record. The scaling procedure gives special conditions between

Table 3 Properties of the ground motions used in NLTH

No.	Earthquake name	Record No. component	PGA (g)	Td (s)	Scale factor
1	Morgan Hill 1984	455-G01230	0.014	55	7.1
2	Düzce 1999	1613–060-E	0.031	120	7.2
3	Taiwan 1999	2753-CHY102E	0.070	30	8.0
4	Taiwan 1999	2759-HWA002N	0.040	224	19.5
5	Taiwan 1999	2929-TN042N	0.026	60	12.3
6	Tottori 2000	3954-SMNH10EW	0.231	300	1.4
7	El Mayor-Cucapah 2010	5993-EML-90	0.053	44	18.9
8	El Mayor-Cucapah 2010	6041-SDR-90	0.016	309	19.4
9	El Mayor-Cucapah 2010	6051-TOR-90	0.059	63	16.0
10	Tottori 2000	6212- HRSH08EW	0.028	155	12.5
11	Tottori 2000	6372- SMN013EW	0.024	164	17.2

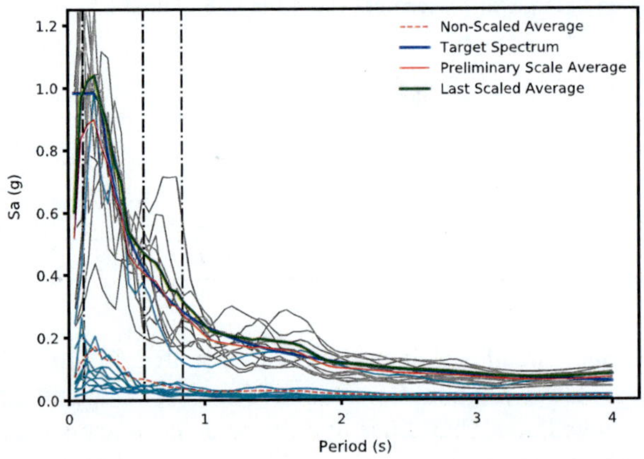

Fig. 4 Amplitude scaling of selected ground motions to given design spectrum

0.2 and 1.5 times of the first natural period of the system which can be seen as shaded area in Fig. 4.

4 Plastic Energy Calculations

Comparison of energy values dissipated in three structural systems are made between nonlinear time history analysis results and proposed plastic energy spectrum. Each structural member in all three systems is defined with plastic hinges where the moment and rotation relation is defined. The area enclosed by moment-rotation curve is accounted as the energy dissipated by given hinge. At the end of each NLTH

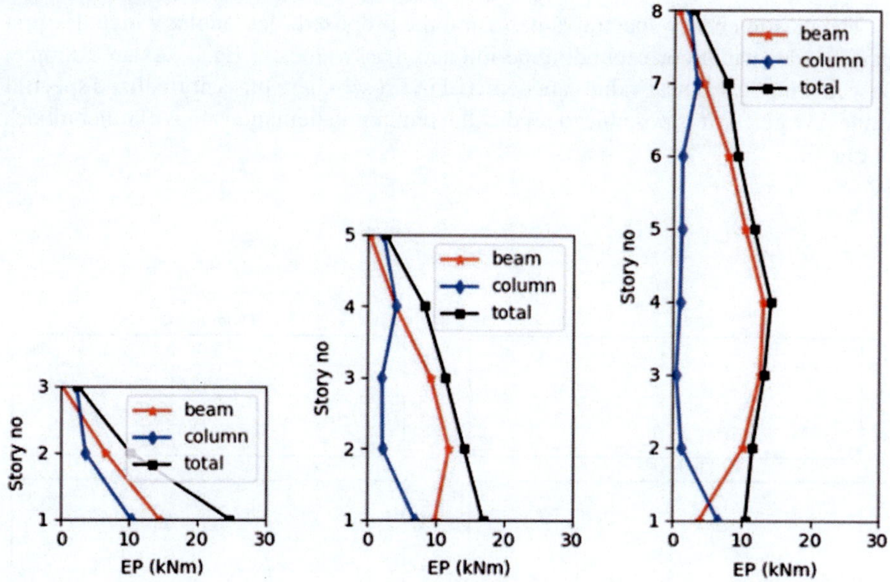

Fig. 5 Distribution of energy dissipation by the structural members over the system height

analysis, areas of moment-rotation curves are grouped as beam and columns at each story. Arithmetic average of dissipated energies at story levels as seen in Fig. 5 indicates that total energy dissipation at the stories significantly changes as the number of stories in the system increases. As a result of increased story number, higher modes show their contribution to structural response.

Since every system is analyzed with 11 different ground motions, there are 11 different dissipated energy values in each plastic hinge. The common practice in earthquake engineering is to estimate mean of each NLTH response at given engineering demand parameter. However, energy is a scalar value, thus neither arithmetic mean nor median value of the response may represent the true behavior of the member. To cover the variation of energy, mean plus one standard-deviation of calculated values is taken into account. This approach was also chosen in development of plastic energy spectrum in Dindar et al. [5]. Dissipated energy values obtained from NLTH analyses are given in Table 4. It is apparent that the increment in story number is proportional with increase of energy dissipation as a result of hinge locations' quantity.

Table 4 Plastic energy values obtained from NLTH analysis (values are in kNm)

GM's	1	2	3	4	5	6	7	8	9	10	11	μ	σ	$\mu + \sigma$
3 stories	3	12	7	32	29	12	51	17	36	15	28	22	14	36
5 stories	21	50	36	46	14	27	64	18	26	31	49	35	16	50
8 stories	101	36	19	42	27	50	22	87	37	65	57	49	26	76

The plastic energy spectra values using the proposed methodology includes first three modes using corresponding period and mass values in Eq. 1. As an example, only first mode spectral values are depicted in Fig. 6 where mass normalized spectral values are given. It is possible to predict the remaining demand values of other modes in Fig. 6.

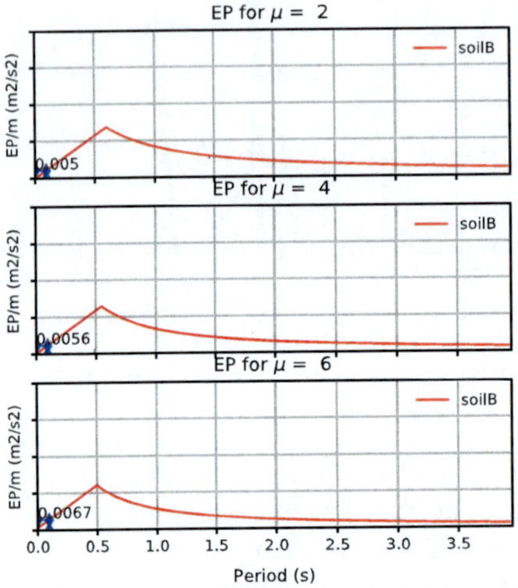

(a) Three story frame system

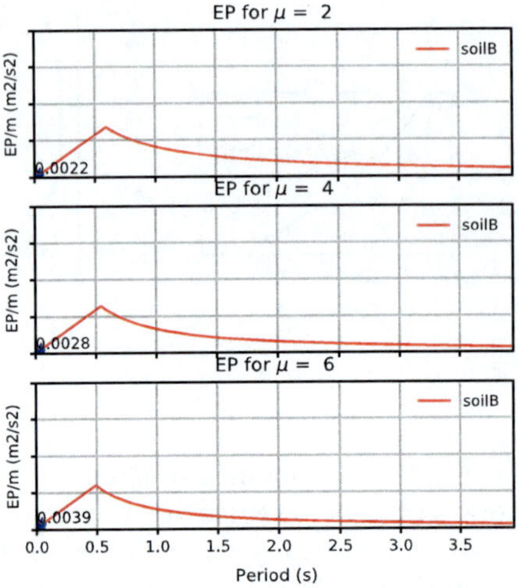

(b) Five story frame system

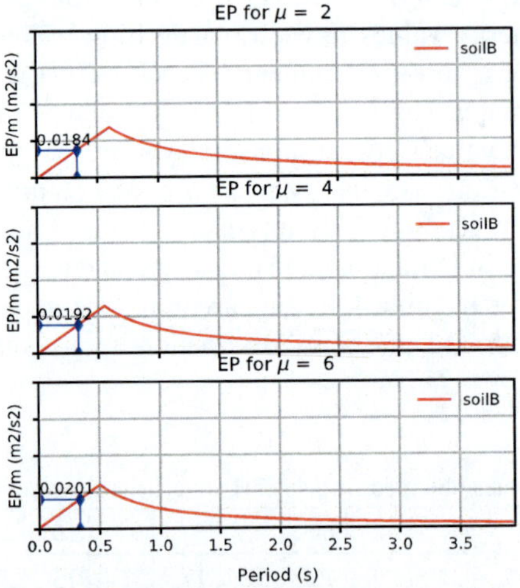

(c) Eight story frame system

Fig. 6 Distribution of energy dissipation by the structural members over the system height

Table 5 Plastic energy values obtained from EP Spectra (values are in kNm)

Mode	3 story			5 story			8 story		
	$\mu = 2$	$\mu = 4$	$\mu = 6$	$\mu = 2$	$\mu = 4$	$\mu = 6$	$\mu = 2$	$\mu = 4$	$\mu = 6$
1st	29.88	31.18	32.64	58.42	60.40	62.87	118.81	117.69	98.95
2nd	0.98	1.10	1.32	2.78	3.05	3.42	5.21	5.61	6.24
3rd	0.11	0.14	0.19	0.53	0.63	0.80	1.08	1.22	1.48
Total	30.97	32.42	34.15	61.73	64.07	67.09	125.10	124.52	106.67

Table 6 Comparison of plastic energy values

Story number	μ	$EP_{spectra}$ (kNm)	EP_{NLTH} (kNm)	Difference (%)
3	3.21	31.84	36	13
5	3.18	63.15	50	−21
8	2.91	124.75	76	−39

Plastic energy demand values for each mode are given in Table 5 at every ductility level. Dominancy of first mode in low and mid-rise systems is significantly apparent in the table in spite of contribution of the higher modes.

Exact spectral demand values are calculated using displacement ductility of three systems. Comparison of both plastic energy demand values reveals that NLTH analyses' results are higher than spectral values for three story system but lower than for five and eight story systems (Table 6).

5 Conclusion

Conventional seismic design codes implicitly consider energy dissipation as a damage measure. Particularly, performance based design requires controlling damage occurrence and propagation in the members at predicted seismic intensity. Even tough, plastic deformations are providing a control mechanism of the member and eventually system performance, seismic demand is obtained either from elastic response spectrum for nonlinear static push-over analysis or from nonlinear time-history dynamic analysis. Both analysis approaches are in different characteristics. Push-over analysis is relatively simpler than the comprehensive time-history analysis. Whereas NLTH is applicable to any structure without concerns limiting push-over analysis. Perhaps, energy concept may form the link lacking between both nonlinear analysis approaches.

This study shows potential of employing plastic energy spectrum in seismic design of engineering structures provideing an approach that does not require a comprehensive analysis compared to NLTH analyses. Despite larger dispersion for five and eight story systems, proposed spectrum yields approximate value for low-rise system.

Considering the growing number of studies aiming to develop a practical energy-based seismic design for next generation codes, the current study might be a valuable step when determining the plastic energy seismic demand.

References

1. Cheng Y, Lucchini A, Mollaioli, F.: Ground-motion prediction equations for constant-strength and constant-ductility input energy spectra. Bull. Earthquake Eng. **18**, 37–55 (2020). https://doi.org/10.1007/s10518-019-00725-x
2. Galé-Lamuela, D., Donaire-Avila, J., Escolano-Margarit, D., et al.: Energy dissipation capacity of rc columns subjected to unidirectional and bidirectional seismic loading. In: Papadrakakis, M., Fragiadakis, M. (eds.) COMPDYN 2019 7th ECCOMAS Thematic Conference on Computational Methods in Structural Dynamics and Earthquake Engineering, pp. 2316–2331. Crete, Greece (2019)
3. Lucchini, A., Mollaioli, F.: Energy-based seismic design methodology. In: 16th World Conference on Earthquake Engineering, Santiago, Chile, 9–13 Jan 2017
4. Mezgebo, M.G., Lui, E.M.: A new methodology for energy-based seismic design of steel moment frames. Earthq. Eng. Eng. Vib. **16**, 131–152 (2017). https://doi.org/10.1007/s11803-017-0373-1
5. Dindar, A.A., Yalçın, C., Yüksel, E., et al.: Development of earthquake energy demand spectra. Earthq. Spectra **31**, 1667–1689 (2015). https://doi.org/10.1193/011212EQS010M
6. Vidic, T., Fajfar, P., Fischinger, M.: Consistent inelastic design spectra: strength and displacement. Earthq. Eng. Struct. Dynam. **23**, 507–521 (1994). https://doi.org/10.1002/eqe.4290230504
7. Fajfar, P., Vidic, T.: Consistent inelastic design spectra: hysteretic and input energy. Earthq. Eng. Struct. Dynam. **23**, 523–537 (1994). https://doi.org/10.1002/eqe.4290230505
8. Dindar AA (2009) Energy-based earthquake response analysis and design of reinforced concrete SDOF columns. Boğaziçi University
9. Ministry of Environment and Urban (2018) Turkish Building Seismic Code
10. Computers and Structures Inc. (2020) SAP2000 Structural Analysis Program

Effects of Pre-existing Damage on Fragility of URM and RC Frame Buildings

P. García de Quevedo Iñarritu, N. Šipčić, M. Kohrangi, and P. Bazzurro

Abstract After the recent earthquake sequences that have hit Italy, New Zealand, and other parts of the world, the evidence that damage in buildings that experienced multiple shocks is, in general, more severe at the end of the sequence than after the mainshock is even more apparent and better documented. However, analytical studies still struggle in providing realistic estimates of how the damage progresses during a sequence of shocks. Predicting damage accumulation is of paramount importance for achieving accurate risk estimates for building stocks in all regions of the world where earthquake sequences are common. This study attempts to provide a framework for generating damage-state-dependent seismic fragility functions for two types of structures: unreinforced masonry buildings (URMs) and reinforced concrete frame buildings (RCFs). We adopted a component-based approach for estimating damage, built on energy-based parameters. In the case of URMs, we consider the energy released through shear damage of masonry components. For RCFs, we utilized the modified Park and Ang damage index. The findings suggest that estimates of maximum global or local deformation experienced during a shock are not fit to model damage progression throughout earthquake sequences. Furthermore, a component-based methodology with local parameters is superior to using global response parameters, such as maximum inter-story drift ratio. However, this study's results suggest that the ground to cover before achieving defensible cumulative damage estimates is still considerable. The need to develop and calibrate demand parameters to capture the progression of damage through all damage states with better precision is needed but not yet within our reach.

P. G. de Quevedo Iñarritu (✉) · N. Šipčić · P. Bazzurro
University School for Advanced Studies IUSS, Pavia, Italy
e-mail: pablo.garcia@iusspavia.it

N. Šipčić
e-mail: nevena.sipcic@iusspavia.it

P. Bazzurro
e-mail: paolo.bazzurro@iusspavia.it

M. Kohrangi
RED Risk Engineering + Development, Pavia, Italy

A. Benavent-Climent and F. Mollaioli (eds.), *Energy-Based Seismic Engineering*, Lecture Notes in Civil Engineering 155,
https://doi.org/10.1007/978-3-030-73932-4_2

Keywords Mainshock–aftershock sequences · Cumulative damage · Component-based fragility assessment · Unreinforced masonry · RC frames

1 Introduction

Current performance-based earthquake engineering (PBEE) (Cornell and Krawinkler 2000) relies on the assumption that the building is in an intact state once the earthquake has occurred, meaning that we assume that the building has been fully repaired since the last shock took place, which is often not the case. Large earthquakes commonly and in a relatively short period can trigger numerous offspring events resulting from the complex stress interaction between fault planes and tectonic plates. These events may cause considerable destructions to the exposed infrastructure (e.g., Moon et al. 2014), which was observed in several sequences such as the 2010–2011 Darfield-Christchurch (Shcherbakov et al. 2012), the 2011 Van in Turkey (Di Sarno et al. 2013), the Great East Japan earthquake (Goda et al. 2013), the 2012 Emilia Romagna in Italy (Penna et al. 2014b) and the 2016 earthquake sequences in central Italy (Sextos et al. 2018).

Ignoring the precursors' and the subsequent events' impact may underestimate the losses, and thus may result into poor and ill-informed decisions in designing insurance products, devising emergency planning strategy, etc. Specifically, the seismic hazard analysis and fragility and vulnerability estimation are the steps that should be modified accordingly to account for these impacts. A seismic hazard model typically adopts declustered earthquake catalogs considering only the largest events in the process and thus disregarding all foreshocks, aftershocks, and triggered events. It was shown, however, that this assumption results in a significant underestimation of the earthquake hazard. This underestimation has been addressed in several previous studies (e.g., Yeo and Cornell 2008; Iervolino et al. 2014; Papadopoulos and Bazzurro 2018). The discussion on how to improve the hazard assessment is, however, outside the scope of the current study.

The development of seismic fragility curves play also a critical role in seismic risk assessment because they represent the probability of attaining different damage states given the ground motion intensity. Most of the fragility functions that are commonly used are developed for intact structures, neglecting that, as mentioned earlier, a building that already experienced damage is probably more vulnerable than an intact one. Therefore, for a realistic estimate of evolving seismic risk during a seismic sequence, suitable modeling of building vulnerability variations should be considered.

Severe cases of damage accumulation in masonry buildings and RC buildings has been observed and reported in several recent earthquake sequences (e.g., Ingham and Griffith 2012; Kam et al. 2011). Sextos et al. (2018) reported that in the 2016 central Italy earthquake sequence, masonry structures suffered significant damage during the first event and quite often experienced an abrupt collapse in a successive earthquake because of the rapid reduction of their residual capacity and their brittle nature.

Similarly, accumulation of damage was reported in RC frame buildings with infills, where infill and column shear failure were observed after the subsequent events. During the ground motion sequences in Christchurch, the unreinforced masonry (URM) buildings were the most damaged among all other construction types. During this sequence, after the first event, a large percentage of masonry buildings were green tagged (i.e., safe to reoccupy), but after the subsequent events, virtually none was determined to be safe, and the majority of the buildings were set to demolition or were destroyed during the aftershock ground shaking (Ingham and Griffith 2012). During the same sequence, cumulative strength degradation of reinforced concrete frames (RCF) was also observed, causing the building's collapse (Kam et al. 2011).

Motivated by these observations, several authors studied the effect pre-existing damage has on buildings' seismic response. Mouyiannou et al. (2014b) proposed and applied a methodology to derive damage-dependent fragility curves for URM buildings adopting calibrated single-degree-of-freedom (SDOF) systems using maximum inter-story drift as a global damage measure. Grimaz and Malisan (2016), using a simplified model for several sets of masonry building cases, studied the impact of cumulative damage on the EMS-98 macro-seismic intensity scale. Burton et al. (2017) used Incremental Dynamic Analysis (IDA) to find both collapse capacity of intact and mainshock-damaged buildings. They studied reduction in the capacity of mainshock-damaged concrete frames with infill whose behavior was simulated using 2D models. Jeon et al. (2015) recognized the IDA approach's limitations in terms of computational effort and proposed a framework where IDA is employed to access mainshock response and cloud approach (using unscaled records) for aftershock response and development of aftershock fragility curves.

Nonetheless, most analytical studies related to aftershock fragility concluded that further damage is experienced especially when the aftershock ground motion intensity is more severe than that previously experienced during the mainshock. These conclusions based on numerical methods are somehow in contradiction with the progressive damage observed, e.g., in the 2016 central Italy earthquake sequence. In these events, at most sites the aftershocks shaking intensity was lower compared to that caused by the mainshock. The relatively low impact of aftershock shaking on the progressive damage of URM and RC buildings in these numerical investigations may be related to (1) the adoption of simplified structural models that are unable to capture damage accumulation and/or (2) improper global damage indices that do not reliably reflect the deterioration of the structural performance.

The current study's primary goal is to provide a framework for developing mainshock-damaged aftershock-fragility functions for typical URM and RC frame building structures found in Italy, representing the damage progression accurately. To better understand and simulate the buildings' progressive damage, we use 3D models of the multi-degree-of-freedom (MDOF) systems accounting for cyclic and in-cycle degradation. Besides, a component-based damage index instead of the widely utilized global ones is adopted to evaluate the progression of global damage for these case study buildings. Nonlinear back-to-back dynamic analyses using the un-scaled ground motion from real earthquake sequences are then performed to generate damage-dependent fragility curves.

2 Structural Modeling of Case Study Buildings

As mentioned earlier, we use 3D models and nonlinear dynamic analysis to derive damage-dependent fragility curves and establish the proposed framework. We have chosen two typical Italian URM and one RC frame as case study buildings. URM buildings were originally designed according to the latest version of the Italian seismic design code NTC08 (RINTC Workgroup 2018) for a site located in L'Aquila, assuming a design ground acceleration of $a_g = 0.347$ g. The two URM building examples include a two-story regular in plan, with symmetry in the X-axis and continuous walls (Building A) and a two-story irregular in plan, without symmetry on any axes and discontinuous walls throughout its length (Building B). Note that both buildings have a total height of 6.0 m, 3.0 m per story. The layout of the two-building examples is shown in Fig. 1. We use TREMURI (Lagomarsino et al. 2013), an equivalent-frame macro-element program capable of performing nonlinear static and dynamic analyses of masonry structures, to create 3D models of masonry buildings. The adopted macro-element model (Gambarotta and Lagomarsino 1997) considers the evolution of flexural and shear cracking for increasing displacement. This macro-element model is specifically developed for simulation of the cyclic and dynamic response of masonry buildings, and it also reproduces the hysteretic response of masonry walls. For this reason, the expected energy dissipation due to material nonlinearity is directly accounted for in the model. Herein, we consider only the in-plane response behavior of the masonry panels, while the out-of-plane failure modes are neglected, which is a limitation that should be addressed in future investigations. To calibrate the macro-element model properties for the specific masonry type building considered in this work, we use the experimental test results (Cattari et al. 2018) that monitor the displacement and the shear strength of the wall response under cyclic loading.

Figure 2a shows the hysteretic behavior (i.e., top displacement versus shear force) of a single wall based on the above-mentioned experimental tests. In this Figure, the analytical model of the wall created in TREMURI is also compared with the test results, verifying the adopted modeling approach's reliability. More details about the specimen and the test results can be found in (Cattari et al. 2018). Fig. 2b shows the two building models' pushover curves in TREMURI, in X and Y directions. These curves show that for Building A, the behavior in both directions is somewhat similar in terms of strength and drift capacities, even if the failure mode is different, due to the different length of the walls in the two directions (more slender walls tend to have a more flexural-like failure). Building B pushover shows the irregularity's effect, as the building behaves very differently in X compared to the Y direction, where it fails more abruptly after peaking under small drifts. Note that in all of the dynamic analyses performed herein, we considered 3% modal Rayleigh damping for both URM buildings.

The RCF building (Building C) was designed according to the past Italian code (DM 96, 1996), and it is located in the moderate seismic zone 2 according to seismic hazard classification of (DM 96). The building has a plan area of 28.0 m × 18.0 m, four stories, and a total height of 14.0 m. To create the building's nonlinear model, we

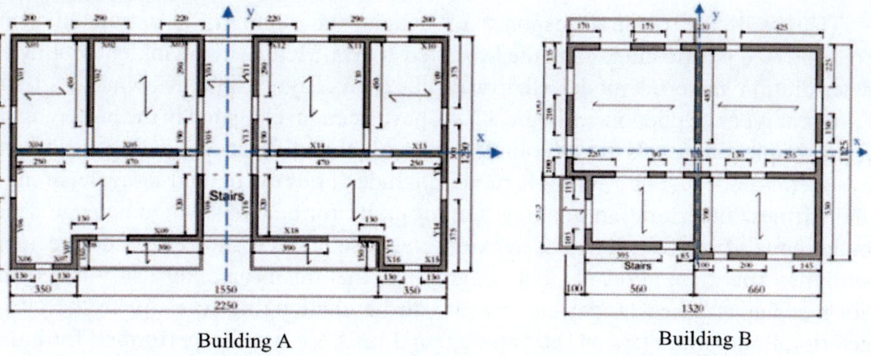

Fig. 1 The layout of the URM case study building examples utilized in this study

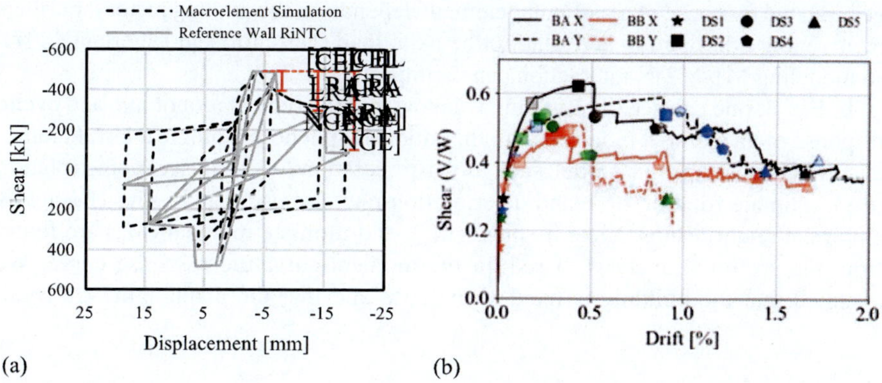

Fig. 2 **a** Comparison of the hysteretic response of a masonry wall specimen with the analytical macro-element modeling in TREMURI (Adopted from Cattari et al. 2018), *Note* (**a–c**) correspond to the loss of 20%, 40% and 80% of the wall total shear strength capacity, respectively; **b** pushover curves with the component-based damage states as solid markers and the global-based ones as transparent markers

used the OpenSees software (Mazzoni et al. 2006). Two main approaches adopted in the numerical modeling of beams and columns in RCF buildings are a fiber-based model (with either distributed plasticity or concentrated plasticity) and a model that uses elastic elements with rotational springs at the ends, which simulate the inelastic response. Using fiber-based models might not be appropriate for the type of analysis conducted herein as cyclic degradation is of high importance, while most of the currently available constitutive models for the prediction of steel and concrete response only account for in-cycle degradation. There are ways to account for this deterioration in fiber-based models, simulating bar slip, rebar buckling, or fracture, but these approaches come with the high computational cost and often numerical instabilities (ATC 2017).

That said, the flexural response of beams and columns is simulated using concentrated plastic hinges and the Modified-Ibarra-Medina-Krawinkler (ModIMK) deterioration material model (Ibarra et al. 2005; Lignos and Krawinkler 2010). Different types of phenomenological laws have been investigated in the past, ranging from simple elastic-plastic to complex strength and stiffness degrading curvilinear hysteretic models. The ModIMK model includes both cyclic and in-cycle strength and stiffness deterioration, making it appropriate for analysis where damage accumulation is of interest. It has been extensively used, and it showed good agreement with the experimental results. One should note that this is an empirically-based and not mechanics-based model and that it can be challenging to estimate the cyclic deterioration parameters as the experimental tests are usually performed for either monotonic or cyclic loading protocols, not both. An additional drawback is that the M1-M2-N interaction is not taken into account. As pointed out in Taucer et al. (1991), the axial load in columns can significantly fluctuate during the ground motion, and, in general, the response of a particular element depends on the loading history. Furthermore, energy dissipation is related to the axial load (Haselton and Deierlein 2007), so modeling M1-M2-N interaction can be important.

In Fig. 3, one can see the Building C layout and the typical monotonic and cyclic response of an element modeled with the ModIMK phenomenological law. Parameters used to describe this model, such as plastic rotation ($\theta_{c,pl}$), post capping rotation (θ_{pc}), ultimate rotation (θ_u), and deterioration rate, are estimated using Haselton's empirical relationships. Yield moment (M_y), and ultimate moment (M_u) are found from the sectional analysis, based on the moment-curvature response curve. We assumed that the building is fixed at the base and that the diaphragms are rigid.

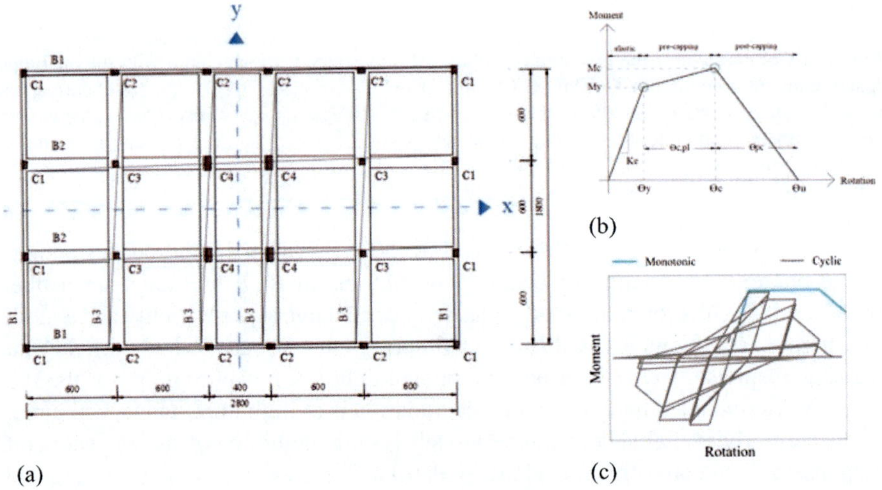

Fig. 3 **a** Layout of the RCF building (Building C); **b** typical monotonic response and **c** comparison of the monotonic and cyclic response of the beam-column element modeled with Modified Ibarra Krawinkler model

To calculate the damping forces, we adopted the Rayleigh model, proportional to the current stiffness matrix. To calculate the gravity loading, we included the load from structural and non-structural elements, and 30% of the live loading assumed as 2 kN/m^2. We performed the Eigen-analysis and found that the fundamental periods are 1.06 and 1.05 s in Y and X directions, respectively.

3 Damage State Definition

Damage states can be thought of as a communication tool that helps us synthetize the extent of damage after the earthquake event, and that can, furthermore, help us evaluate post-earthquake accessibility to the building. Hence, it is intuitive that a meaningful damage state definition is an important task. Furthermore, damage states are used in vulnerability and loss assessment to effectively link the intensity of ground motion to structural and non-structural damage and, consequently, to the repair cost needed to fix that damage. Generally speaking, damage in RC and URM structures is related to inelastic deformations, and hence damage parameters should refer to quantities such as strain, curvature, rotation, displacement, inter-story drift, energy dissipated during the reversed cyclic loading. As the inelastic response of a structure is highly complex and affected by several factors, various damage variables and indices have been proposed in the literature with a fundamental principle to correlate the damage level with parameters whose values can be calculated analytically.

Damage indices could be classified in different ways, but, having in mind that in this study we want to evaluate the effect of seismic sequences on structural response, classification on traditionally used non-cumulative indices and cumulative indices that attempt to include the effects of repeated cyclic loading is of our interest. Non-cumulative damage indices, such as ductility ratio, maximum inter-story drift, rotation, stiffness degradation, and maximum curvature, have been conventionally used due to their simplicity and practicality. For example, experimental studies showed that damage in well-detailed RC members, which are not supposed to experience shear failure or bond-slip, could be represented well with non-cumulative indices (Cosenza and Manfredi 1998). Additionally, the use of non-cumulative indices is justified in cases when structures are subjected to impulse-type or short-duration earthquake ground motions. In other cases, however, cumulative damage measures could be a more rational choice, particularly when we want to represent the progression of damage or when we want to compare the excitations with a different duration as in the case of mainshock only and sequence ground motions. Besides the classification in a cumulative and non-cumulative group, we can also distinguish the local and global damage indices. Local measures are used to quantify individual members' damage, while global ones are related to the entire structure.

Herein we adopt two approaches to define the damage states of the case study buildings subjected to sequences of earthquakes. The first method, referred to as the 'global damage' approach, defines damage states globally in terms of drift (i.e., non-cumulative), while the second approach is 'component-based,' built on energy-based parameters (i.e., cumulative). For URM buildings, global damage states are defined based on the building model's pushover curves (Rota et al. 2010). These global damage states, i.e., five damage states labeled by DS_{G1} to DS_{G5}, are based on the EMS-98 damage scale definitions for masonry buildings (Grünthal 1998). The description of the expected physical damage at each of the global damage states is provided in Table 1, together with the drift limit threshold values corresponding to the predefined points on the pushover curve that are considered as DS threshold criterion. These points are defined based on Tomazêvic (2007) for DS_{G1} to DS_{G3} and based on Penna et al. (2014a) for DS_{G4} and DS_{G5}.

On the other hand, within the component-based approach, we use the parameter α, which describes the energy dissipated by the damage happening in a masonry panel during its shear response, which can be related to cracks in the central region (Brencich et al. 1998). This parameter is found within the nonlinear part of the masonry's constitutive laws as a part of the toughness function. The quantity α is available within any nonlinear analysis conducted in TREMURI, as long as the panel fails in shear (Gambarotta and Lagomarsino 1997). It starts from zero before any loading condition and tends to increase once the panel damages. During any numerical analysis, α increases up to one at the point of peak shear strength (when the toughness function is maximized) and then continues to grow as the capacity degrades. Once the panel achieves a specific displacement limit associated with its failure, the values of α typically jump orders of magnitude. The onset of the four local damage states (herein labeled by DS_{L1} to DS_{L4}) considered for each wall element of the building are assumed to be associated with α values of 1.00, 2.35, 12.35, and 22.00 during the nonlinear analysis. These α threshold values are correlated to different physical damages on single walls based on the results of experimental cyclic loading tests of two wall specimens tested at EUCENTRE (Cattari et al. 2018), shown in Fig. 2. These experimental tests were essentially performed to calibrate the macro-element properties of the specific masonry type considered herein, simultaneously monitoring the displacement, shear strength, and α.

Once the nonlinear dynamic analyses are completed, the structure's global damage state is identified based on the two approaches described above. In the 'global damage' method, the maximum inter-story drift value readily shows the structure's damage state. The procedure is not as simple in the 'component-based' approach, as there is no unique method for transitioning from a component damage state to a global damage state. In this case, the local damage states (DS_{L1} to DS_{L4}) for each masonry wall are determined based on the α threshold values. The aggregation of these local damage measures is then translated into a global damage state following the approach introduced by Augenti et al. (2004). The latter assigns global damage states to damaged buildings according to the preliminary assessment AeDEs forms (Field manual for post-earthquake damage and safety assessment and short-term countermeasures, developed by the Italian Department Civil Protection) (Baggio et al. 2007). This approach seeks to correlate the number of damaged elements (found in a particular damage state) versus the building's global damage states. We first define a parameter that quantifies the percentage of the walls' length in the *j*th local damage state over the total lengths of the walls in the building L_{DSLj}, according to Eq. (1):

$$L_{DSL_j}[\%] = \frac{\sum_n^N L_{DSL_{j,n}}}{L_t} \cdot 100 \quad (1)$$

N is the total number of walls found in the jth local damage state, $L_{DSLj,n}$ is the length of the nth wall found in the jth local damage state, and L_t is the total length of the building's walls. Accordingly, we define the global damage states of this method using a set of criteria defined as a function of L_{DSL_j}. The last column of Table 1 shows the global damage states definition based on the 'component-based' approach.

When it comes to the RC building (Building C), global damage states are defined following Rossetto and Elnashai (2003), who proposed a new damage scale (homogenized RC damage scale or HRCS) that is subdivided into seven damage states

Table 1 Description of the damage states and the corresponding damage onset definitions based on the 'global damage' and 'component-based' approaches

DS_G	Damage description	Damage onset definition[a]	
		Global Damage approach (max inter-story drift thresholds)	Component-based approach Global Damage definition (% of the wall length)
DS_{G1}	Negligible damage: non-structural damage present, typically hairline cracks on very few walls and fall of plaster	The drift at which the first pier yields	$8\% < L_{DSL_1} \leq 15\%$
DS_{G2}	Slight damage: Some diagonal cracks on the middle parts of the structural walls is formed	The drift value at which maximum global shear resistance is reached	$15\% < L_{DSL_1} \leq 55\%$
DS_{G3}	Moderate damage: The number of cracks increases, and their length and become more visible. Most walls are in considerably damaged condition	The drift corresponding with a 20% loss of the structure's peak shear strength	$L_{DSL_2} > 25\%$
DS_{G4}	Heavy damage: The structure is considered to have heavy damage, with failure in some walls and individual pieces	The drift corresponding with a 50% loss of the structure's peak shear strength	$L_{DSL_4} > 0\%$ or $L_{DSL_3} > 25\%$
DS_{G5}	Collapse: The structural collapse, either partial or total	Maximum failure drift	$L_{DSL_4} > 15\%$

[a]The global damage state values for "Building A" and "Building B" are shown on the pushover curves in Fig. 2b

defined in terms of maximum inter-story drift. This scale was calibrated using experimental tests for European RC shear-wall, infilled, and bare models and, therefore, is appropriate for use in the current study. Seven damage states originally proposed by Rossetto with a clear description of structural and non-structural damage were a starting point to define the four damage states used in the current study (Table 2).

On the other hand, we used a modified Park and Ang damage index (MPADI) as defined by Kunnath et al. (1992) to define local damage states, expressed as a linear combination of deformation damage (plastic deformation) and the damage caused by hysteretic energy dissipation.

$$DI_{component} = \frac{\theta_m - \theta_r}{\theta_u - \theta_r} + \frac{\beta}{M_y\theta_u}E_h \tag{2}$$

The application of the Equation above requires determination of the maximum rotation θ_m (from the nonlinear time history), ultimate rotation θ_u and recoverable rotation θ_r, found using Haselton's empirical relations. Yield moment My was estimated based on the sectional analysis, dissipated hysteretic energy E_h based on the area under the moment-rotation curve, and strength degradation parameter β based on the available experimental estimates. We use a β factor of 0.15, a value recommended by Cosenza and Manfredi (1998). The local damage states (DS_{L1} to DS_{L4}) for each RC column element are found following Beck et al. (2002), as indicated in Table 3.

Local damages are then translated into the global one following the methodology proposed by Bracci et al. (1989), summarized by Eqs. (3) and (4) on a story and global level, respectively.

$$D_{story,j} = \frac{\sum_{i=1}^{N} w_i D_i^m}{\sum_{i=1}^{N} w_i D_i} \tag{3}$$

Table 2. Description of the damage states as defined by Rossetto and Elnashai (2003)

Global damage state	Damage description	MIDR (%)
Slight	Fine cracks in plaster partitions/infills, start of structural damage; hairline cracking in beams and columns near joints (<1 mm)	0.32
Moderate	Flexural and shear cracking in most beams and columns. Some yielding in a limited number. Shear cracking and spalling are limited	1.02
Extensive	Loss of bond at lap-splices, bar pull-out, broken ties. The main rebar may buckle, or the element may fail in shear	2.41
Collapse	Shear failure of many columns or impending soft-story failure	4.27

Table 3 Description of the local damage states as defined by Beck et al. (2002)

Local damage state	Description of damage in RC columns	MPADI
DS_{L1} Slight	Widespread light cracking, or a few cracks > 1 mm, ≤2 mm wide; or light shear cracks tending to flatten towards 30°	0.08
DS_{L2} Moderate	Significant cracking, e.g., 900 cracks > 2 mm; 450 cracks > 2 mm; 300 cracks > 1 mm. or yielded, not yet reached ultimate, where ultimate—maximum moment capacity	0.31
DS_{L3} Severe	Exceeded ultimate moment capacity but not yet decreased to 80% of maximum	0.71
DS_{L4} Collapse	Moment capacity decreased below 80% of maximum	1.28

$$D_{global} = \frac{\sum_{story,j=1}^{N} W_{story,j} D_{story,j}^{m}}{\sum_{story,j=1}^{N} W_{story,j} D_{story,j}}. \tag{4}$$

Local indices are weighted as the function of the gravity load supported. The importance factor w_i is calculated as the ratio between the member's tributy area and the total weight. In this manner, more weight is given to the base-story compared to the upper ones. The severity of the damage is taken into account with parameter m, which is usually considered equal to 1. One should note that only vertical elements are considered as we found that most of the dame in a studied building is experienced there. In this manner, we managed to reduce the computational time, which is significant, as the calculation of energy requires integrating the moment-rotation curve.

4 Ground Motion Database

The studies in the literature that developed damage-dependent fragility curves adopted different approaches for selecting the ground motion sequences. In this discussion, we will refer to the mainshock as MS and the aftershock events as AS for simplicity. Ryu et al. (2011) and Raghunandan et al. (2015) used scaled ground motions for MS-AS analyses to develop AS collapse fragility curves, while Jeon et al. (2015) used scaled ground motions for MS analysis and unscaled ground motions for AS analysis to develop AS fragility curves for multiple damage states. To have a realistic understanding of the impact seismic sequences have on the masonry and RC buildings, we compiled a database of 280 real MS-AS motion sequences recorded in 135 MS ground motions worldwide, followed by one or more aftershocks. These ground motion recordings are extracted from the NGA-WEST2 (Ancheta et al. 2014), RESORCE (Akkar et al. 2013), ITACA (Luzi et al. 2017), ESM (Luzi et al. 2016), and Japanese KiK-net (Shin et al. 2017) databases. Because some parts of the MS-AS *PGA* domain were not well populated (see Fig. 4a), we generated an additional

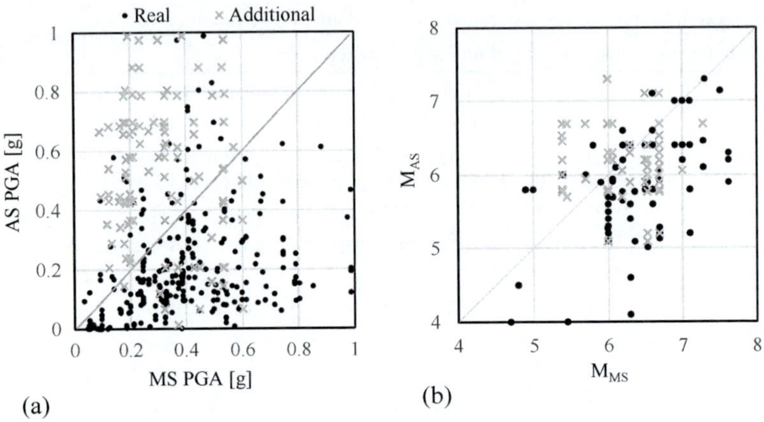

Fig. 4 **a** MS-AS PGA sequence pairs for both the real sequences and the additional random combinations; **b** MS-AS magnitude pairs

ground motion sequence set consisting of 90 pairs of records by randomly pairing the ground motions of some of the MSs included in the original dataset to create new sequences with intensities in the uncovered ranges of the original set. Figure 4a shows the scatter plot of the MS-AS sequence *PGA* pairs for both sets. This Figure shows that, in most of the real MS-AS sequences (black dots), the MS intensity is stronger than its corresponding AS. Figure 4b shows the magnitude pairs for the sequences. Note that because many of the ground motion sequences used here are recorded at multiple locations, the number of data points in Fig. 4b is less than that in Fig. 4a.

5 Development of Damage Dependent Fragility Curves

In the classical seismic assessment, fragility curves are generated assuming the structure is in its intact state (i.e., DS_{G0}), and thus such curves are conditioned on the ground motion intensity level only. However, when the performance of a damaged structure is of interest, the fragility curve should also be conditioned on its current damage state. That said, the probability of exceeding the *i*th global damage state (DS_{Gi}) due to AS ground motion, given that the structure is in the *j*th damage state (DS_{Gj}, where $i \geq j$), can be defined by Eq. (5):

$$P\left[ds > DS_{Gi} | IM_{AS}, DS_{Gj}\right] = \Phi\left[\frac{\ln(IM_{AS}) - \mu_{\ln(IM_{AS})}}{\sigma_{\ln(IM_{AS})}}\right], \tag{5}$$

$\Phi[\bullet]$ denotes the cumulative Gaussian distribution function, while $\mu_{\ln(PGA_{AS})}$ and $\sigma_{\ln(PGA_{AS})}$ are the fragility curves' parameters in terms of logarithmic mean and standard deviation of IM_{AS}, respectively. The structural response evaluated by the cloud analysis (Cornell et al. 2002) is used to derive these parameters using the method of moments (Baker 2015). The building's damage state is identified based on both the 'global damage' approach and the 'component-based' approach after every mainshock and aftershock analysis. Generally, displacement-based methods, such as using MIDR, were found to be less sensitive to the evolution of damage on the sequential analyses. Very rarely would there be an additional level of damage after the aftershock, which does not reflect the observations from real life sequences. Based on the results found, the increase of drift in both masonry and RC buildings has been observed mainly when the AS intensity is higher than the MS one. Even in these cases, the increase was observed only when the damage experienced during the MS is significant. These results are in sync with the findings of other analytical studies, but they, nonetheless, conflict with the observations. That said, the progression of damage is more apparent in the component-based approach, e.g., Building A changed its damage state in 30% of all cases, where only 29% of the cases with changed damage state were caused by sequences where the AS ground motion intensity is higher than that of the MS. On the other hand, the global approach leads to a 19% of cases showing an increase of damage state at the end of the sequence, out of which 95% of these changes happen when the intensity of the AS ground motion is larger than that of the MS. When Building C is concerned, in component-based approach damage state is changed in 19% of all cases, 21% of which occur when $S_a(T_{av})$ is lower during the aftershock event. In global approach, however, building changes its damage state in 12% of the cases, only 2% of which correspond to the cases when aftershock intensity is lower compared to the mainshock one.

In Fig. 5, we compare the damage-dependent fragility curves obtained for both masonry buildings using a component-based approach. Each panel shows the fragility curves for a final jth global damage state given the building was in the ith damage state after the mainshock, where of course $j > i$. When comparing the fragility curves of the intact building (in DSG0) with the damaged ones, the reduction in the median capacity becomes more evident as the level of the initial damage is increasing. For instance, Fig. 5f shows the fragility curves at the fifth global damage state (DSG5) conditioned on the building being in DSG0, DSG1, DSG2, DSG3, and DSG4. The median collapse capacity (namely that for DSG5), expressed in terms of PGA, of building A in its intact damage state is about seven times the median collapse capacity of the same building when the initial damage state is DSG4 (PGA_{AS}, DSG5|DSG4 = 0.13g vs. PGA_{AS}, DSG5|DSG0 = 0.76g). We can also see that building B has a relatively lower capacity compared to building A due to its in-plan irregularity. That said, a very small PGA increase of about 0.1 g was enough to cause changes on the structure's damage states.

In Fig. 6 we compare the damage-dependent fragility curves obtained for RC frame building using a component-based approach. Similarly, as in the masonry case, one can observe that the reduction in the capacity increases with the amount of initial damage experienced. One should note that there was not sufficient data to fit the fragility curve for DSG_4 conditioned on DSG_2, and it is hence omitted.

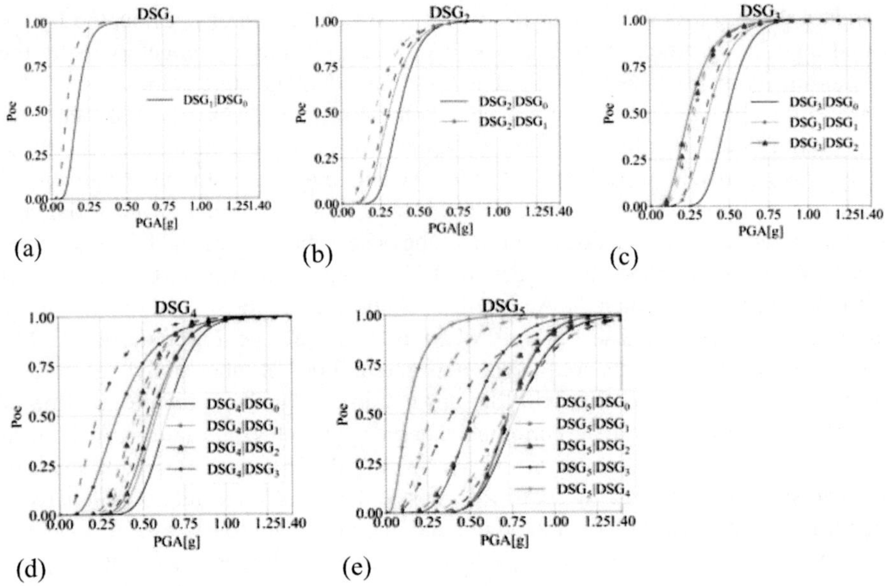

Fig. 5. Damage-dependent fragility curves for Building A (solid) and Building B (dashed) with the final damage state of **a** Damage State 1: DSG1; **b** Damage State 2: DSG2; **c** Damage State 3: DSG3; **d** Damage State 4: DSG4; **e** Damage State 5 (collapse): DSG5

Our findings indicate a non-negligible difference in the fragility of intact and damaged structures that can significantly affect the loss estimates. The figures above confirm the engineering expectation that when the building experiences heavy damage due to MS ground motion, it is more likely that a progression of damage will be observed in the aftershocks, while such damage evolution is less likely when the structure is left at a lower damage state (such as DSG_1 or DSG_2) after the MS motion. In the former case, the reduction in the structural capacity is rather significant and, therefore, the relatively low IM level of the AS may inflict further damage to the building or even cause its collapse.

6 Conclusions

In most risk assessment studies, the same fragility functions are used for intact and damaged structures, despite the evidence that pre-existing damage can have a non-trivial effect on the response. In other words, it is assumed that a building is in an intact state for all shocks, an assumption implying that immediate repair takes place after every earthquake, which is clearly untenable. In this context, we investigated the effects that pre-existing damage has on the structural response of the damaged

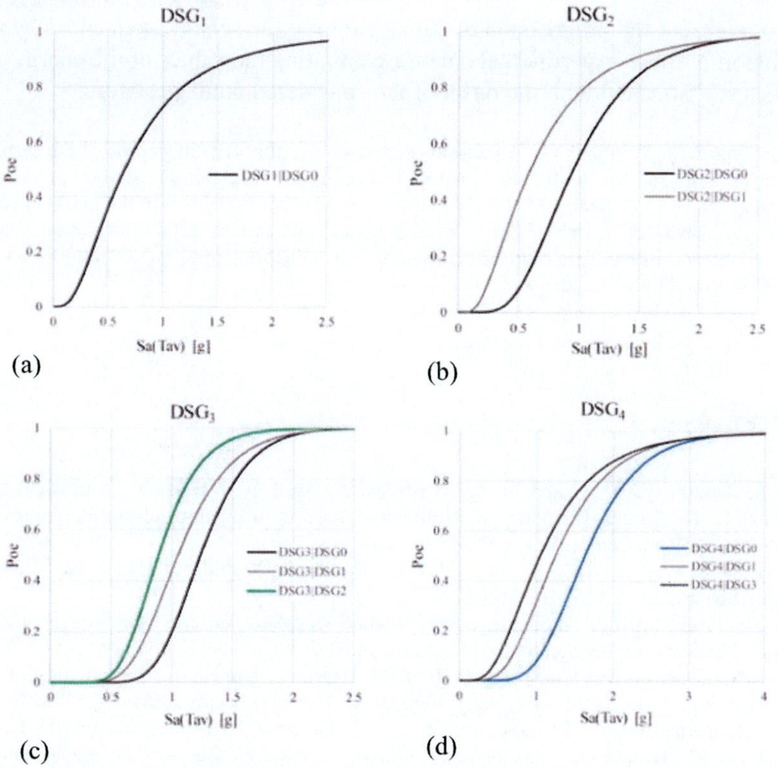

Fig. 6 Damage-dependent fragility curves for Building C with the final damage state of **a** Damage State 1: DSG1; **b** Damage State 2: DSG2; **c** Damage State 3: DSG3; **d** Damage State 4: DSG4 (collapse)

building using two different methodologies. First, we used MIDR as a global, non-cumulative, displacement-based parameter, typically used in studies with a similar objective. We found that damage increases only when the AS intensity is greater than the MS one or when significant damage is observed during the MS. This result is not in line with empirical observations. In addition, we also used a component-based approach, combined with the energy-based, cumulative, local damage parameters. This methodology proved as a better tool for capturing the accumulation of damage during the seismic sequences, indicating a significant decrease in the capacity that should not be omitted in the seismic assessment of buildings.

Although this study sheds some light into the issue of assessing damage accumulation in buildings, we certainly believe that there is still a lot of research ground to be covered in this subject on at least two fronts. First, more advanced modeling techniques, including out-of-plane failure mechanism for masonry walls, modeling shear, axial and joint failure in RC buildings, including moment-axial load interaction, and investigating the effect of infills are some aspects that can significantly impact the results. Second, to develop reliable and defensible energy-based damage parameters able to represent the progression of damage during the seismic sequences, we need the support of more experimental data for calibrating the values of the energy-based parameters associated with the onset of the considered damage states.

Acknowledgment We thank Dr. Athanasios Papadopoulos for providing some of the earthquake sequence records and for the fruitful discussions occurred throughout this study. We would also like to thank Dr. Andrea Penna, Dr. Maria Rota and Dr. Martina Mandirola of the masonry research group at EUCENTRE for sharing the models, software, and experimental information used here. Finally, we would like to thank Dr. Ettore Faga for his valuable inputs and discussions about the assessment on the URM buildings.

References

Akkar, S., Sandıkkaya, M.A., Şenyurt, M., Azari Sisi, A., Ay, B.Ö., Traversa, P., Douglas, J., Cotton, F., Luzi, L., Hernandez, B., Godey, S.: Reference database for seismic ground-motion in Europe (RESORCE). Bull. Earthq. Eng. (2013)

Ancheta, D., Stewart, S., Silva, C., Wooddell, G., Kottke, B., Kishida, D.: NGA-West2 database. Earthq. Spectra **30**(3), 989–1005 (2014)

ATC. (2017) Recommended modeling parameters and acceptance criteria for nonlinear analysis in support of seismic evaluation, retrofit, and design. NIST GCR 17-917-45

Augenti, N., Cosenza, E., Dolce, M., Manfredi, G., Masi, A., Samela, L.: Performance of school buildings during the 2002 Molise, Italy, Earthquake. Earthq. Spectra **20**(S1), S257–S270 (2004)

Baggio, C., Bernardini, A., Colozza, R., Coppari, S., Corazza, L., Della Bella, M., Di Pasquale, G., Dolce, M., Goretti, A., Martinelli, A., Orsini, G., Papa, F., Zuccaro, G.: Field manual for post-earthquake damage and safety assessment and short term countermeasures (AeDES). In: European Commission, Joint Research Centre Institute for the Protection and Security of the Citizen, European Communities (2007)

Baker, J.W.: Efficient analytical fragility function fitting using dynamic structural analysis. Earthq. Spectra **31**(1), 579–599 (2015)

Beck, J.L., Porter, K.A., Shaikhutdinov, R., Au, S.K., Moroi, T., Tsukada, Y., Masuda, M.: Impact of seismic risk on lifetime property values. Final Report for CUREE-Kajima Phase IV Project, Consortium of Universities for Research in Earthquake Engineering, Richmond, CA (2002)

Brencich, A., Gambarotta, L., Lagomarsino, S.: A macroelement approach to the three-dimensional seismic analysis of masonry buildings. In: 11th European Conference on Earthquake Engineering, pp. 6–11 (1998)

Burton, H., Sreekumar, S., Sharma, M., Sun, H.: Estimating aftershock collapse vulnerability using mainshock intensity, structural response and physical damage indicators. Struct. Saf. **68**, 85–96 (2017). https://doi.org/10.1016/j.strusafe.2017.05.009

Cosenza, E., Manfredi, G.: A seismic design method including damage effect. In: Proceedings of the 11th European Conference on Earthquake Engineering, Paris, 6–11 Sept 1998

Cattari, S., Camilletti, D., Lagomarsino, S., Bracchi, S., Rota, M., Penna, A.: Masonry Italian code-conforming buildings. Part 2: nonlinear modelling and time history analysis. J. Earthq. Eng. (2018). https://doi.org/10.1080/13632469.2018.1541030.

Cornell, C.A., Jalayer, F., Hamburger, R.O., Foutch, D.A.: (2002) Probabilistic basis for 2000 SAC federal emergency management agency steel moment frame guidelines. J. Struct. Eng. ASCE **128**(4), 526–533 (2002)

Cornell, C.A., Krawinkler, H.: Progress and challenges in seismic performance assessment. PEER Center News **3**(2), 1–2 (2000)

Di Sarno, L., Yenidogan, C., Erdik, M.: Field evidence and numerical investigation of the Mw = 7.1 October 23 Van, Tabanlı and the M = 5.7 November earthquakes of 2011. Bull. Earthq. Eng. **11**(1), 313–346 (2013)

Haselton, C.B., Deierlein, G.G.: Assessing seismic collapse safety of modern reinforced concrete frame buildings. PEER Report 2007/08, Pacific Engineering Research Center, University of California, Berkeley, Berkeley, CA, 274p (2007)

Gambarotta, L., Lagomarsino, S.: Computational models for the seismic response of damaging structures. In: Abrams, D.P., Calvi, S. (eds.) Proceedings of the US-Italian Workshop on Seismic Evaluation and Retrofit. US National Center for Earthquake Engineering Research, Buffalo, New York, pp. 45–66 (1997)

Geotechnical Extreme Events Reconnaissance Association (GEER): Engineering reconnaissance following the October 2016 Central Italy Earthquakes, Version 2, Report No. GEER-050D (2017)

Goda, K., Pomonis, A., Chian, S.C. et al.: Ground motion characteristics and shaking damage of the 11th March 2011 M w9.0 Great East Japan earthquake. Bull. Earthq. Eng. **11**, 141–170 (2013). https://doi.org/10.1007/s10518-012-9371-x

Grimaz, S., Malisan, P.: How could cumulative damage affect the macroseismic assessment? Bull. Earthq. Eng. (2016)

Grünthal, G. (ed.): European Macroseismic Scale 1998 (EMS-98). Centre Europèen de Géodynamique et de Séismologie, Luxembourg (1998)

Ibarra, L.F., Medina, R.A., Krawinkler, H.: Hysteretic models that incorporate strength and stiffness deterioration. Earthq. Eng. Struct. Dynam. **34**, 1489–1511 (2005)

Iervolino, I., Giorgio, M., Polidoro, B.: Sequence-based probabilistic seismic hazard analysis. Bull. Seismol. Soc. Am. **104**(2), 1006–1012 (2014)

Ingham, J., Griffith, M.: The performance of unreinforced Masonry Buildings in the 2010/2011 Canterbury Earthquake Swarm. In: 15th World Conference on Earthquake Engineering, Lisboa, Portugal (2012)

Jeon, J.-S., DesRoches, R., Lowes, L.N., Brilakis, I.: Framework of aftershock fragility assessment–case studies: older California reinforced concrete building frames. Earthq. Eng. Struct. Dyn. **44**, 2617–2636 (2015). https://doi.org/10.1002/eqe.2599

Kam, W. Y., Pampanin, S., Elwood, K.: Seismic performance of reinforced concrete buildings in the 22 February Christchurch (Lyttelton) earthquake. Bull. N.Z. Soc. Earthq. Eng. **44**(4), 239–278 (2011)

Kunnath, S.K., Reinhorn, A.M., Lobo, R.F.: IDARC Version 3.0: a program for the inelastic damage analysis of reinforced structures. Technical report NCEER-92-0022. Buffalo, NY: State University of New York at Buffalo (1992)

Lagomarsino, S., Penna, A., Galasco, A., Cattari, S.: TREMURI program: an equivalent frame model for the nonlinear seismic analysis of masonry buildings. Eng. Struct. **56**, 1787–17993 (2013)

Lignos, D.G., Krawinkler, H.: Deterioration modelling of steel components in support of collapse prediction of steel moment frames under earthquake loading. J. Struct. Eng. **137**, 1291–1302 (2010)

Luzi, L., Pacor, F., Puglia, R.: Italian Accelerometric archive v 2.3. Istituto Nazionale di Geofisica e Vulcanologia, Dipartimento della Protezione Civile Nazionale (2017). https://doi.org/10.13127/ITACA.2.3

Luzi, L, Puglia, R, Russo, E., ORFEUS WG5: Engineering strong motion database, version 1.0. Istituto Nazionale di Geofisica e Vulcanologia, Observatories & Research Facilities for European Seismology (2016). https://doi.org/10.13127/ESM

Mazzoni, S., McKenna, F., Scott, M.H., et al.: Opensees Command Language Manual, p. 264. Pacific Earthquake Engineering Research Center, Berkeley, CA (2006)

Moon, L., Dizhur, D., Senaldi, I., Derakhshan, H., Griffith, M., Magenes, G., Ingham, J.: The Demise of the URM building stock in Christchurch during the 2010–2011 Canterbury Earthquake sequence. Earthq. Spectra **30**, 253–276 (2014). https://doi.org/10.1193/022113EQS044M

Mouyiannou, A., Rota, M., Penna, A., Magenes, G.: Implications of Cumulative Seismic Damage on the Seismic Performance Of Unreinforced Masonry Structures. Bulletin of the New Zealand Society for Earthquake Engineering, New Zealand (2014)

Papadopolous, A.N., Bazzurro, P.: Towards incorporation of seismicity clustering in probabilistic seismic loss estimation. In: Eleventh US National Conference on Earthquake Engineering, Los Angeles, California, US, 25–29 June 2018

Penna, A., Lagomarsino, S., Galasco, A.: A nonlinear macroelement model for the seismic analysis of masonry buildings. Earthq. Eng. Struct. Dynam. **43**(2), 159–179 (2014)

Penna, A., Morandi, P., Rota, M., Manzini, C.F., Da Porto, F., Magenes, G.: Performance of masonry buildings during the Emilia 2012 earthquake. Bull. Earthq. Eng. **12**(5), 2255–2273 (2014)

Polese, M., Marcolini, M., Zuccaro, G., Cacace, F.: "Mechanism based assessment of damage-dependent fragility curves for RC building classes. Bull. Earthq. Eng. **13**(5), 1323–1345 (2015)

Raghunandan, M., Liel, A.B., Luco, N.: Aftershock collapse vulnerability assessment of reinforced concrete frame structures. Earthquake Eng. Struct. Dynam. **44**(3), 419–439 (2015)

RINTC Workgroup: Results of the 2015–2017 Implicit seismic risk of code conforming structures in Italy (RINTC) project. ReLUIS report, Rete dei Laboratori Universitari di Ingegneria Sismica (ReLUIS), Naples, Italy (2018). Available at https://www.reluis.it/

Rossetto, T., Elnashai, A.: Derivation of vulnerability functions for European-type RC structures based on observational data. Eng. Struct. **25**, 1241–1263 (2003). https://doi.org/10.1016/S0141-0296(03)00060-9

Rota, M., Penna, A., Magenes, G.: A methodology for deriving analytical fragility curves from masonry buildings based on stochastic nonlinear analyses. Eng. Struct. **32**(5), 1312–1323 (2010)

Ryu, H, Luco, N., Uma, S.R., Liel, A.B.: Developing fragilities for mainshock-damaged structures through incremental dynamic analysis. In: Proceedings of the 9th Pacific Conference on Earthquake Engineering, Auckland, Paper No. 225, p. 8 (2011)

Sextos, A., Risi, R, Pagliaroli, A., Foti, S., Passeri, F., Ausilio, E., Cairo, R., Capatti, M., Chiabrando, F., Chiaradonna, A., Dashti, S., de Silva, F., Dezi, F., Durante, M., Giallini, S., Lanzo, G., Sica, S., Simonelli, A.L. and Zimmaro, P. (2018). Local site effects and incremental damage of buildings during the 2016 Central Italy Earthquake sequence. Earthq. Spectra **34**. https://doi.org/10.1193/100317EQS194M

Shcherbakov, R., Nguyen, M., Quigley, M. (2012) Statistical analysis of the 2010 M W 7.1 Darfield Earthquake aftershock sequence. NZ J. Geol. Geophys. **55**(3), 305–311. https://doi.org/10.1080/00288306.2012.676556

Shin, A.O.I., Kunugi, T., Adachi, S., Nakamura, H., Morikawa, N., Suzuki, W., Fujiwara, H.: K-NET and KiK-net Strong-motion Data. National Research Institute for Earth Science and Disaster Resilience, Japan (2017)

Taucer, F., Spacone, E., Filippou, F.: A fiber beam-column element for seismic response analysis of reinforced concrete structures (1991)

Tomazêvic, M.: Damage as a measure for earthquake-resistant design of masonry structures: Slovenian experience. Canad. J. Civil Eng. (2007)

Yeo, G.L, Cornell, C.A.: A probabilistic framework for quantification of aftershock ground-motion hazard in California: methodology and parametric study. Earthq. Eng. Struct. Dynam. **38**, 45–60 (2008)

Energy Dissipation Capacity of RC Columns Subjected to Dynamic Biaxial Seismic Loadings on a Shake Table

Amadeo Benavent-Climent, David Escolano-Margarit, and Leandro Morillas

Abstract Knowing the (ultimate) energy dissipation capacity of structural members is a key aspect of seismic design based on the energy-balance, since it is the main parameter for verification. The gap in knowledge about the energy dissipation capacity of structural members is considerable, particularly for reinforced concrete (RC) structural elements. In this study, the energy dissipation capacity of three RC columns subjected to dynamic seismic loadings on a shake table is investigated. The columns are tested within a structure that represents a scaled portion of a three-story prototype building. Two of the columns reached failure and the third one was severely damaged. It is found that the total ultimate energy dissipation capacity of the two columns that failed is very similar, although they followed different loading paths and dissipated very different amounts of energy in the X and Y directions. Finally, it is shown that the chord rotation (at least alone) is not a good indicator of damage.

Keywords Energy based design · Reinforced concrete · Shake table · Ultimate energy dissipation capacity

1 Introduction

The seismic design approach based on the energy-balance verifies safety by comparing the energy dissipation demand on the structural system and the individual members with their capacities. There is a large gap in our knowledge of the energy dissipation capacity of structural components, particularly in the case of reinforced

A. Benavent-Climent (✉) · D. Escolano-Margarit
Department of Mechanical Engineering, Universidad Politécnica de Madrid, 28006 Madrid, Spain
e-mail: amadeo.benavent@upm.es

D. Escolano-Margarit
e-mail: d.escolano@upm.es

L. Morillas
Department of Structural Mechanics and Hydraulic Engineering, School of Architecture, Universidad de Granada, Campo del Príncipe, 18071 Granada, Spain
e-mail: lmorillas@go.ugres

A. Benavent-Climent and F. Mollaioli (eds.), *Energy-Based Seismic Engineering*, Lecture Notes in Civil Engineering 155,
https://doi.org/10.1007/978-3-030-73932-4_3

concrete RC structures [1]. Characterizing the ultimate energy dissipation capacity is cumbersome and challenging, as it can be affected by the history of displacement reversals (loading path), which in turn is influenced by the characteristics of the ground motion and of the structure [2–4]. Besides its key role in the energy-based approach, knowing the ultimate energy dissipation capacity of structural members is important for further reasons. One is that structural damage is related not only to the maximum apparent deformation demand, but also to the cumulative plastic deformations—which, multiplied by the forces, give the hysteretic energy dissipation demand. The cumulative inelastic cyclic demand component can prove more significant than the displacement component in many cases [5]. It has moreover been shown by uniaxial cyclic testing on RC columns [6, 7] that the displacement component controls damage for low drifts (below about 1%), but for larger drifts the energy component becomes equally or more important than the displacement component. The need to combine apparent maximum deformation with dissipated energy to quantify damage is acknowledged in the well-known damage index proposed by Park and Ang [2] for uniaxial loading given by Eq. (1).

$$DI_{uniaxial} = \frac{\delta_M}{\delta_u} + \frac{\int dE}{\left(V_y \delta_u / \beta\right)} \tag{1}$$

Here, V_y is the strength at yielding, δ_M is the maximum deformation under the earthquake, δ_u is the ultimate deformation under monotonic loading (calculated considering all possible failure modes under repeated cyclic loadings), dE is the incremental dissipated hysteretic energy, and β an empirical parameter. Under uniaxial loading, Eq. (1) assumes that failure occurs when $DI_{uniaxial} \geq 1$. Making $DI_{uniaxial}=$ 1 in Eq. (1), and solving for $(\int dE)_{DI=1}$ gives the dissipated energy:

$$(\int dE)_{DI=1} = \frac{V_y}{\beta}(\delta_u - \delta_M) \tag{2}$$

It is worth recalling that in Eq. (1), V_y, δ_u and β are independent of the loading history. Equation (1) was not intended to be used to estimate the ultimate hysteretic energy dissipation capacity. Therefore,$(\int dE)_{D=1}$ given by Eq. (2) cannot be interpreted as the ultimate hysteretic energy dissipation capacity of the member, since in the case of monotonic loading up to $\delta_M = \delta_u$, Eq. (1) yields that the dissipated energy is zero, while it is in fact $V_y(\delta_u - \delta_y)$. Here, δ_y is the displacement at yielding.

Park and Ang's damage index has been extended to biaxial loadings by several authors [8]. Rodrigues et al. [7] assessed different formulae proposed in the literature with the results of quasi-static tests and concluded that the bidirectional index given by Eq. (3) was the most accurate.

$$DI_{biaxial} = DI_{uniaxial,X} + DI_{uniaxial,Y} - \lambda \cdot \min\left\{DI_{uniaxial,X}, DI_{uniaxial,Y}\right\} \tag{3}$$

where $DI_{uniaxial,X}$ and $DI_{uniaxial,Y}$ are calculated using Eq. (1) for each direction, X and Y, and λ is a constant empirical parameter. Similarly to the uniaxial index, $DI_{biaxial} \geq 1$ indicates failure.

This work focuses on the hysteretic energy dissipation capacity until failure (ultimate energy dissipation capacity) of RC columns subjected to biaxial cyclic loadings. This topic has been addressed in previous studies, although the available test results are very limited and, in most cases, correspond to columns subjected to quasi-static cyclic loads and not always up to failure. A comprehensive review of tests conducted until the end of the twentieth century can be found in Lehman et al. [9]. A summary of shake table tests involving yielding columns can be found in Hachem et al. [10], although very few of them are bidirectional. These experiments suggest that dynamic tests contain many features that cannot be detected in quasi-static experiments applying prescribed displacement histories [10]. Past studies agreed in that: (i) strength and stiffness degradation is more pronounced for columns under biaxial lateral loading than under uniaxial loads [11]; and (ii) damage is highly dependent on the loading path and history [12].

Acun and Sucuoglu [13] analyzed the experimental data obtained from 23 column specimens tested under constant-amplitude and variable-amplitude displacement cycles. Based on these analyses, they developed a model for predicting the cyclic energy dissipation capacity under constant-amplitude inelastic displacement cycles; it was used to propose an analytical procedure for estimating the energy dissipation under variable-amplitude displacement cycles.

Rodrigues et al. [6, 7] tested 24 RC columns under uniaxial and biaxial horizontal quasi-static loadings and found that for biaxial loading, specific damage occurs for drift demands that are 50–75% lower than those corresponding to uniaxial demand, while the plastic hinge length L_p remains approximately the same. They also found that the maximum strength in one specific direction was from 8 to 30% lower in the case of the specimens subjected to biaxial loading as opposed to the corresponding uniaxial tests. The ultimate ductility was also lower, and the strength degradation more pronounced, under biaxial loading than under uniaxial loading. These authors likewise found that the amount of dissipated energy—expressed in terms of equivalent global damping—depends on the loading paths.

In this study, an RC structure was subjected to bidirectional seismic loading on a shake table that reproduced two horizontal components of a historic (near-fault) ground motion. The RC structure was intentionally designed with a significant in-plan irregularity which, combined with the bidirectional action, produced important biaxial bending demands on the columns. Three RC columns that reached or approached failure are analyzed from the energy point of view. The columns are located in the upper part of a specimen that represents a scaled portion of the three-story prototype building. Results are not conclusive because the number of specimens tested is very limited, but they provide quantitative data on the energy dissipation capacity of RC columns under realistic seismic loadings and point out that attention must be paid to the cumulative hysteretic energy when evaluating damage —the maximum displacement alone is not a good indicator of the level of damage and the proximity to failure.

2 Experimental Study

2.1 Description of the Overall Test Specimen and Set-Up

Experimental results are obtained from shake table tests involving a 2/5 scale RC waffle-flat-plate structure (Fig. 1) conducted at the Laboratory of Structural Dynamics of the University of Granada (Spain). The RC specimen represents a portion of a prototype 3-story building equipped with hysteretic dampers. The prototype RC structure (Fig. 2) is composed of 3 × 3 bays; the floors are waffle slabs supported on a quincunx grid of columns. The RC structure is designed to sustain vertical gravity load only: a dead load of 3.13 kN/m^2 and live load of 2 kN/m^2. The characteristic design values of the material properties are $f_{c,k} = 25$ MPa for concrete C25/30, and $f_{y,k} = 500$ MPa for B500 steel rebars.

The tested specimen is a portion of the prototype building consisting of 1.5-bay and 1.5 story-height. In order to preserve the boundary conditions of the specimen in relation to the prototype structure, the specimen is extracted by assuming negligible or zero bending moments under horizontal loading at mid-span and mid-height sections of the slabs and columns, respectively. The effects of the rest of the structure on the specimen are introduced by pinned joints at these zero-bending-moment sections. Reactive steel blocks are located on the top columns and on the waffle-flat-plate to reproduce the mass of the upper part of the structure and to satisfy scaling similitude laws. Figure 3 offers a schematic representation of the overall set-up and instrumentation on the top floor, where the three columns investigated in this study are located.

The specimen is equipped with six hysteretic dampers (three dampers per story) designed to withstand the seismic loading and to control lateral displacements under the design and the maximum credible earthquakes. The experimental instrumentation included uniaxial accelerometers, displacement transducers (LVDTs and laser), load cells, and more than 400 strain gauges. In order to record the relative displacement or lateral drift of the structure, LVDTs and laser displacement transducers were attached to auxiliary rigid frames. Strain gauges were installed in rebars at the end sections of columns and on the relevant sections of the waffle slab. The top sections of the upper columns of the specimen were instrumented with load cells that measured the portion of the inertial forces sustained by each column in each horizontal direction, as detailed next.

2.2 Detailed Description of the Columns Investigated

This study focuses on the seismic performance of the three columns located on the top story (columns C1, C2 and C3 in Fig. 3b). All columns have the same size (120 × 120 mm^2) and reinforcement consisting of four ϕ 6 mm steel rebars as longitudinal reinforcement, and ϕ 4 mm closed stirrups with 30 mm spacing as

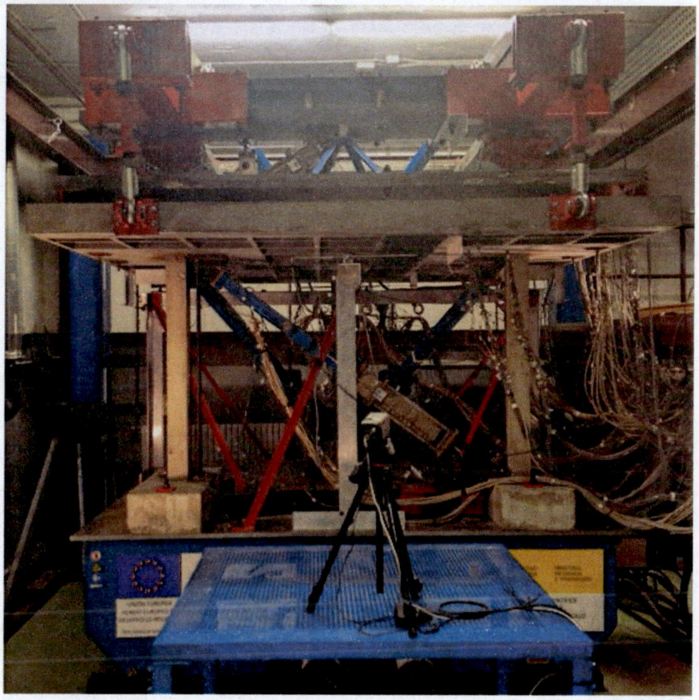

Fig. 1 Test overview

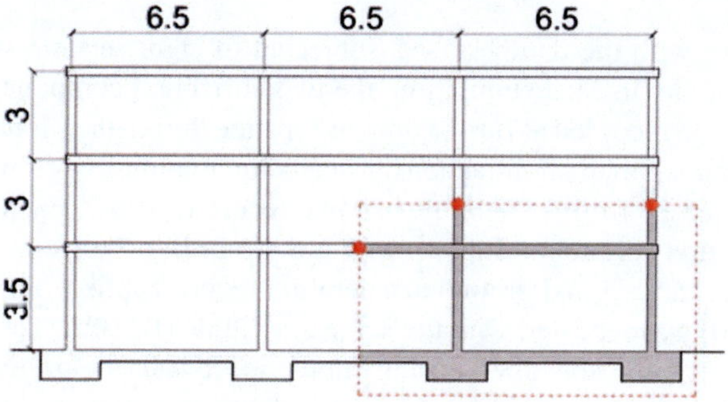

Fig. 2 Prototype building

transverse reinforcement. Figure 4a shows the geometry and detailing of one of the columns. In order to reproduce ordinary construction techniques, the flexural reinforcement overlapped with the reinforcement coming from the story below. The yield stress of the steel obtained from coupon tension tests was 543 MPa for the longitudinal bars and 656 MPa for the stirrups. The concrete strength obtained from compression tests was 44 MPa. The end top section reinforcement was welded to a

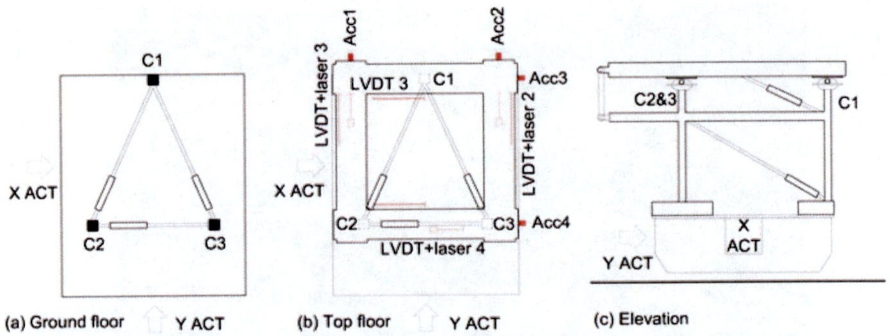

Fig. 3 Overall experimental setup: **a** ground floor in plan; **b** top floor in plan; **c** elevation

10 mm plate connected to the experimental set-up (Fig. 4b). As seen in Fig. 4b, the steel blocks settle on the top plate of the column through a polished stainless-steel sphere. Once the mass was stabilized, four load cells were post-tensioned against the sphere following the principal directions of loading by means of auxiliary L-shaped steel plates. After post-tensioning, the auxiliary L plates were clamped to the top plate at the end of the column. Figure 5 provides a detailed photograph of the final set-up.

2.3 *Seismic Simulations*

The RC structure with the dampers was subjected to eight seismic simulations of increasing amplitude. In each simulation, the two horizontal components of ground motion acceleration recorded at Bar-Skupstina Opstine during the Montenegro earthquake (1979) were applied simultaneously, scaled in amplitude by a factor that was successively increased from 15% of the original record to 190%. Each simulation is identified in the first column of Table 1 with a B (from Bar-Skupstina) followed by the percentage of the original ground acceleration record applied. The test at 130% amplitude (B130) was repeated with the same amplitude (B130b). The total amount of energy E input to the specimen during simulation B100 (the original record not scaled in amplitude), expressed in terms of equivalent velocity $V_E = \sqrt{2E/M}$ where M is the total mass, was $V_E = 156$ cm/s. The seismic tests were performed on a bidirectional MTS 3×3 m^2 shake table. During these simulations, the ultimate capacity of the dampers was reached, and the testing continued in order to investigate the capacity of the RC structure after the failure of the dampers. The Bar-Skupstina Opstines records represents an impulsive near-fault ground motion.

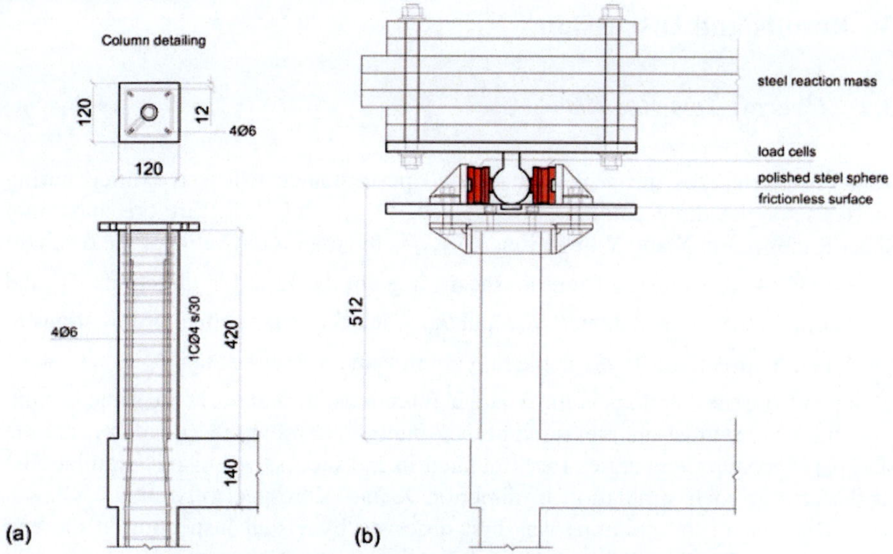

Fig. 4 Detailed column (**a**) and instrumentation (**b**); dimensions in mm

Fig. 5 Experimental set-up overview

3 Results and Discussion

3.1 Overall Test Results

Table 1 summarizes the peak demand and performance values measured during the tests for the three columns investigated. $V_{x,max}$ and $V_{y,max}$ are the maximum shear forces in the X and Y directions. $V_{r,max}$ is the maximum value of the resultant $V_r = \sqrt{V_x^2 + V_y^2}$ obtained from the shear forces in the X and Y directions, V_x and V_y, acting at the same instant. $\theta_{x,max}$ and $\theta_{y,max}$ are the maximum chord rotations in the X and Y directions. $\theta_{r,max}$ is the maximum value of the resultant $\theta_r = \sqrt{\theta_x^2 + \theta_y^2}$ obtained from the rotations in the X and Y directions, θ_{x} and θ_{x}, at the same instant. ε_s is the maximum strain measured at the longitudinal rebars. W_{px} and W_{py} are the dissipated hysteretic energies (accumulated in the successive seismic simulations) at the end of each simulation in direction X and Y, respectively. $W_p = W_{px} + W_{py}$. Damage in the columns was first observed by visual inspection at the end of simulation B100 and increased in following tests. During simulations B15 to B50, the chord rotations remained below the elastic limits, the maximum strains in longitudinal rebars (3004 με) hardly exceeding the yield strain (2586 με), and the plastic energy dissipated by the columns W_p was negligible. Simulations B100 to B190 exerted severe damage on the columns, as indicated by the maximum strains in rebars (well above 2586 με), chord rotations (>1%) and—especially—the amount of energy dissipated, which grew exponentially in successive tests. As will be detailed in sections below, columns C1 and C2 attained their ultimate energy dissipation capacity and failed, while column C3 was severely damaged but did not fail. The three columns exhibited flexural yielding; shear failure was not observed in any case.

3.2 Force and Displacement Demand

Figure 6 shows the chord rotations in X and Y directions throughout simulations B50 to B190, for the three columns. The results for test B15 and B25 are not included because the deformations were very small, and the columns remained elastic. Also shown in Fig. 6 are the limiting values established by FEMA-356 2000 [14] to characterize the boundaries of the seismic performance levels (SPL). Three limits are plotted. The first one is the rotation limit of the SPL of Immediate Occupancy 1% (green line). The second is the limit of the SPL of Life Safety 2% (orange line). The third is the limit of the SPL of Near Collapse 4% (red line). It is worth mentioning that these limits differ from those established earlier, in the document Vision 2000 [15], for the SPLs of Fully Operational (0.2%), Life Safety (1.5%) and Near Collapse (2.5%). Two possible shapes for the interaction curves in the X and Y directions are plotted in Fig. 6 for each SPL. The first shape is a square, and the second one a circumference. The circumference is considered more realistic than the square; some

Table 1 Peak demand and response values

Test	Col	V_x (kN)	V_y (kN)	V_r (kN)	θ_x (%)	θ_y (%)	θ_r (%)	W_{px} (kNm)	W_{py} (kNm)	W_{pt} (kNm)	εs με
15	C1	2.29	0.98	2.42	0.43	0.22	0.45	0.00	0.00	0.00	565
	C2	0.98	2.03	2.25	0.10	0.17	0.17	0.00	0.00	0.00	1464
	C3	0.40	0.26	0.41	0.10	0.31	0.32	0.00	0.00	0.00	1943
	C1	2.45	1.28	2.67	0.59	0.27	0.50	0.00	0.00	0.01	830
25	C2	1.00	2.22	2.40	0.20	0.24	0.26	0.00	0.00	0.00	1912
	C3	0.62	0.43	0.62	0.20	0.39	0.39	0.00	0.00	0.00	2159
	C1	3.09	1.61	3.44	0.78	0.35	0.81	0.02	0.01	0.03	2202
50	C2	1.03	2.46	2.56	0.48	0.44	0.51	0.00	0.01	0.01	2727
	C3	1.20	0.94	1.22	0.48	0.56	0.57	0.01	0.01	0.01	3004
	C1	6.27	3.12	6.28	1.32	0.68	1.35	0.18	0.00	0.18	6891
100	C2	2.19	4.66	4.69	1.02	0.84	1.03	0.07	0.17	0.24	3668*
	C3	2.64	1.77	2.72	1.02	1.16	1.16	0.13	0.10	0.23	8768*
	C1	7.03	3.35	7.10	1.87	0.81	1.90	0.55	0.06	0.61	14,607
130	C2	3.10	5.90	6.18	1.30	1.09	1.32	0.21	0.44	0.65	6525*
	C3	3.74	2.28	3.76	1.30	1.27	1.35	0.30	0.27	0.57	12,529*
	C1	6.86	3.21	6.94	2.05	0.82	2.09	0.91	0.13	1.04	14,476
130b	C2	3.27	6.32	6.57	1.40	1.13	1.42	0.36	0.70	1.06	8705*
	C3	4.04	2.14	4.05	1.40	1.29	1.44	0.45	0.43	0.89	12,958*
	C1	7.55	3.81	7.55	2.45	1.03	2.50	1.53	0.25	1.79	5339*
160	C2	3.67	7.24	7.52	1.68	1.38	1.73	0.65	1.14	1.79	4881*
	C3	4.83	2.17	4.93	1.68	1.52	1.71	0.64	0.67	1.31	12,856*

(continued)

Table 1 (continued)

Test	Col	V_x (kN)	V_y (kN)	V_r (kN)	θ_x (%)	θ_y (%)	θ_r (%)	W_{px} (kNm)	W_{py} (kNm)	W_{pt} (kNm)	εs $\mu\varepsilon$
	C1	8.29	3.86	8.31	3.48	1.39	3.48	2.35	0.49	2.84	19,047*
190	C2	4.56	7.96	8.05	1.97	1.77	2.05	1.17	1.74	2.91	4832*
	C3	5.26	3.30	5.49	1.97	1.82	1.97	0.90	0.99	1.90	12,605*

* max strain of surviving gauges

experimental studies [13] suggest using parabolic curve. The following observations can be drawn from Fig. 5:

- Test B50: All columns performed in the SPL Immediate Occupancy domain in both directions. The columns remained elastic, since the chord rotations were smaller than the yielding rotations estimated with the equations of EN-1998–3 [16] ($\theta_y = 1.06\%$).
- Test B100: Column C1 performed within the Immediate Occupancy domain, with local excursions that exceeded by about 20% the limit of this SPL in the X direction. Columns C2 and C3 performed within the Immediate Occupancy domain.
- Tests B130 and B130b: Column C1 performed within the Life Safety domain reaching the limit of this SPL. C2 and C3 performed mainly in the Immediate Occupancy domain, with local excursions that exceeded by about 15% this SPL while remaining far from the boundary of the SPL of Life Safety.
- Tests B160: Column C1 performed mainly in the Life Safety domain but endured local displacement excursions exceeding the limit of this SPL by about 15%. Columns C2 and C3 performed within the Life Safety domain yet approaching the boundary of this SPL.
- Test B190: Column C1 clearly exceeded the limit of the Life Safety SPL and approached the limit of Near Collapse. Columns C2 and C3 remained within the limits of the Life Safety SPL, yet very close to the boundary.

Meanwhile, Fig. 7 shows the evolution of the bending moments in the X and Y directions, M_x and M_y, throughout simulations B50 to B190. Simulations B15 and B25 are not included because the corresponding moments were very small. Figure 7 also shows the M_x–M_y interaction envelope (dashed line) obtained using a fiber model with commercial software SAP-2000. The axial force ratio considered on the columns was $n_o = 0{,}035$. It must be noted that the intersection of this interaction curve with the X and Y axes gives the nominal bending moment capacity under quasi-static uniaxial bending. This interaction envelope ignores the fact mentioned in the introduction [7]: that under bidirectional loading, the maximum strength in one specific direction is lower (from 8 to 30%) than that obtained from uniaxial tests. Therefore, to derive a more realistic interaction envelope, the dash line of Fig. 7 should shrink between 8 and 30%. In either case—using the nominal interaction envelope or a more realistic one—it is clear from Fig. 7 that columns C1 and C2 reached their bending capacities. In contrast, column C3 did not reach the nominal interaction envelope (dash line) but did reach the counterpart interaction envelope obtained by shrinking the nominal one by 30%. It should also be noted that, as expected, column C1 concentrated the displacement demand in the X direction due to the torsional effects on the structure.

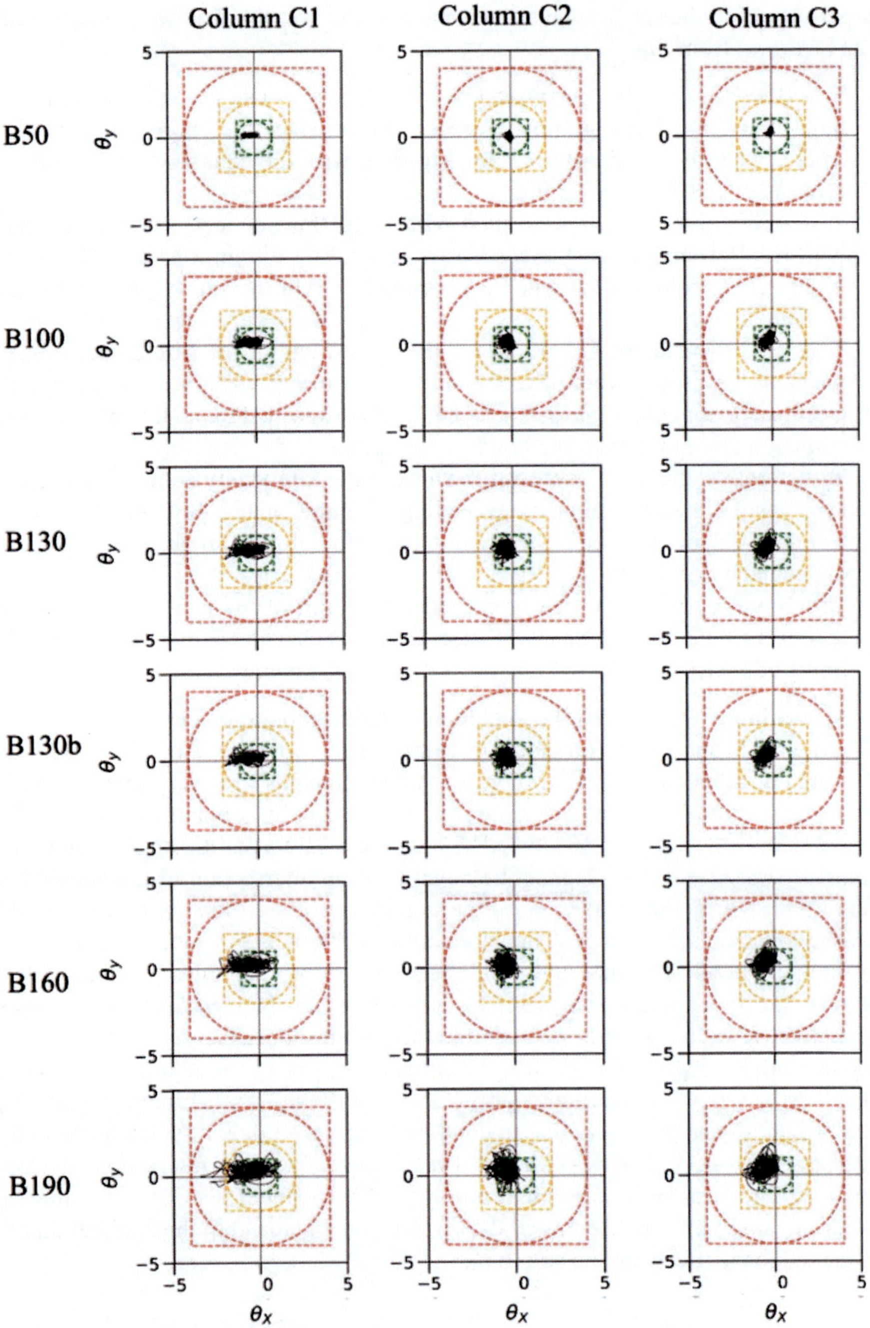

Fig. 6 Chord rotations in columns (in %)

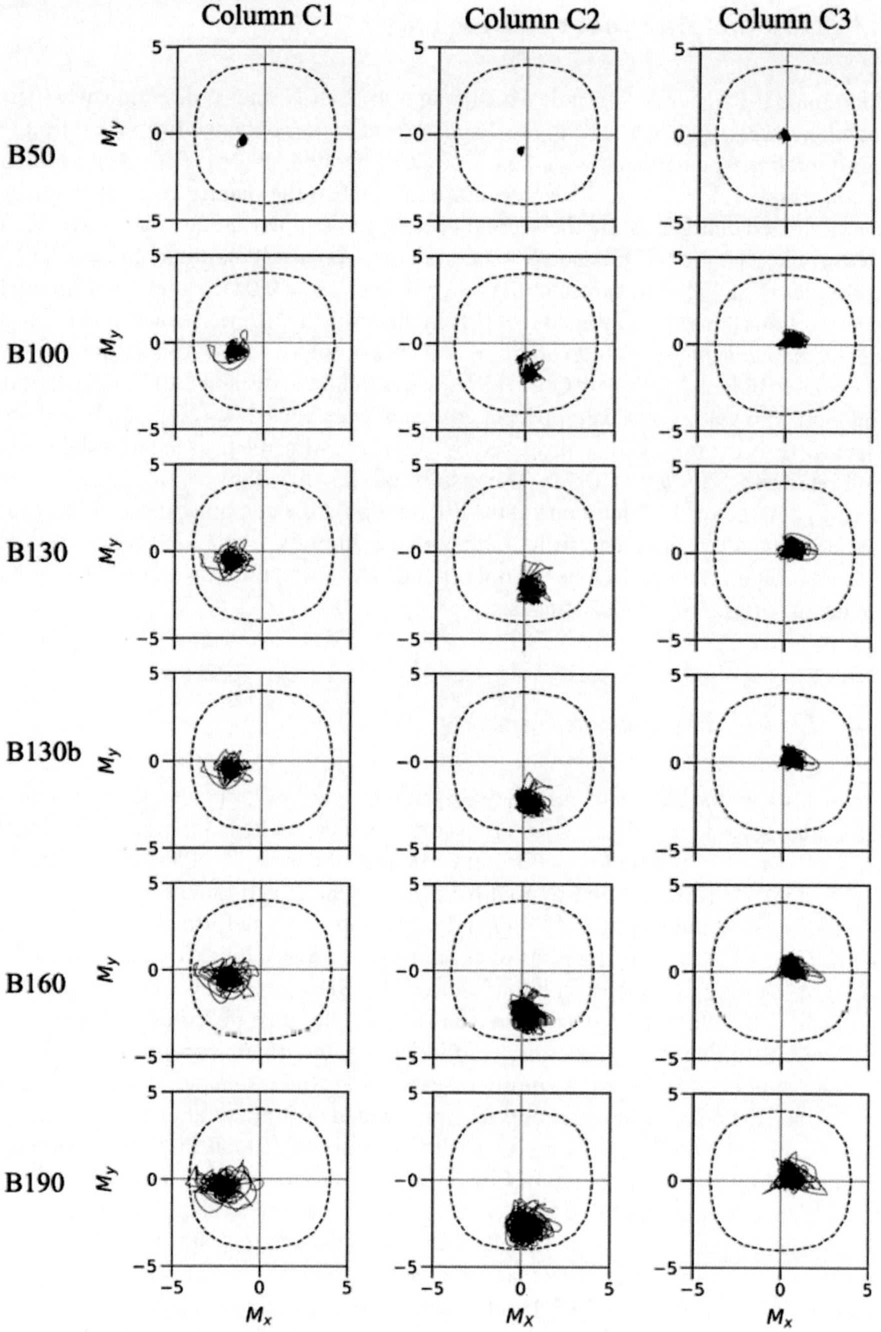

Fig. 7 Bending moments in columns (in kNm)

3.3 Park and Ang Index of Damage

The uniaxial Park and Ang index of damage in each X and Y direction was first calculated with Eq. (1). The ultimate displacement δ_u was obtained from the ultimate chord rotation θ_u calculated with the formulae of EN1998–3 [16]. The value adopted for the empirical parameter β is very important when the damage is controlled by the dissipated energy, i.e. by the second term of Eq. (1). Different values have been used in the past for RC elements. A typical value $\beta = 0.05$ is often adopted [17]. Rodrigues et al. [7] used values in the range $0.038 \leq \beta = 0.047$. Other experimental investigations [18–22] employed $\beta = 0.1$. In this study, $DI_{uniaxial}$ is computed for two representative values $\beta = 0.05$ and $\beta = 0.10$. Next, the Park and Ang damage index extended to biaxial loading by Qiu et al. [8] was calculated by means of Eq. (3). In this equation, two values of the empirical parameter λ were used: $\lambda = 0.5$ and $\lambda = 0.85$. The former is the initial value proposed by Qiu et al. [8] while the second is the value calibrated by Rodrigues et al. [7]. The results are shown in Table 2. According to the $ID_{biaxial}$ obtained with at least one of the different possible combinations of values for β and λ, it can be concluded from Table 2 that columns C1 and C2 reached failure, whereas column C3 came close but did not fail. This interpretation is consistent with the damage observed at end of tests.

3.4 Energy Dissipation Capacity

Figure 8 shows the history of cumulative hysteretic energy dissipated by each column during the seismic simulations. The two columns that reached collapse (C1 and C2) are seen to have dissipated very different amounts of energy along the X (red line in Fig. 8) and Y (yellow line in Fig. 8) directions, but the total dissipated energy (black line in Fig. 8) is quite similar. As seen in Fig. 6, column C1 deformed mainly around the X axis, while the deformations of column C2 were on a direction that formed an angle of approximately 45° with the X direction. Since column C3 approached but did not reach collapse, the dissipated energy is smaller than for columns C1 or C2.

Figure 8 furthermore shows, with solid circles, the points corresponding to the first attainment of the rotation limit corresponding to the Operational SPL (green circle), and the points when the columns first attained the limit rotation corresponding to the Life Safety SPL. It can be seen that the amount of total dissipated energies corresponding to the limit points of Life Safety SPL (yellow circles) are very different for column C1 and for column C2. As the amount of dissipated energy is an indicator of damage, and its closeness to the ultimate energy dissipation capacity indicates the proximity to failure, this means that the rotation angle (at least alone) is not a good indicator of the level of damage. In other words, column C1 attained the limit rotation corresponding to the Life Safety SPL when only about one third of its ultimate energy dissipation capacity was consumed, while column C2 attained the same limit when more than two thirds of its ultimate energy dissipation capacity were exhausted.

Table 2 Park and Ang index of damage

Test	Col	ID_{uniX} $\beta = 0.05$	ID_{uniX} $\beta = 0.1$	ID_{uniY} $\beta = 0.05$	ID_{uniY} $\beta = 0.1$	ID_{biax} $\beta = 0.05$ $\lambda = 0.5$	ID_{biax} $\beta = 0.05$ $\lambda = 0.85$	ID_{biax} $\beta = 0.1$ $\lambda = 0.5$	ID_{biax} $\beta = 0.1$ $\lambda = 0.85$
15	C1	0.05	0.05	0.03	0.03	0.07	0.06	0.07	0.06
	C2	0.01	0.01	0.02	0.02	0.03	0.02	0.03	0.02
	C3	0.01	0.01	0.04	0.04	0.05	0.04	0.05	0.04
	C1	0.07	0.08	0.03	0.03	0.09	0.08	0.09	0.08
25	C2	0.03	0.03	0.03	0.03	0.04	0.03	0.04	0.03
	C3	0.03	0.03	0.05	0.05	0.06	0.05	0.06	0.05
	C1	0.10	0.10	0.05	0.05	0.12	0.11	0.13	0.11
50	C2	0.06	0.06	0.06	0.06	0.09	0.07	0.09	0.07
	C3	0.06	0.06	0.07	0.07	0.10	0.08	0.10	0.08
	C1	0.19	0.22	0.09	0.09	0.23	0.20	0.26	0.23
100	C2	0.14	0.15	0.13	0.16	0.20	0.16	0.23	0.18
	C3	0.15	0.17	0.16	0.17	0.23	0.18	0.26	0.20
	C1	0.32	0.40	0.11	0.12	0.37	0.33	0.46	0.42
130	C2	0.19	0.23	0.20	0.27	0.30	0.23	0.38	0.31
	C3	0.21	0.25	0.20	0.24	0.31	0.24	0.37	0.29
	C1	0.40	0.53	0.12	0.14	0.46	0.41	0.61	0.56
130b	C2	0.23	0.28	0.25	0.36	0.36	0.28	0.50	0.40
	C3	0.24	0.31	0.23	0.29	0.36	0.28	0.46	0.36
	C1	0.54	0.77	0.17	0.21	0.62	0.56	0.88	0.80
160	C2	0.31	0.41	0.35	0.52	0.50	0.39	0.72	0.58
	C3	0.31	0.41	0.29	0.39	0.45	0.35	0.60	0.47
	C1	0.79	1.15	0.25	0.32	0.92	0.83	1.31	1.20
190	C2	0.42	0.60	0.49	0.75	0.70	0.55	1.05	0.84
	C3	0.38	0.52	0.38	0.53	0.57	0.44	0.79	0.61

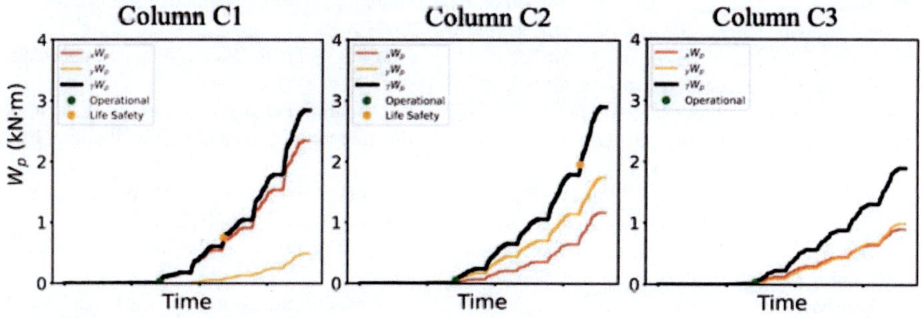

Fig. 8 History of cumulative hysteretic energy dissipated

4 Conclusions

This study investigated experimentally the energy dissipation capacity of three ductile RC columns (flexural yielding) subjected to biaxial seismic loadings on a shake table. Two of them attained their ultimate energy dissipation capacity and failed. The third one suffered severe damage but did not reach failure. The main findings can be summarized as follows.

1. The two columns that reached failure followed very different loading paths and dissipated very different amounts of energy along the X and Y directions, but the total amount of dissipated energy was almost the same. This means that, at least in these tests, the ultimate energy dissipation capacity of the columns was not affected by the loading path.
2. The chord rotation (at least alone) is a poor indicator of the level of damage on the columns; it is not correlated with the amount of dissipated energy to ultimate energy dissipation capacity.

Acknowledgements This research was funded by Spain´s Ministry of Economy and Competitivity, research project reference MEC BIA2017 88814 R, and received funds from the European Union (Fonds Européen de Dévelopment Régional).

References

1. Fardis, M.N.: From Force- to displacement-based seismic design of concrete structures and beyond. In: Pitilakis, K. (eds.) Recent Advances in Earthquake Engineering in Europe. ECEE 2018. Geotechnical, Geological and Earthquake Engineering, p. 46. Springer, Cham (2018)
2. Park, Y., Ang, A.H., Wen, Y.K.: Seismic damage analysis of reinforced concrete buildings ASCE. J. Struct. Eng. **111**(4), 740–757 (1985)
3. Erberik, A., Sucuoglu, H.: Seismic energy dissipation in deteriorating systems through low-cycle fatigue. Eartqhq. Eng. Struct. Dyn. **33**, 44–67 (2004)
4. Benavent-Climent, A.: An energy-based damage model for seismic response of steel structures. Earthq. Eng. Struct. Dynam. **36**(8), 1049–1064 (2007)
5. Kunnath, S.K., Chai, Y.H.: Cumulative damage-based inelastic cyclic demand spectrum. Earthq. Eng. Struct. Dynam. **33**, 499–520 (2004)
6. Rodrigues, H., Varum, H., Arede, A., Costa, A.: A comparative analysis of energy dissipation and equivalent viscous damping of RC columns subjected to uniaxial and biaxial loading. Eng. Struct **35**, 149–164 (2012)
7. Rodrigues, H., Arede, A., Varum, H., Costa, A.: Damage evolution in reinforced concrete columns subjected to biaxial loading. Bull. Earthq. Eng. **11**, 1157–1540 (2013)
8. Qui, F., Li, W., Pan, P., Qian, J.: Experimental tests on RC columns under biaxial quasi-static loading. Eng. Struct. **24**, 419–428 (2002)
9. Lehman, D.E., Mohele, J.P.: Seismic performance of well-confined bridge columns. In: PEER-1998/01, Berkeley, Univeersity of California, Berkeley (1998)
10. Hachem M.M., Mahin S.A., Mohele J.P.: Performance of circular reinforced concrete bridge columns under bidirectional earthquake loading. In: PEER 2003/06-UBC/ENG-9374, University of California, Berkeley (2003)

11. Kitajima, K.: Response characteristics of reinforced concrete columns under bi-directional earthquake motions. In: Proceedings of 10th World conference on Eaerthquake Engineering, Balkema, Rotterdam, pp. 3019–3024 (1992)
12. Bousias, S.N., Verzelleti, G., Fardis, M.N., Magonette, G.: RC columns in cyclic biaxial bending and axial load. In: 10th World Conference on Earthquake Engineering, pp. 3041–3046 (1992).
13. Acun, B., Sucuoglu, H.: Energy dissipation capacity of reinforced concrete columns under cyclic displacements. ACI Struct. J. **109**(4), 531–540 (2012)
14. FEMA-356: Prestandard and commentary for the seismic rehabilitation of buildings. Federal Emergency Management Agency, Washington, D.C., USA (2000)
15. SEAOC: Performance based seismic engineering of buildings. VISION 2000 Committee, Structural Engineers Association of California, Sacramento, CA, (1995)
16. European Committee for Standardization (CEN): Eurocode 8: Design of Structures for Earthquake Resistance—Part 3: Assessment and Retrofitting of Buildings, European Standard EN 1998–3:2005. Brussels: European Committee for Standardization (2005)
17. Bento, R.: Assessment of the seismic behavior of reinforced concrete framed structures: an approach based on damage indices. IST Lisbon (1996)
18. Lee, C.S., Honh, H.P.: Statics of inelastic responses of hysteretic systems under bidirectional seismic excitations. Eng. Struct. **32**, 2074–2086 (2010)
19. Benavent-Climent, A., Donaire-Avila, J., Oliver-Sáiz, E.: Seismic performance and damage evaluation of a waffle-flat plate structure with hysteretic dampers through shake-table tests. Earthq. Eng. Struct. Dyn. **47**, 1250–1269 (2018)
20. Benavent-Climent, A., Donaire-Avila, J., Oliver-Sáiz, E.: Shaking table tests of a reinforced concrete waffle–flat plate structure designed following modern codes: seismic performance and damage evaluation. Earthq. Eng. Struct. Dyn. **45**, 315–336 (2016)
21. Benavent-Climent, A., Morillas, L., Escolano-Margarit, D.: Seismic performance and damage evaluation of a reinforced concrete frame with hysteretic dampers through shake-table tests. Earthquake Engng Struct. Dyn. **43**, 2399–2417 (2014)
22. Benavent-Climent, A., Escolano-Margarit, D., Morillas, L.: Shake-table tests of a reinforced concrete frame designed following modern codes: seismic performance and damage evaluation. Earthq. Eng. Struct. Dyn. **43**, 791–810 (2014)

Spatial Distribution of Hysteretic Energy in Reinforced Concrete Moment Resisting Frames

M. Altug Erberik and Mahyar Azizi

Abstract Nowadays, structural engineers perceive that conventional force-based seismic design method is not still the unique way of designing structures subjected to ground motions. The reason is that it does not consider inelastic displacement, plastic structural behavior and duration of seismic motion. At the present time, there are new and popular alternatives like displacement-based method, in which the aforementioned issues are mostly handled. Energy-based approach is another convenient tool to examine the seismic response of structures under seismic action and probably the best way to include duration of ground motion within the analysis. In this approach, the energy input to the structure should be dissipated through inelastic action (hysteretic energy) and damping. Hence it is an important challenge to obtain the distribution of hysteretic energy within the building in order to develop energy-based design and analysis tools. Such studies have been conducted for steel frames previously, but not extensively for RC frame structures. Accordingly, this study is focused on the story-wise and component-wise distribution of hysteretic energy in RC moment resisting frames. For this purpose, RC frames with different number of stories and bays are designed according to the 2018 Turkish Seismic Code. Then the designed frames are modeled by using lumped plasticity approach. The developed models are subjected to a set of strong ground motion records and the distributions of hysteretic energy for each frame and analysis are obtained. The results indicate that it is possible to set up some rules for the hysteretic energy distribution in RC frames that can be used in energy-based design and analysis procedures.

Keywords Energy-based design · Hysteretic energy · RC frame · Story-wise distribution · Component-wise distribution

M. A. Erberik (✉)
Middle East Technical University, Ankara 06800, Turkey
e-mail: altug@metu.edu.tr

M. Azizi
Yüksel Proje, Ankara 06610, Turkey

A. Benavent-Climent and F. Mollaioli (eds.), *Energy-Based Seismic Engineering*, Lecture Notes in Civil Engineering 155,
https://doi.org/10.1007/978-3-030-73932-4_4

1 Introduction

In the conventional force-based design approach, the basic aim is to provide enough strength capacity in component and system levels by employing design base shear force. The design calculations are based on linear elastic analysis, in which inelastic action is accounted for by using force reduction factors. However, numerous past studies have shown that force-based approach has many pitfalls and it is not a good candidate to make the comparison between inelastic seismic demand and capacity in a realistic manner. In order to overcome this issue, there are two alternatives: displacement-based approach and energy-based approach. Displacement based approach has been very popular for the last two decades and it has been implemented to most of the current seismic codes as an alternative design tool since target demand of structure and performance of seismic action are quantifiable in this approach. On the other hand, energy-based approach is rather old when compared to displacement-based counterpart, as it has been first proposed by Housner in 1956 [1] who simply stated that input energy of an earthquake should be less than energy dissipation capacity of the structure. After this date, numerous studies have been conducted on energy-based design and analysis procedures, but practical and widely accepted energy-based methodologies have not been developed for real-life structures yet. The main reason is the complexity of energy-based parameters since energy is simply the product of force and displacement, two basic parameters that are used in alternative design approaches. Hence it is not easy to propose practical calculation procedures that can be easily conceived by practicing engineers. On the other hand, it should also be noted that the use of energy parameters means total control on force and displacement terms. Hence one can monitor the full inelastic seismic response and compare it with the corresponding seismic capacity throughout the complete duration of ground motion. Moreover, energy-based design and analysis procedures have an indisputable superiority when one has to deal with base-isolated structures or structures with external damping mechanisms.

This study aims to contribute to the development of practical energy-based design and assessment methodologies to be implemented in the future generations of seismic codes and standards. For this purpose, code-compliant reinforced concrete (RC) moment resisting frames (MRFs) are employed in this study to investigate the spatial distribution of hysteretic energy through members and stories when the planar frame models are subjected to a series of ground motion records. The design of the frames is carried out by using the latest version of Turkish Building Seismic Code, which can be abbreviated as TBSC-18 [2]. The reason for selecting RC frames in this study is two folds: First, most of the related past studies have been carried out on steel frame structures. There are few studies that are focused on the energy-based response of RC frame structures [3–8]. Second, in Turkey, RC frame buildings constitute the majority of the building stock, therefore development of energy-based design and assessment procedures for this construction type seems to be more crucial.

2 Energy-Based Seismic Response

Energy-based approach is focused on the premise that energy demand during seismic action can be quantified and energy supply of the structure can be provided. In order to obtain the energy terms, equation of motion for an inelastic damped Single-Degree-of-Freedom (SDoF) system should be integrated with respect to the relative displacement

$$\int m\ddot{u}du + \int c\dot{u}du + \int f_s du = -\int m\ddot{u}_g du \tag{1}$$

where m, c and f_s are the mass, damping coefficient and restoring force of the SDoF oscillator, respectively, and $\ddot{u}_g$ is the ground motion acceleration. For an inelastic SDoF system, restoring force is a function of relative displacement u, i.e. $f_s = f_s(u)$. Hence it should be represented by a hysteresis model that simulates the cyclic force-displacement response of the inelastic SDoF system.

The above equation is defined as "the relative energy balance equation" since the ground excitation term is on the right-hand side of the formulation. This physically represents a SDoF system with a fixed base subjected to an equivalent lateral earthquake force $m\ddot{u}_g$. On the other hand, if this term is taken to the left-hand side of the equation, then Eq. (1) can be written as

$$\int m\ddot{u}_t du + \int c\dot{u}du + \int f_s du = 0 \tag{2}$$

where $\ddot{u}_t$ denotes the total acceleration. This version is known as "the absolute energy balance equation" and it physically represents a SDoF system with a moving base subjected to the ground excitation $m\ddot{u}_g$ from the base. This model considers both the relative deformation of the frame and the rigid body translation due to ground displacement u_g.

The differences between these two different interpretations of energy response has been discussed in literature in a detailed manner. Starting with the early works of Uang and Bertero [9], the results of numerical analysis reveal that there is not a significant difference in seismic energy response when these two approaches are compared within the period range 0.3–5.0 s. This is practically the range of periods in which most of the structural systems reside. Bruneau and Wang [10] stated that the relative energy equation is physically more meaningful since all the internal forces are computed by using the relative displacements and velocities. In addition, Kalkan and Kunnath [11] claimed that the response statistics obtained by using the relative energy approach is more rational when near-fault ground motion records are used in the analyses. Since the RC frame models considered in this study have periods within the range 0.3–5.0 s, both energy approaches can be used since they are deemed to yield similar results in terms of energy response. However, relative energy approach seems to be more practical since it considers relative kinematic terms that can be directly extracted from the output data of the used finite element structural analysis program. Hence relative energy approach, i.e. Equation (1) is selected in this study. This equation can also be written by using the abbreviations of the relative energy terms as

$$E_K + E_D + E_A = E_I \tag{3}$$

where E_K, E_D, E_A and E_I stand for the relative terms of kinetic energy, damping energy, absorbed energy on the left-hand side and input energy on the right-hand side of the equation. Absorbed energy is composed of two sub-terms: recoverable elastic strain energy (E_S) and irrecoverable hysteretic energy (E_H). When a structural system behaves beyond the linear elastic range under ground motion excitation, energy input to the structure should be dissipated by both E_D and E_H (or in other words, inelastic displacement). At the end of ground motion duration, since E_K and E_S die out, the summation of E_D and E_H gives E_I. This shows the importance of E_H distribution within the structure during seismic action since it is directly related with the deformations and damage levels of the members. In the case of Multi Degree of Freedom (MDoF) systems, the same energy balance equation can be used by replacing single-valued kinematic terms with vectors and other terms (m, c and f_s) with matrices.

There are many studies that have been focused on the estimation of E_I in the literature, so it is simple to estimate the energy demand of a given structure. However, there exist fewer research on the determination of E_H since it is a more complex energy parameter and it is not an easy task to determine the energy dissipation characteristics of structural members. On the other hand, it is crucial to determine the spatial distribution of E_H within the structure under different ground motion excitations in order to estimate energy demand and then compare it with energy supply. For this purpose, in this study, multi-story RC MRFs are modeled and analyzed by using nonlinear time-history analysis (NLTHA), as it is explained in the following sections.

3 Code-Compliant RC MRF Buildings

In this study, 3, 5, 7 and 9 story code-compliant RC MRF building models with different numbers of bays (for 5-story models only) are selected to examine the E_H response by using NLTHA. The following sub-sections summarize the details in design and numerical modeling of the frame building models.

3.1 Seismic Design of the RC Frame Buildings

The building models are designed and detailed to satisfy the requirements according to Turkish Standards TS-500 [12] and TBSC-18. The reason for selecting code-compliant buildings is to examine the distribution of inelastic behavior in well-designed RC frame buildings so that energy-based design principles can be proposed by taking the force-based design rules as reference for comparison.

All RC frame buildings are constructed as 3-D numerical models with elastic section properties as dictated by the seismic code. The plan and elevation views of one of the selected RC frame buildings are shown in Fig. 1. In total, six frame models are considered with different number of stories and bays: RCFS3B3 (3 stories and 3 bays), RCFS5B2 (5 stories and 2 bays), RCFS5B3 (5 stories and 3 bays), RCFS5B4 (5 stories and 4 bays), RCFS7B3 (7 stories and 3 bays) and RCFS9B3 (9 stories and 3 bays). Span length in all of the models is considered as 6 m whereas story height is taken as 3 m. Applied dead loads and live loads are taken from the standard TS-498 [13]. In accordance with the design regulations, concrete and steel grades are selected as C25 ($f_{ck} = 25$ MPa) and S420 ($f_{yk} = 420$ MPa).

Designed frame buildings are assumed to be located at a site with peak ground acceleration (PGA) equal to 0.4 g and site class ZD (stiff clay and medium compact sand). The design of the frames is carried out by using the conventional force-based design approach in TBSC-18 for a seismic hazard level with a return period of 475 years and Life Safety performance level. Frames are designed as high ductile with structure system behavior factor R = 8 and overstrength factor D = 3 according to the seismic code.

The dynamic properties of the frame buildings are determined by using cracked section rigidities for beams and columns of the frame models as enforced by the TBSC-18. All beams are designed as T-sections with constant dimensions of 25 cm × 45 cm and 100 cm effective flange width. The amount of beam reinforcement varies

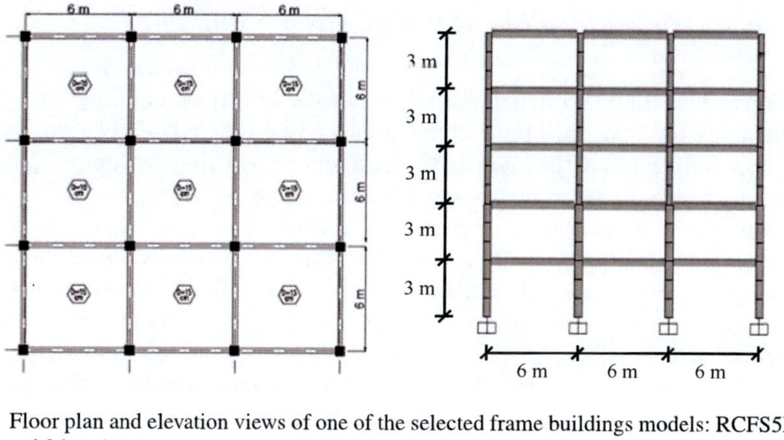

Fig. 1 Floor plan and elevation views of one of the selected frame buildings models: RCFS5B3 (5 stories and 3 bays)

Table 1 Beam member design details

Beam section (cm × cm)	Beam Rebar (top)	Beam Rebar (bottom)	Flange reinf	Transverse reinf. (confinement zone)	Transverse reinf. (central zone)
25 × 45	3ϕ16	3ϕ16	6ϕ10	ϕ10/10	ϕ10/15
	3ϕ18	4ϕ16	6ϕ10	ϕ10/10	ϕ10/15
	4ϕ18	4ϕ16	6ϕ10	ϕ10/10	ϕ10/15

Table 2 Column member design details

Column section (cm × cm)	Column Rebar	Transverse reinf. (confinement zone)	Transverse reinf. (central zone)
35 × 35	8ϕ20, 8ϕ25	ϕ10/10	ϕ10/20
40 × 40	8ϕ20, 8ϕ25	ϕ10/10	ϕ10/20
50 × 50	12ϕ20	ϕ10/10	ϕ10/20
60 × 60	16ϕ20	ϕ12/10	ϕ12/20
70 × 70	20ϕ20	ϕ12/10	ϕ12/20

depending on the design load on the member as shown in Table 1. For column design, square sections are used for which dimensions and reinforcement details vary from story to story for all frames. Member section details used in the design of the frames are presented in Table 2. Slab thickness is taken as 15 cm for all frame buildings.

All design checks have been performed for the considered frame buildings, including the capacity design principles (especially strong column-weak concept) and it has been verified that strength and deflection requirements are all satisfied.

3.2 *Numerical Modeling of the RC Frame Buildings*

Numerical modeling and dynamic analyses of the RC frame buildings are carried out by using the SAP2000 finite element analysis software [14]. During dynamic analysis, nonlinear materials properties are assigned by using the lumped plasticity approach. This is achieved by placing nonlinear hinges at two ends of linear elastic frame members. Plastic hinge length (L_p) is assumed as half of the section depth in the considered direction as recommended by TBSC-18. The selection of the lumped plasticity approach is rational for MRFs since maximum moment and in turn concentration of plastic deformation usually take place at the end zones of beam and column members.

Different hysteresis models can be employed to simulate the cyclic nonlinear behavior at hinge elements. In this study, Takeda hysteresis model [15] is used since it is known to simulate the nonlinear cyclic behavior of RC frame members with sufficient accuracy and reasonable simplicity. Since the frame members are expected to be in high ductility class due to design requirements, premature failure modes like shear and bond slip are eliminated. In addition, design calculations show that the columns are subjected to low-to-medium ranges of axial load. Hence flexural behavior seems to be dominant under seismic action for all frame members. These points verify the use of Takeda model in this study to simulate the global behavior of building frame models under seismic action.

Takeda hysteresis model uses a trilinear backbone curve for force-displacement (or moment-curvature) relationship with three limit states: tensile cracking (C), yielding of longitudinal reinforcement (Y) and ultimate capacity (U) in both positive and negative directions (Fig. 2). There are 16 rules in this hysteresis model to determine the instantaneous stiffness at a loading or unloading step. The details of these rules can be obtained from [15]. Takeda model has already been implemented in SAP2000 software.

In order to use Takeda model in nonlinear hinges, moment-curvature information of all beam and column end sections are determined and 3 limit states (C, Y and U) are defined in terms of moment-curvature pairs to form the backbone curve of the selected model. The limit state values for all members and frames are provided in Azizi [16].

It should also be mentioned that only the planar interior frames are considered for dynamic analysis instead of 3-D numerical models used in design due to symmetrical plan layout in both orthogonal directions. Rayleigh damping is used in analysis for a damping ratio of 5% in which the first two natural frequencies of the frame models are considered.

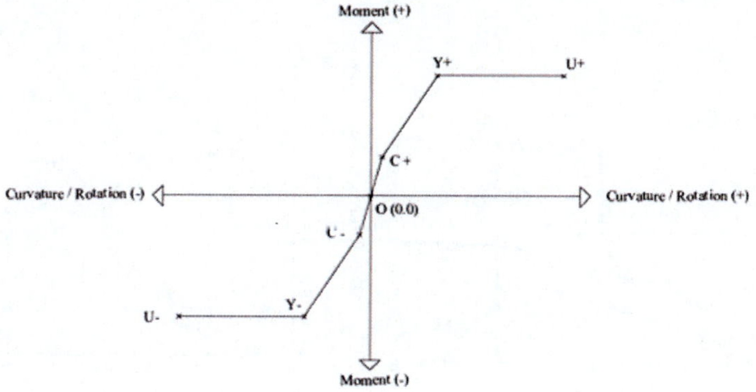

Fig. 2 Backbone curve and limit states for Takeda hysteresis model

4 Dynamic Analyses Applied to the Model Frames

The selected frame models are analyzed by using NLTHA. For this purpose, 20 strong ground motion records are employed in two groups: local records and global records. Local set is composed of 10 records from past major earthquakes that occurred in Turkey (labelled as L1-L10) whereas global set has the same number of records which had been recorded during major earthquakes in different parts of the world (labelled as G1-G10). In the selection of ground motion records, the main criterion is to have ground motion variability in terms of duration, intensity and frequency content. The major characteristics of the ground motion records are provided in Table 3.

Before performing dynamic analysis, all ground motions are scaled for each frame building model according to the ordinate of the 5% damped target design spectrum at the fundamental period of that building in order to impose the same acceleration demand from different records at design earthquake level.

In this study, 6 different frame models are subjected to 20 ground motion records, which means in total 120 NLTHA are conducted and energy parameters given in Eq. 3 are calculated by considering the nonlinear response history of each frame subjected to a specific ground motion record. But before conducting the NLTHA, a benchmark case study is carried out in order to verify the energy calculations within the software. For this purpose, the 10 story RC frame developed by Zhu [3] is modeled in SAP2000 and then the benchmark frame is subjected to Nahanni, Canada (1985) ground motion record. The energy time history results obtained by Zhu in the original research are compared with the ones obtained in this study. As seen in Fig. 3, the energy values and the trends are very close to each other.

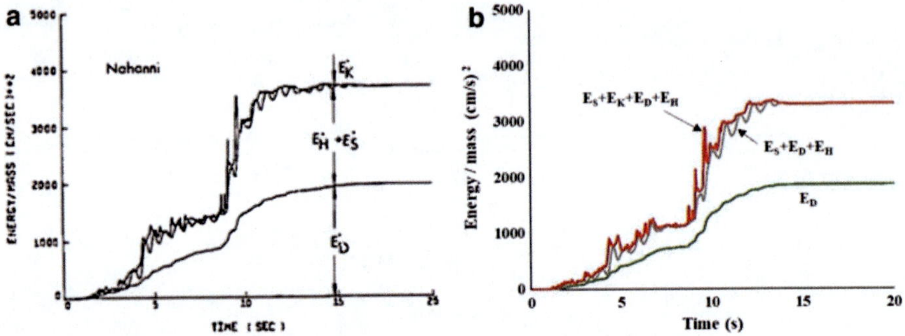

Fig. 3 Time history of energy components for 10 story RC frame under Nahanni (1985) record **a** by Zhu [3], **b** by the verification study

Table 3 Characteristics of the selected global and local ground motion records

Label	Event	Country	Year	Location	M	PGA (g)	PGV (cm/s)
L1	Horasan	Turkey	1983	Horasan	6.7	0.13	36.9
L2	Erzincan	Turkey	1992	Erzincan	7.3	0.47	92.1
L3	Dinar	Turkey	1995	Dinar	6.1	0.32	40.6
L4	Kocaeli	Turkey	1999	Yarımca (NS)	7.8	0.32	79.6
L5	Kocaeli	Turkey	1999	Yarimca (EW)	7.8	0.23	84.7
L6	Kocaeli	Turkey	1999	Duzce	7.8	0.34	60.6
L7	Duzce	Turkey	1999	Duzce (NS)	7.3	0.41	65.8
L8	Duzce	Turkey	1999	Duzce (EW)	7.3	0.51	86.1
L9	Bingol	Turkey	2003	Bingol	6.4	0.51	34.5
L10	Ceyhan	Turkey	1998	Ceyhan	6.2	0.23	29.8
G1	Imperial Valley	USA	1979	El Centro Array #5	6.5	0.37	95.9
G2	Montenegro	Yugoslavia	1979	Ulcinj	7.0	0.24	47.1
G3	Loma Prieta	USA	1989	Hollister	7.0	0.37	62.8
G4	Manjil	Iran	1990	Abhar	7.3	0.21	55.4
G5	Cape Mendocino	USA	1992	Petrolia	7.0	0.66	89.5
G6	Northridge	USA	1994	Slymar	6.7	0.37	118.9
G7	Northridge	USA	1994	Jensen F.P	6.7	0.42	106.2
G8	Kobe	Japan	1995	JMA	6.9	0.83	90.7
G9	Chi-Chi	Taiwan	1999	TCU074	7.6	0.60	74.6
G10	Tabas	Iran	1978	Tabas	7.3	0.24	47.1

5 Energy-Based Results from Dynamic Analyses

This section is devoted to the presentation of energy-based results from nonlinear dynamic analyses conducted for the proposed 6 RC frame models under the selected 20 ground motion records. Although all of the results in terms of energy parameters have been obtained from the NLTHA, only the ones related with the distribution of hysteretic energy within the frames are discussed in this section as the main focus of this study. For detailed discussion regarding the variation of input energy and the ratio of hysteretic to input energy, one can refer to Azizi [16]. In this section, the energy results are presented in terms of the story-wise and member-wise distributions of E_H demand within the given frame.

5.1 Story-Wise Distribution of Hysteretic Energy

Dissipated E_H at column and beam ends are summed for each story and then divided by the total E_H of the frame to find the normalized E_H demand of each story. The distribution of story-wise E_H to total E_H ratio (i.e. $E_{SH}/\Sigma E_H$) over the height of the frame under global ground motion set are shown in Fig. 4, in which MG represents mean value of the results. It is worth stating that the results obtained for 5-story frame models with different number of bays are very close to each other. Hence, model RCFS5B3 is used to represent the results of all 5-story frame models. The mean values of the $E_{SH}/\Sigma E_H$ ratio for global and local records are provided in Table 4.

From evaluation of results obtained for the story-wise distribution of E_H, the following comments can be concluded:

- General trend of the results shows that distribution of E_H over the height of the structure depends on both ground motion characteristics and structural properties.
- For model RCFS3B3, ratios for nearly all ground motions seem to be close to each other whereas the ratio values have more scatter for models with higher number

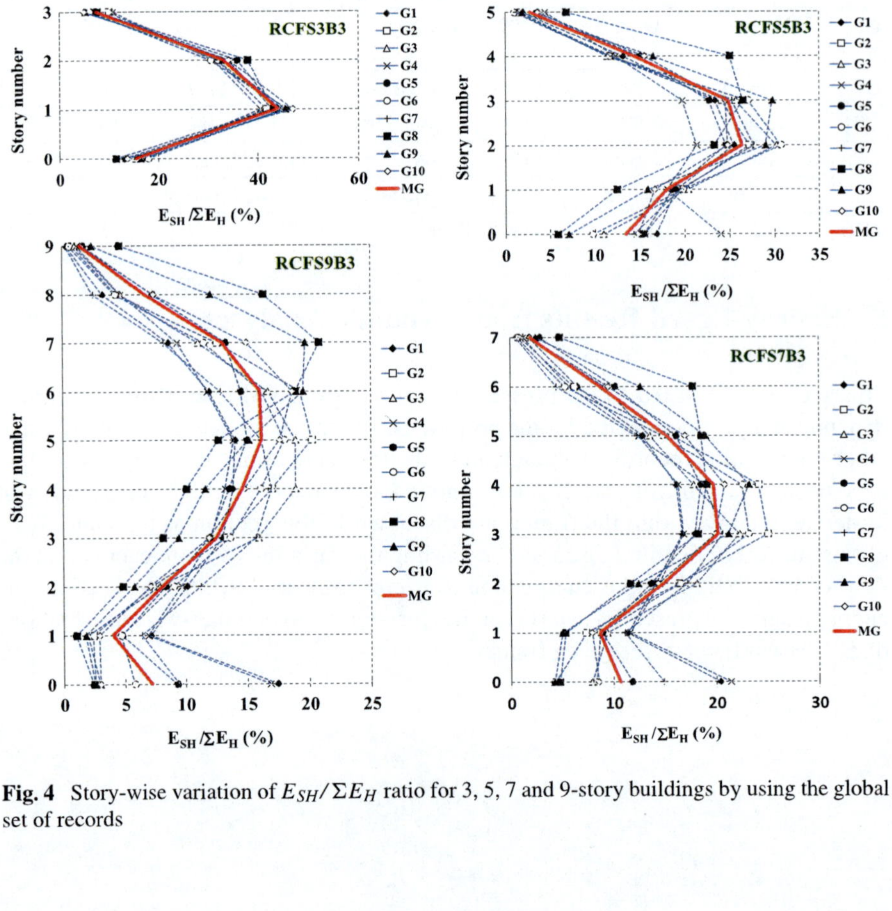

Fig. 4 Story-wise variation of $E_{SH}/\Sigma E_H$ ratio for 3, 5, 7 and 9-story buildings by using the global set of records

Table 4 Mean values of $E_{SH}/\Sigma E_H$ (%) for global and local record sets

	RCFS3B3		RCFS5B3		RCFS7B3		RCFS9B3	
Story	Global	Local	Global	Local	Global	Local	Global	Local
9	–	–	–	–	–	–	1.4	1.1
8	–	–	–	–	–	–	6.7	4.9
7	–	–	–	–	1.9	2.3	13.0	10.2
6	–	–	–	–	8.8	7.8	16.1	13.8
5	–	–	2.8	3.7	15.2	13.9	16.2	15.9
4	–	–	14.5	15.0	19.7	18.7	14.6	15.5
3	7.4	8.7	24.9	23.7	20.2	19.5	12.5	13.7
2	33.2	33.5	26.4	25.1	14.9	15.1	8.2	9.5
1	44.0	43.5	18.0	17.4	8.6	9.5	4.0	5.3
Base	15.4	14.3	13.4	15.1	10.6	13.3	7.2	10.0

of stories. This shows that story-wise distribution of E_H is less sensitive to ground motion characteristics for low-rise buildings.

- According to Table 4, 75% of E_H is dissipated in the 2nd and 3rd stories for RCFS3B3. This value decreases to 45% for RCFS5B3, 25% for RCFS7B3 and 15% for RCFS9B3. In addition, E_H in the base columns decreases from 15% for RCFS3B3 to 9% for RCFS9B3. It means that as number of stories increases, E_H propagates from the lower stories to the middle and upper stories.

5.2 Member-Wise Distribution of Hysteretic Energy

The dissipated E_H at each beam and column end is divided by the total E_H to obtain the ratio of member-wise (beam or column) E_H to total E_H ($E_{MH}/\Sigma E_H$) for all ground motions as a measure. The percentages of total E_H dissipated by the base columns, story columns and story beams for all records are presented in Fig. 5 for model frame RCFS5B5 only, since the results for the remaining 3, 7 and 9-story frame models have similar trends. In addition, the mean values of E_H percent dissipated by beam and column members for global and local ground motion sets are provided in Table 5.

Evaluating the member-wise distribution of E_H results given in Fig. 5 and Table 5, the following comments can be concluded:

- Calculated mean values of E_H dissipated at column and beam ends for all frame models indicate that 78% of total E_H is dissipated by beam hinges whereas 9% is dissipated by column hinges in the stories and 13% is dissipated by the hinges at the base columns. It is also observed that at a beam-column joint, 10% of the E_H is dissipated by columns whereas 90% is dissipated by the beams.

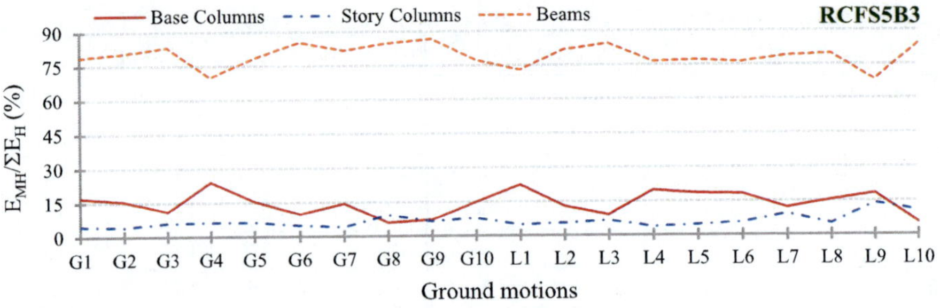

Fig. 5 Member-wise E_H distribution for frame model RCFS5B3

Table 5 Mean $E_{MH}/\Sigma E_H$ values (%) for structural members under global and local records

	Base columns		Story columns		Beams	
Frame	Global	Local	Global	Local	Global	Local
RCFS3B3	15.4	14.3	16.4	18.9	68.2	66.8
RCFS5B2	12.8	14.2	5.7	6.5	81.6	79.3
RCFS5B3	13.4	15.1	5.9	7.0	80.7	77.9
RCFS5B4	13.9	15.5	6.1	7.2	80.1	77.3
RCFS7B3	10.6	13.3	8.3	8.7	81.0	78.0
RCFS9B3	7.2	10.0	7.7	7.0	85.1	83.0

- As stated before, strong column-weak beam criterion (i.e. sum of column moment capacities at a joint is 20% more than the sum of beam moment capacities) has been ensured in the design of frame models in the context of TBSC-18. Accordingly, the total column moment capacity to total beam moment capacity ratios at the joints of frame model RCFS3B3 are 1.7 on the average whereas the same ratios get values around 2.0 for the other frame models. These values seem to be directly reflected on E_H ratios as seen on Table 5. The mean $E_{MH}/\Sigma E_H$ value for story columns in frame model RCFS3B3 is approximately 18%. However, for the other frames, this ratio is observed to take values around 7%. Hence it can be clearly stated that the ratio of dissipated E_H by columns to dissipated E_H by beams is sensitive to the ultimate moment capacities of column and beam sections at joints.
- Although the column moment capacity to beam moment capacity ratio is more than 1.2 for all frame models according to the force-based design requirements as stated above, the columns did not behave all in the linear elastic range and they exhibited hysteretic energy due to inelastic deformation. It could be concluded that the strong column-weak beam requirement (i.e. ratio of 1.2 in many seismic codes) does not guarantee the elastic behavior of columns but it induces a ductile beam-column failure mechanism.

- Referring to Table 5, it is observed that as the frame model becomes more flexible (i.e. number of stories increases), dissipated E_H is transferred from the base columns to the beams.

Mean values of $E_{MH}/\Sigma E_H$ are presented at each nonlinear hinge for frames RCFS3B3, RCFS5B3, RCFS7B3 and RCFS9B3 in Fig. 6. The following comments are based on these E_H distributions:

- Comparison of the $E_{MH}/\Sigma E_H$ ratios shows that the percentage of E_H dissipated by exterior and interior members are not the same.
- The values indicate that the $E_{MH}/\Sigma E_H$ ratios for exterior beam hinges are greater than the ones for interior beam hinges. The inverse trend occurs for columns, i.e. the $E_{MH}/\Sigma E_H$ ratios for exterior column hinges are less than the ones for interior columns hinges. This means that the E_H demand of exterior beams and interior columns are more critical than the E_H demand of the interior beams and exterior columns of the same story in RC MRFs.
- If one finds the difference of $E_{MH}/\Sigma E_H$ ratios between exterior and interior beam hinges for all stories and takes their average value, it can be observed that the E_H dissipated by exterior beam hinges is 30% more than the E_H dissipated by the interior beam hinges for RCFS3B3, 15% for RCFS5B3, 8% for RCFS7B3 and 4% for RCFS9B3. This means that as the number of stories increases, the difference in $E_{MH}/\Sigma E_H$ ratios between exterior and interior beam hinges decreases and the distribution becomes quite regular.
- Since, the amount of E_H dissipated by columns is small and majority of the E_H is dissipated by beams, it is possible to ignore the difference in $E_{MH}/\Sigma E_H$ ratios between exterior and interior column hinges.
- Overall, member-wise distribution of E_H is directly influenced by the moment capacities of beam and column sections at a joint whereas the dependence is slight for different ground motion sets. This shows that it is possible to propose practical energy-based design rules to control the distribution of inelastic action within a frame structure.

6 Conclusions

This study is a preliminary attempt to promote energy-based design and assessment approaches to next generation seismic codes. In order to achieve this task, seismic response in terms of energy should be examined thoroughly. This has somewhat been accomplished for the input energy demand since there are many past studies in the literature focusing on the estimation of the parameter E_I. However, research about the distribution of E_H demand within a structure has not been studied to the same extent, especially for RC frame buildings. Hence this study plans to give contribution to the findings regarding the story-wise and member-wise distributions of E_H in RC frame buildings. This study contains some assumptions and simplifications as stated in the previous sections. The following conclusions can be stated without ruling out these limitations:

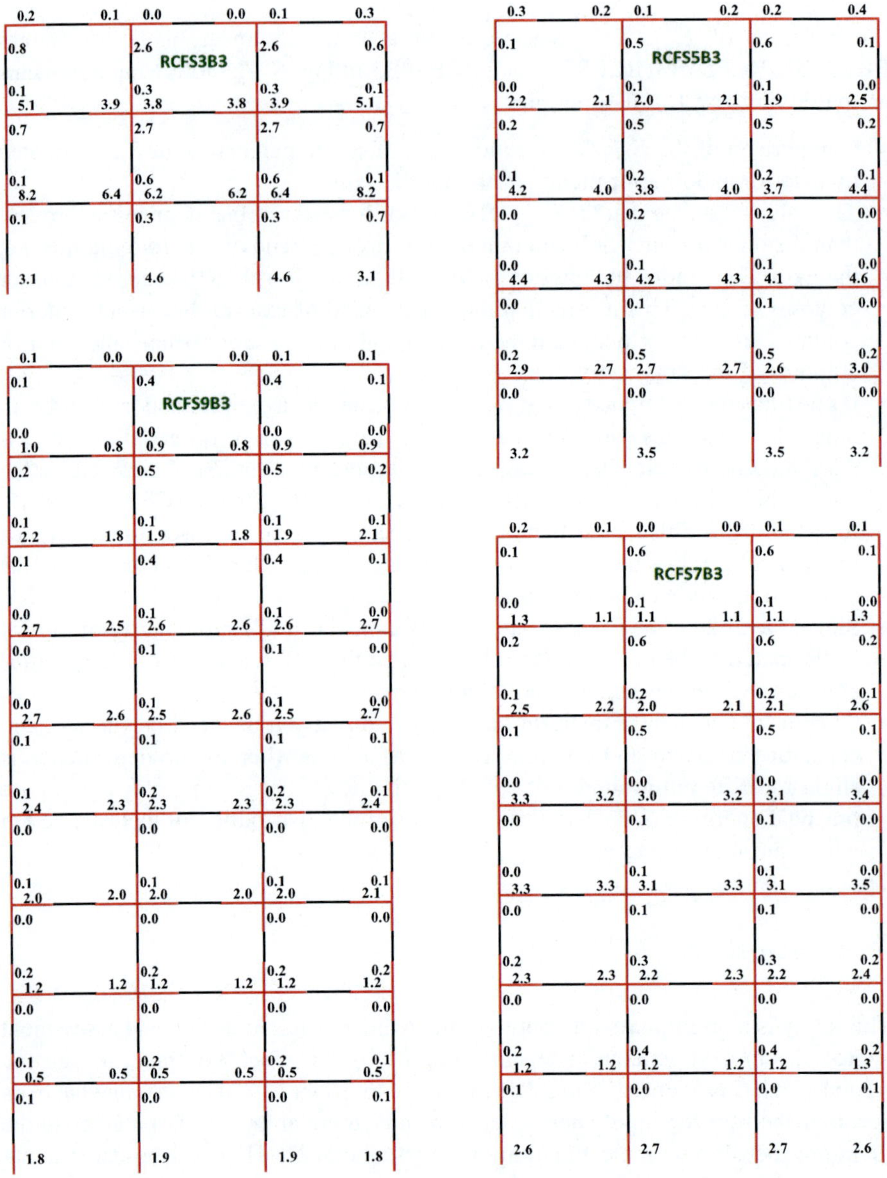

Fig. 6 Mean values of $E_{MH}/\Sigma E_H$ at the joints of all frame models with three bays under global ground motion set

- Considering the general trend of the story-wise distribution of E_H results, it is observed that distribution of E_H over the height of structure depends on both ground motion characteristics and structural properties. The dependency of the $E_{SH}/\Sigma E_H$ ratio to ground motion characteristics become more obvious as the number of stories increases (i.e. structure becomes more flexible).
- The $E_{SH}/\Sigma E_H$ ratio of ground story and lower stories decrease as the number of stories increases. Hence E_H demand shifts from lower stories to upper ones as structure becomes more flexible (i.e. going from 3-story to 9-story frame models). This reveals that the second mode of the structures should also be considered in an energy-based design or assessment methodology (during estimation of the story-wise distribution of hysteretic energy) for mid-rise and high-rise structures.
- The results of this study show that in a well-designed RC moment-resisting frame, approximately, 70–85% of the E_H is dissipated by beams, 8–18% is dissipated by story columns and 7–15% is dissipated by base columns. Consequently, it seems that in a ductile RC moment-resisting frame, majority of E_H is dissipated by beams, which is a verification of the intended behavior in force-based capacity design of RC frame structures. It also seems that the percent of E_H dissipated by columns or beams strongly depends on the ultimate moment capacities of columns and beams sections at the joints of the frame.
- Although strong column-weak beam criterion is considered in design of frame models, the columns exhibited inelastic behavior. This observation indicates that assigned safety factor of 1.2 for the ratio of column moment capacity to beam moment capacity does not guarantee elastic behavior for columns and it causes a ductile beam-column failure mechanism. This is not surprising since the frame models are designed for Life Safety performance level, for which controlled damage is allowed. The important point is that the percentage of inelastic action is very limited in columns when compared to beams. This is a verification in the force-based seismic design process for ductile behavior.
- Member-wise distribution of E_H in the same story shows that E_H is distributed uniformly between interior members. This may also be verified for exterior members. However, comparing interior and exterior members together indicates that $E_{MH}/\Sigma E_H$ values at the ends of members are not equal. In addition to this, the E_H demand of exterior beam hinges are generally more than interior beam hinges whereas the E_H demand of interior column hinges are more than exterior column hinges. This difference in E_H demand between interior and exterior members becomes more pronounced in low rise RC frame building. Hence it can be stated that as number of stories increases, E_H is distributed more uniformly in the same story.
- The dynamic analysis results obtained in this study reveal that the spatial distribution of E_H within a frame is highly affected by the number of stories whereas it does not seem to be influenced by the number of bays.
- The aforementioned results regarding story-wise and member-wise distributions of E_H can assist to estimate the role of each member to dissipate a certain amount of energy in an energy-based design methodology and the capacities of members can be arranged in accordance with this demand. So, it can be finally stated that

energy-based parameters are promising in order to estimate the distribution of energy demand in a RC frame structure. This leads to the motivation that simple yet robust energy-based approaches can be developed and implemented to the future releases of seismic codes if the energy dissipation capacities of the members can be determined in a satisfactory manner.

References

1. Housner, G.W.: Limit design of structures to resist earthquakes. In: Proceedings of the 1st World Conference on Earthquake Engineering, pp. 5.1–5.13. California USA (1956)
2. Disaster and Emergency Management Presidency (AFAD): Turkish Building Seismic Code, TBSC-18. Republic of Turkey, Ministry of Interior, Ankara Turkey (2018)
3. Zhu, T.J.: Inelastic response of reinforced concrete frames to seismic ground motions having different characteristics. Doctoral dissertation, McMaster University, USA (1989)
4. Benavent-Climent, A., Zahran, R.: An energy-based procedure for the assessment of seismic capacity of existing frames: application to RC wide beam systems in Spain. Soil Dyn Earthquake Eng **30**, 354–367 (2010)
5. Guan, M., Du, H.: Energy-based seismic performance of reinforced concrete frame structures. Mag. Concr. Res **65**(8), 494–505 (2013)
6. Merter, O., Ucar, T.: Design of RC frames for pre-selected collapse mechanism and target displacement using energy–balance. Sadhana **39**(3), 637–657 (2014)
7. Benavent-Climent, A., Escobedo, A., Donaire-Avila, J., Oliver-Saiz, E., Ramírez-Márquez, A.L.: Assessment of expected damage on buildings subjected to Lorca earthquake through an energy-based seismic index method and nonlinear dynamic response analyses. Bull Earthquake Eng **12**, 2049–2073 (2014)
8. Tu, B.B., Zhao, D.: Distribution of accumulated irrecoverable hysteretic energy in MDOF structures. Multidiscipline Model. Mater. Struct. **14**(2), 202–215 (2018)
9. Uang, C.M., Bertero, V.V.: Evaluation of seismic energy in structures. Earthquake Eng. Struct. Dyn. **19**(1), 77–90 (1990)
10. Bruneau, M., Wang, N.: Some aspects of energy methods for the inelastic seismic response of ductile SDOF structures. Eng. Struct. **18**(1), 1–12 (1996)
11. Kalkan, E., Kunnath, S.K.: Relevance of absolute and relative energy content in seismic evaluation of structures. Adv Struct. Eng. **11**(1), 17–34 (2008)
12. Turkish Standards Institute: Requirements for design and construction of reinforced concrete structures, TS-500. Turkish Standards Institute, Ankara, Turkey (2003)
13. Turkish Standards Institute: The calculation values of loads used in designing structural elements, TS-498. Turkish Standards Institute, Ankara, Turkey (1997)
14. CSI: SAP2000 Version 20.1.0, Linear and nonlinear static and dynamic analysis and design of three-dimensional structures: Basic analysis reference manual. Computers and Structures, Inc. Berkeley, California (2018)
15. Takeda, T., Sozen, M.A., Nielsen, N.N.: Reinforced concrete response to simulated earthquakes. J. Struct. Div. **96**(12), 2557–2573 (1970)
16. Azizi, M.: Energy based evaluation of RC framed structures. Master Thesis, Middle East Technical University, Ankara, Turkey (2019)

Reinforced Concrete Columns: Insight on Energy-Based Assessment from Biaxial Tests with Different Load Paths

Paolo Franchin, Andrea Lucchini, and José Miranda Melo

Abstract Results from a recently concluded experimental campaign on non-conforming reinforced concrete cantilever columns failing in both flexure and flexure-shear modes under five different load paths are used to discuss whether an energy-based failure criterion would be advantageous with respect to a deformation-based one. Based on the limited evidence of these 18 tests it appears that the relation between energy and damage, compared to that of drift ratio and damage, is influenced to a lesser extent by load path, but this only holds in some cases and cannot be generalized. In particular, the occurrence of any phenomenon that interrupts the stable energy dissipation mechanism through inelastic flexure leads to failure prior to the attainment of the energy threshold. This can occur in many practical situations indicating that it may not be feasible to use energy as a useful measure in the seismic assessment of existing frames.

Keywords Energy capacity · Deformation capacity · Biaxial response · Seismic damage prediction

1 Introduction

As early as the 1960s, the possibility of expressing in terms of energy the verification inequality comparing demand with capacity, was put forward by John Blume in a paper at the 2nd World Conference on Earthquake Engineering in Tokyo [1]. This was also included in a very influential report the following year, but confined to an annex [2], apparently as a trade-off on a first name struggle (academics …). That same report contains the concept of capacity design, even if the name with which it became famous was invented later elsewhere [3]. One could speculate about the

P. Franchin (✉) · A. Lucchini
Sapienza University of Rome, Rome, Italy
e-mail: paolo.franchin@uniroma1.it

J. M. Melo
CONSTRUCT-LESE, Faculty of Engineering, University of Porto, Porto, Portugal

A. Benavent-Climent and F. Mollaioli (eds.), *Energy-Based Seismic Engineering*, Lecture Notes in Civil Engineering 155,
https://doi.org/10.1007/978-3-030-73932-4_5

future of energy-based design and assessment (or rating, as Blume called it at the time), were this approach not relegated to an annex in that report. In any case, the history is known, and earthquake-resistance design remained force-based.

In 1997, the first of a series of influential international workshops on performance-based earthquake engineering was held in Bled. In the summary of those days, contained in the Resolutions at the beginning of the proceedings, a clear path was traced towards developing and then adopting methods that could grant a better control of actual performance of structures under seismic actions [4]. In there, deformation-controlled design was identified as the better option. It was also recognized the need to educate the technical community and to only gradually implement the research developments into normative documents. It was envisaged that three stages would have taken place, starting with the enhancement of conventional force-based approach with deformation verifications and the direct deformation-based design approach added only as an option, to familiarize code users. This was the time when the first generation of the structural Eurocodes was being drafted, and indeed Eurocode 8 experimentally introduced displacement-based assessment into its Part 3 [5]. This contained the first generation of deformation acceptance criteria of members in terms of chord-rotation [6]. In the second stage, it was thought that the force-based and the direct deformation-based design approaches could co-exist with incentives provided for the use of the latter. We can see that we are entering into this stage, at least in Europe, only now, more than two decades later, with the second generation Eurocode 8 presenting updated deformation criteria for reinforced concrete [7–9] and steel [10–13], and allowing their use in displacement-based design in its main part [14], not just for assessment as it was the case before [5]. Finally, it was expected that a direct deformation-based design approach would be eventually adopted in a third stage, that appears relatively distant in the future at the time of writing.

What about energy then? While displacements and deformations are certainly better correlated than forces to actual damage, it was also clear that duration and the associated number of cycles played an important role. Thus, again in [4], it was recommended that cumulative damage/dissipated energy should be considered for structures with rapidly deteriorating components, and in case of long duration ground motions. The solution envisaged at the time was that energy or some other characterization of duration effects should be included in establishing deformation targets in order to account for the effects of damage accumulation on deformation capacities (equivalent ductility concept). The new Eurocode 8 [14] has not adopted an explicitly energy-based approach to the design of conventional structures, opting for accounting for the duration effects through the deformation capacity (with the notable exception of structures equipped with energy-dissipating devices, to which an energy-based approach is dedicated).

This short paper humbly aims at contributing to the discussion about the actual need and practical feasibility of introducing an energy-based verification format also for more conventional structures, such as for instance, RC moment-resisting frames. It does so by looking at deformations and energy as experimentally derived from a campaign on 18 columns tested to collapse under different uniaxial and biaxial load paths, recently concluded by the authors and colleagues [15].

2 Brief Illustration of Tests Carried Out

As pointed out in [15], the proportion of cyclic tests carried out on RC members that is not uniaxial is almost negligible. In almost forty years since 1984 (the first multi-axial test campaign retrieved in the literature during the study in [15]) only 113 such tests have been apparently performed, of which at a closer scrutiny only 87 are truly bi- or triaxial tests. Most test focus on "conforming" members, with deformed bars. None considers non-symmetric load paths. In general, as it is most often the case, the same load paths were used to test different specimens. For this reason, the authors and colleagues carried out an experimental campaign to investigate the response of non-conforming members with light transverse reinforcement, focusing on the load path influence.

All specimens have the same geometry: cantilever column, square cross section of 300 mm side, four 16 mm diameter longitudinal bars corresponding to a geometric reinforcement ratio ρ_s of 0.89%, and height equal to 1500 mm, for an aspect ratio L_V/h of 5. Four combinations of two levels of axial force (150 kN and 450 kN, corresponding approximately to a normalized axial force of 10% and 20%, respectively) and two levels of transverse reinforcement (6 mm diameter hoops at 150 and 250 mm spacing, respectively, with 90° hooks) are considered. They are denoted in the following L10, VL10, L20 and VL20, where L and VL stand for "low" and "very low" transverse reinforcement ratio (all values are representative of Southern European old RC frames).

Five load paths are considered, both uniaxial (U) and biaxial (B). After the test campaign started, it was decided to include also a monotonic (UM) and a cyclic anti-symmetric (UCA) one, but these were not carried out for the L10 type specimen. The monotonic test was included because this reference response is most often not established experimentally but inferred somehow from the cyclic one, and the anti-symmetric was included because it can be representative of actual load paths under severe ground motions, e.g., near-field ones. The latter is also referred to as "collapse-consistent", and it was used before only in tests of structural steel columns. The resulting 18 tests are indicated in Table 1. For each specimen and load path, the table reports the failure mode, adopting the common distinction between flexure (F) and flexure-shear or ductile shear (FS). Details can be found in [15].

The tests in Table 1 are arranged by column in an ideal transition from the least energy demand (uniaxial monotonic) to the largest one (biaxial circular, BC), passing

Table 1 Failure mode by load path for the four combinations of geometry and axial load

Specimen	UM	UCA	UCS	BE	BC
L10	–	–	F	F	F
VL10	FS	FS	F	F	F
L20	FS	FS	FS	FS	FS
VL20	FS	FS	FS	FS	FS

through uniaxial cyclic antisymmetric, uniaxial cyclic symmetric (this is the standard protocol of most tests) and biaxial elliptic (BE). As it can be seen, the failure mode is influenced by both axial force and load path. In general, the larger axial force (labelled 20) leads to ductile shear failure due to the increased plastic shear demand, while, given the slenderness of the specimens, has negligible effect on the shear strength. At the lower level, the load path influences the failure mode.

Three values of the drift ratio θ (or chord rotation, for these cantilever columns) have been identified, among other quantities, in each test: at peak lateral force, at "ultimate" (minus 20% post-peak strength) and at axial failure. Figure 1 illustrates their evaluation in four of the 18 cases, showing the force-drift ratio loops and envelopes in the North–South direction, which is the direction along which the first increment of displacement is applied in the case of the BC tests, and the main direction of loading for the BE tests. Details are given in [15], but it is apparent how a larger energy demand (L20-UCA $\rightarrow$VL10-UCS $\rightarrow$VL10-BC) implies lower values of the drift ratio even when the failure mode does not change, as in the VL10-BC test as compared to the VL10-UCS one. In the following section, these drift ratios are presented for all tests and compared with the associated strain energy values.

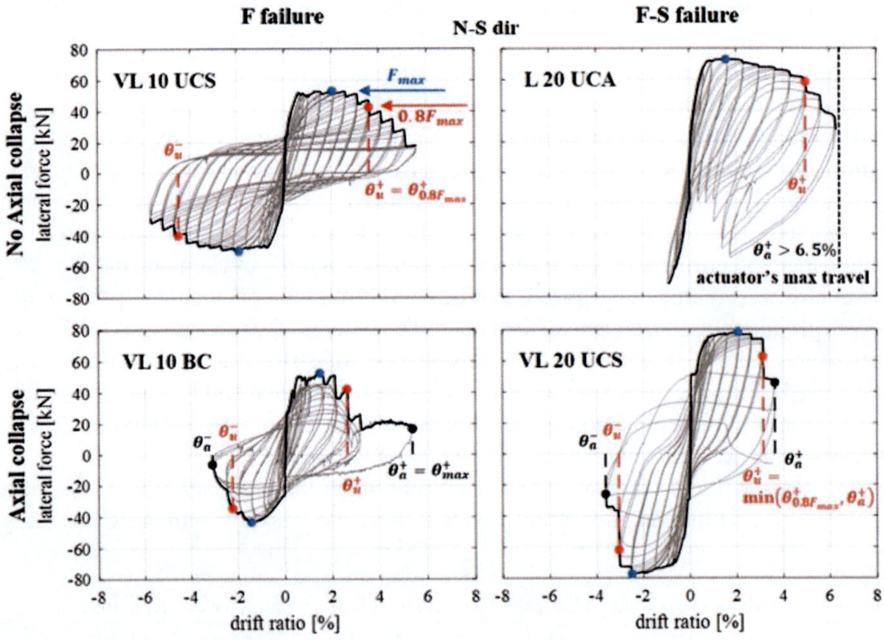

Fig. 1 Definition of peak strength, ultimate rotation and axial failure points on the force-drift ratio loops of the tested columns

3 Results and Discussion

It is of interest to understand the relation between load path and failure deformation, on the one hand, and energy on the other. As shown in Fig. 1 with reference to some tests and summarized for all 18 specimens in Fig. 2, there is a clear tendency of the ultimate drift ratio θ_u to decrease going from the least to the most energy demand. Figure 2 also shows that θ_u values have an average around 4% (the starting value of the plastic part of chord rotation capacity at ultimate, θ_u^{pl}, in Eurocode 8 [14] is 3.9% for code-conforming members), and can reduce to half that value in biaxial conditions (BC tests). Yielding is between 0.5 and 1%.

What value of ultimate rotation should then be used, given that the actual load path is not known in advance? How valid are "interaction" formulas [8] like Eq. (1) that define an elliptical domain using results from uniaxial tests to establish uniaxial capacities in two orthogonal directions (denoted as North–South and East–West in the formula)? Would be an energy-based failure criterion more useful?

$$\left(\frac{\theta_{NS}}{\theta_{u,NS}}\right)^2 + \left(\frac{\theta_{EW}}{\theta_{u,EW}}\right)^2 = 1 \quad (1)$$

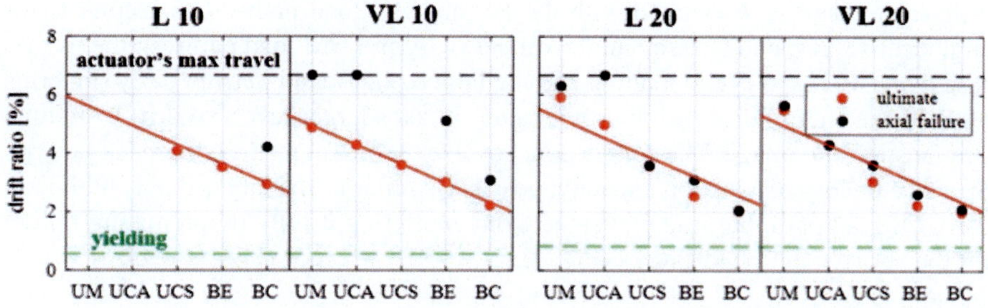

Fig. 2 Drift ratio versus load path for the four considered combinations of geometry and axial load

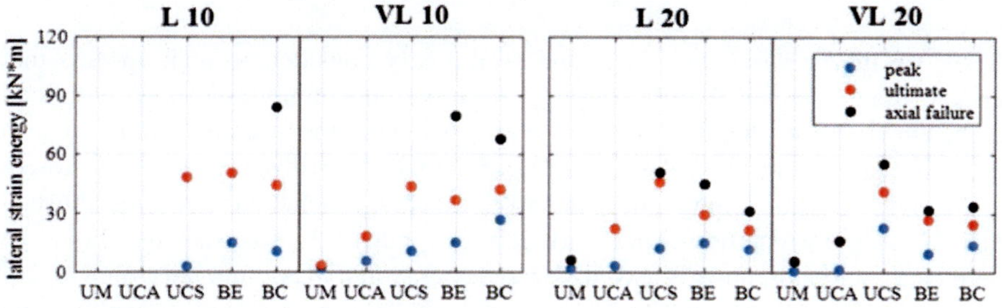

Fig. 3 Strain energy value at the three key points (peak, ultimate, axial failure) versus load path, for the four considered combinations of geometry and axial load

Figure 3, similarly to Fig. 2, reports for each of the four combinations of axial force and transverse reinforcement, and for the different load paths, the value of the lateral strain energy demand E up to time t, as obtained by integration of the force–displacement $(F - s)$ histories in the two orthogonal directions NS and EW:

$$E(t) = \int_{0}^{s_{NS}(t)} F_{NS} ds_{NS} + \int_{0}^{s_{EW}(t)} F_{EW} ds_{EW} \tag{2}$$

at each of the three previously defined points. Colors are the same as in Fig. 3.

The point at "ultimate", i.e. the strain energy value corresponding to the attainment of θ_u in each test, is also shown on energy vs maximum in-cycle drift plots of Fig. 4, where drift is shown in the N–S direction (note that colors indicate the load path in this figure). The results in these figures highlight how energy values are much closer to each other than drift ratios, with values between 40 and 50 kNm, corresponding to drifts that range between about 2 and almost 4%, when specimen failing in flexure are considered (L10 and VL10 for load paths UCS, BE and BE, see also Table 1).

When the stable dissipating mechanism provided by inelastic flexure is instead interrupted, as it happens when the compression chord fails due to concrete crushing and bar buckling, or the shear reinforcement fails preventing the internal couple from working, then failure happens at largely different values of energy. It may be useful to observe that this can occur in two apparently different ways. The first is for low drift values, and is associated with the BE and BC load paths. The second is for large drift values, and is associated with the UCA and UM load paths. In the former case damage due to cyclic loading and the larger axial load interrupt early (in terms of drift) the energy dissipation, in the latter, it takes larger levels of drift to attain a similar state of damage. The conclusion, however, is the same: the column has always the same energy dissipation capacity, possibly similar to the larger one exhibited in the UCS load path, but it is prevented from exploiting it fully by premature failure for the above mentioned causes (Fig. 4).

4 Conclusions

Based on the previous very brief and preliminary considerations, it appears that energy alone does not free the engineer from the uncertainty associated with load path. On the other hand, this result does not come unexpected, given that also in deformation-controlled design and assessment one still needs to check force-controlled i.e. brittle failure modes in terms of forces, and that it is a well-established fact since the formulation of the Park and Ang index [16], that damage should be defined as compound measure of dissipated energy and maximum deformation.

Looking at the tests and trying to cast these onto practical situations that can arise, one could say that energy may work as a load-path independent measure to

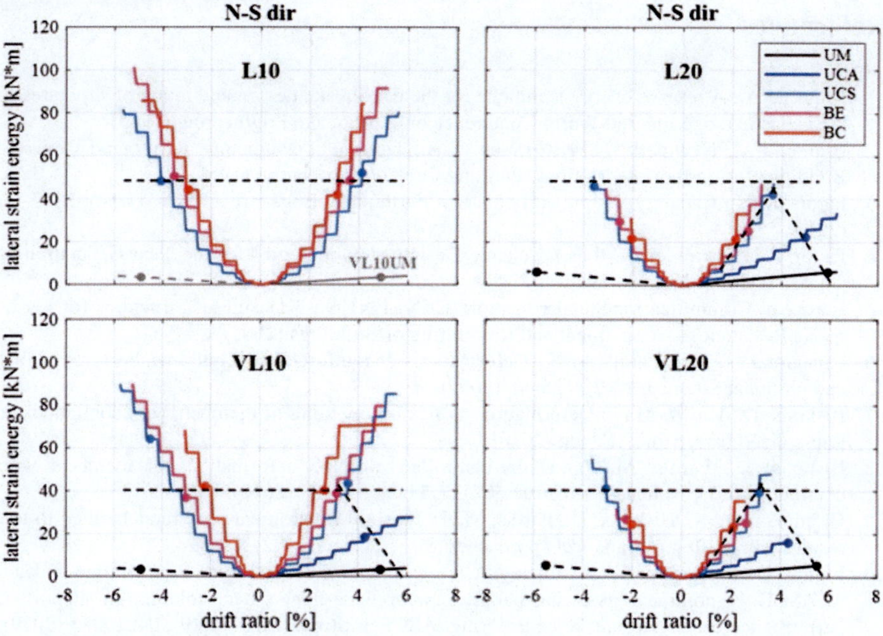

Fig. 4 Key points (ultimate) on the strain energy diagrams for every load path and the four considered combinations of geometry and axial load

express performance when response is not appreciably anti-symmetric (like in the UCA or UM cases), i.e. in far-field situations, or in other words when directivity effects are not expected to play a major role, and for code-conforming members, where shear-failure should be prevented through correct capacity design. It probably cannot offer more than deformation criteria used jointly with force-checks in the assessment of existing structures. Whether switching to energy for conventional RC frames is warranted based on the apparently limited scope remains an open question.

Finally, it may be more important to increase the number of multi-axial tests in order to: (a) increase the experimental support for interaction rules such as the elliptical one referenced before, to increase confidence in verifications based on nonlinear static analysis performed independently in two orthogonal horizontal directions; (b) calibrate robust coupled, i.e. at least biaxial, phenomenological models for use in response-history analysis.

Acknowledgements The tests in [12] have been carried out within a collaboration between University of Rome La Sapienza, the Faculty of Engineering at the University of Porto, and University College London. The Colleagues António Arêde and Humberto Varum from Porto, and Tiziana Rossetto from London, are gratefully acknowledged. The opinions expressed in this paper are only those of the authors.

References

1. Blume, J.A.: A reserve energy technique for the earthquake design and rating of structures in the inelastic range. In: 2nd World Conference on Earthquake Engineering (1960)
2. Blume, J.A., Newmark, N.M, Corning, L.H.: Design of Multistory Reinforced Concrete Buildings for Earthquake Motions. Portland Cement Association (1961)
3. Fardis, M.N.: Capacity design: early history. Earthquake Eng. Struct. Dyn. **47**(14), 2887–96 (2018)
4. Fajfar, P., Krawinkler, H. (eds.): Seismic Design Methodologies for the Next Generation of Codes. Taylor & Francis (1997)
5. European Committee for Standardization (CEN) EN1998-3 Design of Structures for Earthquake Resistance—Assessment and Retrofitting of Buildings (2005)
6. Panagiotakos, T.B., Fardis, M.N.: Deformations of reinforced concrete members at yielding and ultimate. Struct. J. **98**(2), 135–48 (2001)
7. Biskinis, D., Fardis. M.N.: Deformations at flexural yielding of members with continuous or Lap-spliced bars. Struct. Concr. (2010)
8. Biskinis, D., Fardis, M.N.: Flexure-controlled ultimate deformations of members with continuous or lap-spliced bars. Struct. Concr. **11**(2), 93–108 (2010)
9. Grammatikou, S., Biskinis, D., Fardis, M.N.: Flexural rotation capacity models fitted to test results using different statistical approaches. Struct. Concr. **19**(2), 608–24 (2018)
10. Lignos, D.G., Hartloper, A.R., Elkady, A., Deierlein, G.G., Hamburger, R.: Proposed updated to ASCE 41 nonlinear modeling parameters for wide-flange steel columns in support of performance-based earthquake engineering. ASCE J. Struct. Eng. **145**(9), 04019083 (2019). https://doi.org/10.1061/(ASCE)ST.1943-541X.0002353
11. Lignos, D.G., Hartloper, A.R.: Steel column stability and implications in the seismic assessment of steel structures according to Eurocode 8 part 3. Stahlbau **89**(1), 16–27 (2020). https://doi.org/10.1002/stab.201900108
12. Lignos, D.G., Krawinkler, H.: Deterioration modeling of steel components in support to collapse prediction of steel moment frames under earthquake loading. ASCE J. Struct. Eng. **137**(11), 1291–1302 (2011). https://doi.org/10.1061/(ASCE)ST.1943-541X.0000376
13. El Jisr, H., Elkady, A., Lignos, D.G.: Composite steel beam database for seismic design and performance assessment of composite-steel moment-resisting frame systems. Bull. Earthquake Eng. **17**(6), 3015–3039 (2019). https://doi.org/10.1007/s10518-019-00564-w
14. European Committee for Standardization (CEN): Eurocode 8: Design of structures for earthquake resistance—Part 1-1: general rules and seismic action. Draft CEN/TC250/SC8/N969 (2020)
15. Lucchini, A., Melo, J.M., Arêde, A., Varum, H., Franchin, P., Rossetto, T.: Load path effect on the response of slender lightly-reinforced square RC columns under biaxial bending. J. Struct. Eng., submitted (2021)
16. Park, Y.-J., Ang, A.H.-S.: Mechanistic seismic damage model for reinforced concrete. J. Struct. Eng. **111**(4), 722–39 (1985)

Evaluation of Different Approaches to Estimate Seismic Input Energy and Top Displacement Demand of Moment Resisting Frames

Furkan Çalım, Ahmet Güllü, and Ercan Yüksel

Abstract Energy-based seismic design (EBSD) has many advantages over the conventional design approaches (force- and displacement-based) provided by the current seismic design codes since it accounts for duration, frequency content and pulse-type of the ground motion. Cumulative damage potential of the earthquake excitation is also taken into consideration by this approach. In order to obtain a satisfactory design, energy capacity of a structure should exceed the energy demand from an earthquake in EBSD. The structural damage indicators like inelastic top displacement demand (δ_{top}) of MDOF system, which can be used as a crucial input data for general nonlinear static analysis procedures, are claimed to be accurately predicted using the input energy. Primarily, seismic input energy per unit mass (E_I/m) imparted into a MDOF frame system during an earthquake is determined by only using modal properties of the system and input energy time series of earthquake ground motion on its equivalent SDOF systems. Using the determined seismic input energy per unit mass, δ_{top} is predicted by literature equations. Effects of spectral matching on the success of seismic input energy and inelastic displacement demand estimations were also investigated. Evaluation of the predictions of δ_{top} was achieved by comparing with the results of nonlinear time history analyses (NLTHA) on two distinct three-story moment resisting frames. It was observed that the relative differences between analyses results and the predicted values were calculated as 18% (input energy prediction) and 30% (top displacement prediction) for the original records whereas they were 12 and 20% for the spectrally matched ground motions.

Keywords Energy-based seismic design · Energy balance equations · Seismic input energy · Inelastic top displacement demand

F. Çalım · E. Yüksel
Istanbul Technical University, Istanbul, Turkey
e-mail: calimf@itu.edu.tr

E. Yüksel
e-mail: yukselerc@itu.edu.tr

A. Güllü (✉)
Istanbul Gedik University, Istanbul, Turkey
e-mail: ahmet.gullu@gedik.edu.tr

A. Benavent-Climent and F. Mollaioli (eds.), *Energy-Based Seismic Engineering*, Lecture Notes in Civil Engineering 155,
https://doi.org/10.1007/978-3-030-73932-4_6

1 Introduction

The existing design philosophies, namely force-(FBSD) and displacement-based seismic design (DBSD), are mainly used in the current seismic design codes. However, it is mentioned that the seismic damages in the structural members are not only related with the peak responses. So, FBSD and DBSD approaches have difficulty in accounting for the duration-dependent cumulative damage potential of the seismic excitation [1].

A relatively new approach named energy-based seismic design (EBSD) have been much sought after in recent years. EBSD approach is claimed to be a more rational and effective alternative for FBSD and DBSD approaches due to its ability to account for the cumulative damage potential of a ground motion by considering the parameters like total duration and frequency content of the excitation as well as hysteretic behavior of structural members [2–5].

Initially seismic input energy (E_I) demand of an earthquake should be determined in EBSD, [5–10]. Hereafter, input energy response of a MDOF system is generally estimated using equivalent SDOF systems (E-SDOF), Fig. 1.

The relative seismic input energy of a lumped mass SDOF system can be derived from the equation of motion. Multiplying both sides of the equation of motion by the velocity response and integrating it over the earthquake duration yields the energy balance equation, Eq. (1). The left-hand side of the equation represents the kinetic energy (E_k), damping energy (E_d), and strain energy (E_s), respectively whereas the right-hand side of the equation represents the input energy (E_I) for a SDOF system [11]. Energy responses of MDOF systems can also be obtained by converting the terms of the equation into matrices and vectors.

$$m\int_0^t \ddot{u}(t)\cdot\dot{u}(t)dt + c\int_0^t \dot{u}(t)\cdot\dot{u}(t)dt + k\int_0^t u(t)\cdot\dot{u}(t)dt = -m\int_0^t \ddot{u}_g(t)\cdot\dot{u}(t)dt \quad (1)$$

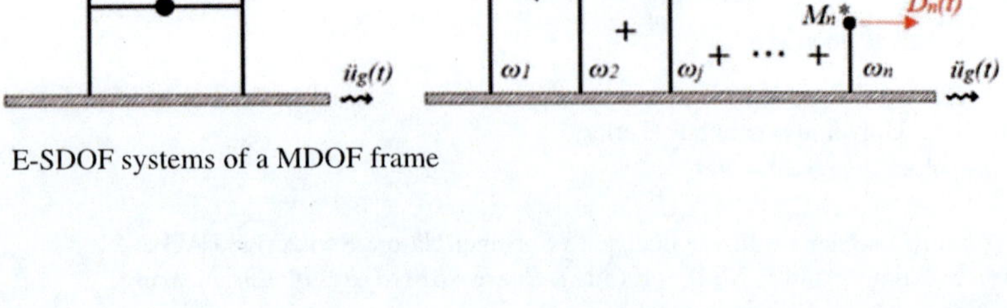

Fig. 1 E-SDOF systems of a MDOF frame

Kalkan and Kunnath [12] suggested an equation to estimate the input energy demand of MDOF systems. The input energy responses of E-SDOF systems are combined using the modal participation factor (Γ), Eq. (2). Here, $\dot{D}i(t)$ is the velocity response of the *ith* mode E-SDOF system. It was mentioned that using up to the first three modes is usually enough to satisfactorily predict the seismic energy demand of MDOF systems. If the mode shapes are not normalized with respect to modal mass, Eq. (3) can be used for the calculation of the E_I time series for MDOF systems instead [13].

$$E_I(t)_{MDOF} = \sum_{i=1}^{n} \left[-\int_0^t \Gamma_i^2 \cdot \ddot{u}_g(t) \cdot \dot{D}_i(t)dt \right] \tag{2}$$

$$E_I(t)_{MDOF} = \sum_{i=1}^{n} \left[-\int_0^t M_i^* \cdot \ddot{u}_g(t) \cdot \dot{D}_i(t)dt \right] \tag{3}$$

The maximum top story displacement (δ_{top}) of a MDOF structure caused by a seismic excitation is crucially important for the evaluation of its seismic performance and imperative for some general nonlinear static analysis procedures [14]. It was also mentioned that predicting the structural damage indicators such as maximum inter-story drift ratios and top displacements based on energy-based ground motion parameters yields more efficient results when compared with the strength-based parameters [15].

A unitless parameter ζ was defined by Teran-Gilmore [16] in order to indicate the relationship between input energy per unit mass and maximum displacement of a SDOF structure, Eq. (4) It was stated that the input energy results in a more effective representation in the seismic demands compared to other energy-based parameters depending on hysteretic energy.

$$\zeta = \frac{\sqrt{E_I/m}}{\omega \cdot \delta} \tag{4}$$

The modified parameter ζ´ was defined in order to study on the correlation between mass normalized input energy and top displacement of a MDOF structure by Mollaioli et al. [17], Eq. (5) where ω_1 is angular frequency of the predominant mode.

$$\zeta' = \frac{\sqrt{E_I/m}}{\omega_1 \cdot \delta_{top}} \tag{5}$$

In this study, predictions for the seismic input energy demand and top displacement of nonlinear MDOF frame systems using both original and scaled earthquake records are presented. The validation of the predictions is done by performing nonlinear time history analyses (NLTHA).

2 Numerical Models

2.1 Selected Benchmark Frames

The analyses were performed for three-story, steel, 3-D moment resisting skeleton frames called *BF1* and *BF2*, Fig. 2. The frames have previously been designed and constructed with a nearly 1/3 geometric scale and experimentally tested on a shake table [18–20].

Both frames have a story height of 0.9 m and one-bay in each direction with a bay width of 1.0 m. The frame *BF1* is formed with the concept of weak column, strong beam while the frame *BF2* is formed considering strong column—weak beam principle. In *BF1*, columns were selected as 40 × 40 × 3 square steel hollow sections (SHS) and beams were selected as 40 × 60 × 3 SHS. Whereas in *BF2*, columns were 40 × 60 × 3 SHS and beams were 40 × 40 × 3 SHS. The benchmark frames were designed such as they have similar vibrational periods [18].

Primarily, modal properties of the frames were calculated. The natural periods of vibration and effective modal masses as proportions of total mass are presented in Table 1 for *BF1 and BF2,* respectively. Since the earthquake excitations were defined

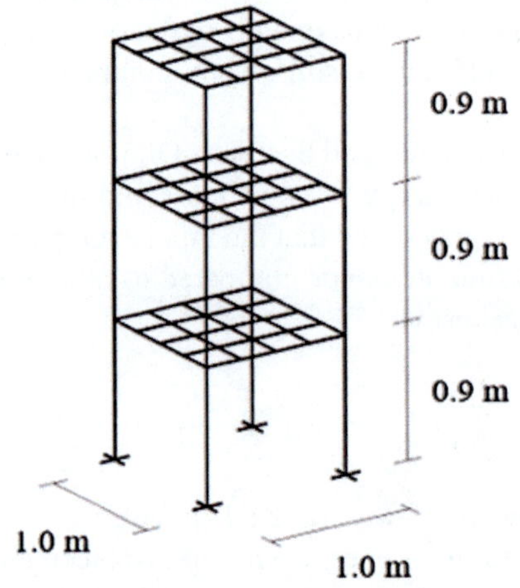

Fig. 2 Geometry of the selected benchmark frames

Table 1 Modal properties of the benchmark frames

Mode	Period (s)		Effective modal mass		Cumulative	
	BF1	BF2	BF1	BF2	BF1	BF2
1	0.477	0.475	0.8825	0.8252	0.8825	0.8252
2	0.159	0.134	0.0974	0.1347	0.9799	0.9599
3	0.101	0.071	0.0117	0.0401	0.9916	1.0000

only in one direction for the NLTHA, the modes in which the vibration of interest is dominant are shown.

2.2 *Ground Motion Selection*

Horizontal acceleration histories of 11 earthquakes were selected in order to perform NLTHA. Spectral matching process was carried out for the selected ground motion records (GMs) using SeismoMatch [21] software. Target design spectrum was constructed for earthquake level-2 (recurrence period of 475 years), local site class of ZD (gravel or clay layers) and location of Istanbul Technical University campus (Latitude: 41.10°, Longitude: 29.02°) based on Turkish Building Earthquake Code [22]. By scaling the original acceleration histories, their response spectra were matched with the target spectrum, Fig. 3. The ground motion parameters are presented in Table 2 for the selected original and matched GMs.

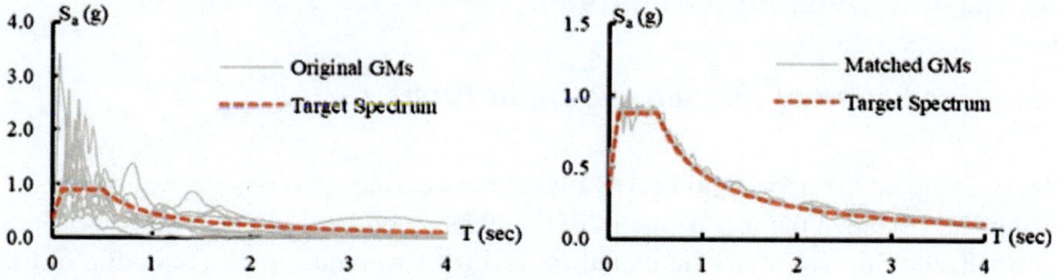

Fig. 3 Matching of selected GMs with the target spectrum

Table 2 Selected ground motion records

GM #	Event	Year	M_W	Original GMs		Matched GMs	
				PGA (g)	PGV (cm/s)	PGA (g)	PGV (cm/s)
1	Chi-Chi	1999	7.7	0.361	21.548	0.547	54.362
2	Duzce	1999	7.2	0.806	65.883	0.654	47.855
3	Friuli	1976	6.5	0.351	22.019	0.524	47.344
4	Hollister	1961	5.6	0.195	12.355	0.500	30.138
5	Imperial Valley	1979	6.5	0.315	31.496	0.374	41.390
6	Kobe	1995	6.9	0.345	27.678	0.417	43.880
7	Kocaeli	1999	7.6	0.349	62.182	0.621	60.818
8	Landers	1992	7.3	0.780	31.598	0.397	43.054
9	Loma Prieta	1989	6.9	0.367	44.695	0.460	49.730
10	Northridge	1994	6.7	0.568	51.827	0.358	42.907
11	Trinidad	1983	5.8	0.194	8.463	0.423	33.196

2.3 Nonlinear Time History Analyses

In order to perform the nonlinear time history analyses and to compute the total seismic input energy imparted into the benchmark frames by GMs, Perform 3D software [23] was used since input energy can be calculated precisely by the program as it provides an energy balance between internal (strain, kinetic, inelastic, and damping) energies and external (input) energy.

Beams and columns of the selected benchmark frames were modelled as nonlinear frame elements using the plastic hinge theory. While the mid-sections of the structural members were assumed to behave elastic under the seismic excitation, lumped plasticity was defined by assigning inelastic fiber hinges at both ends.

An inelastic steel material having a trilinear stress-strain relationship with a strength loss of 25% and without cyclic degradation was defined for inelastic regions of the structural members.

3 Seismic Demand Estimations

3.1 Prediction of the Seismic Input Energy

The seismic input energy imposed by the selected earthquake records were predicted using Eq. (3) since the mode shapes of the considered benchmark frames were not normalized with respect to the modal mass. First two modes in the direction of the applied seismic action were used in the analysis since 98% and 96% of the total mass was achieved for the frames *BF1* and *BF2*, respectively. Effective modal masses to be used in the formula were calculated as 2.601 t and 0.287 t for the first two modes of *BF1* whereas they were calculated as 2.491 t and 0.407 t for the first two modes of *BF2*.

Input energy (E_I) time series obtained as a result of NLTHA and the contribution of each mode in the input energy time series using the modal procedures are presented in Figs. 4 and 5 for original and matched GMs, respectively. Since the benchmark frames are low-rise, the effects of the second mode are negligible for both of the frames.

Maximum E_I values obtained through NLTHA, Eq. (3), and their relative differences are represented in Table 3 for the original GMs and Table 4 for the matched GMs. It was observed that the inelastic input energy demands can be predicted accurately using the elastic input energy demands (associated with the natural modes) by using the equal energy principle, especially for the scaled GMs. The arithmetic means of the relative differences were computed to be 18 and 12% for the original and matched GMs, respectively. Higher discrepancy was observed for some records with relatively higher PGV values.

Table 3 Seismic input energy estimations for the original GMs

GM #	BF1			BF2		
	$(E_I)_{NLTHA}$ (kNm)	(E_I) Eq. (3) (kNm)	Relative Diff. (%)	$(E_I)_{NLTHA}$ (kNm)	(E_I) Eq. (3) (kNm)	Relative Diff. (%)
1	0.233	0.202	13.5	0.248	0.194	21.9
2	3.993	1.750	56.2	3.733	1.678	55.0
3	1.650	1.368	17.0	1.484	1.308	11.8
4	0.332	0.310	6.6	0.342	0.296	13.4
5	1.187	1.273	7.3	1.107	1.249	12.9
6	1.447	1.176	18.7	1.240	1.094	11.8
7	0.543	0.633	16.4	0.576	0.606	5.2
8	0.769	0.729	5.2	0.670	0.652	2.7
9	0.926	0.796	14.0	0.845	0.767	9.2
10	2.608	1.901	27.1	2.530	1.809	28.5
11	0.065	0.075	16.2	0.061	0.072	18.9
Mean			18.0			17.4

Table 4 Seismic input energy estimations for the matched GMs

GM #	BF1			BF2		
	$(E_I)_{NLTHA}$ (kNm)	(E_I) Eq. (3) (kNm)	Relative Diff. (%)	$(E_I)_{NLTHA}$ (kNm)	(E_I) Eq. (3) (kNm)	Relative Diff. (%)
1	1.004	0.885	11.9	0.959	0.849	11.5
2	1.398	0.970	30.6	1.357	0.922	32.1
3	2.698	2.365	12.3	2.381	2.263	5.0
4	2.605	2.212	15.1	2.493	2.112	15.3
5	2.398	2.553	6.4	2.492	2.455	1.5
6	2.764	2.540	8.1	2.590	2.426	6.3
7	1.889	2.015	6.7	1.896	1.930	1.8
8	2.094	2.260	7.9	1.889	2.162	14.4
9	1.265	1.158	8.5	1.229	1.102	10.3
10	2.203	1.977	10.2	2.076	1.882	9.4
11	1.516	1.739	14.7	1.273	1.658	30.2
Mean			12.0			12.5

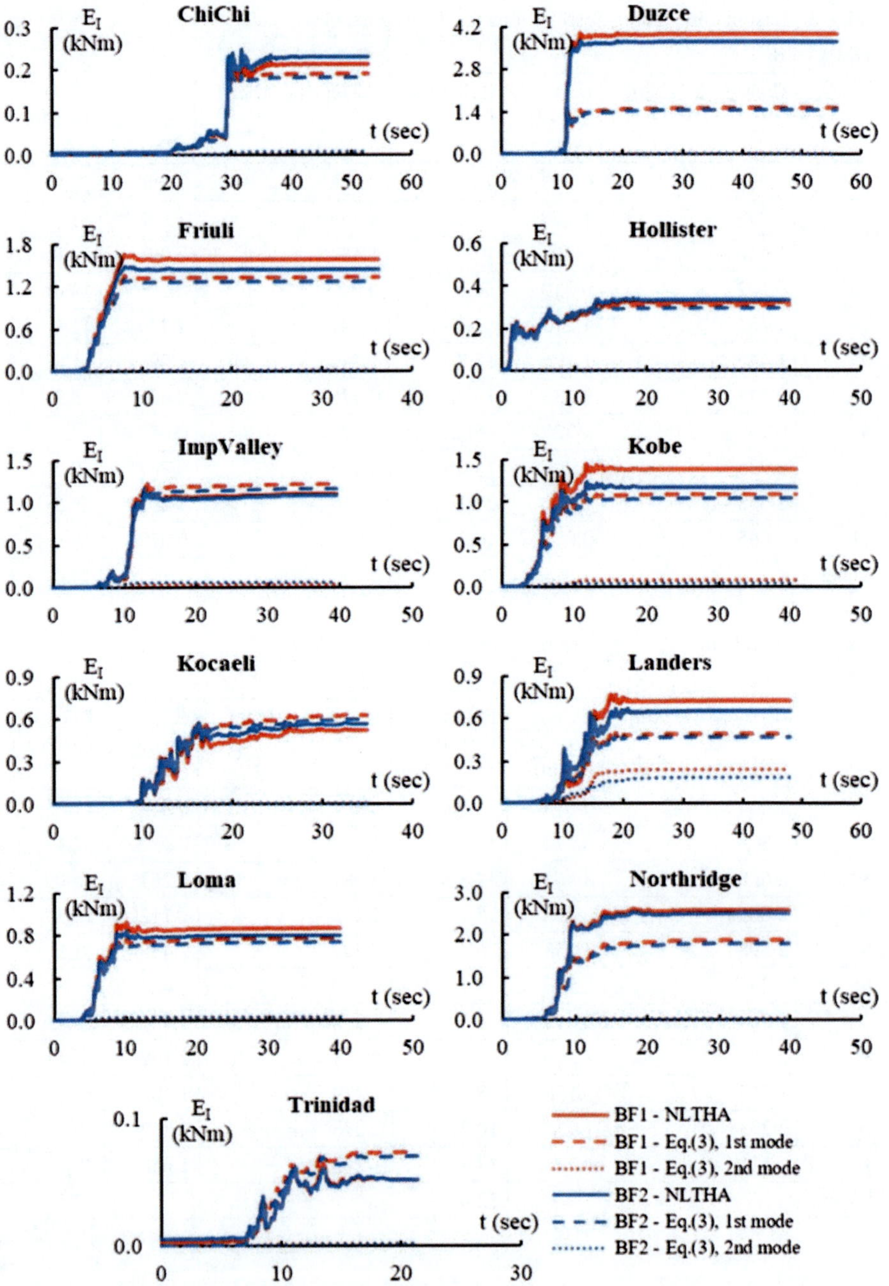

Fig. 4 Input energy time series for the original GMs

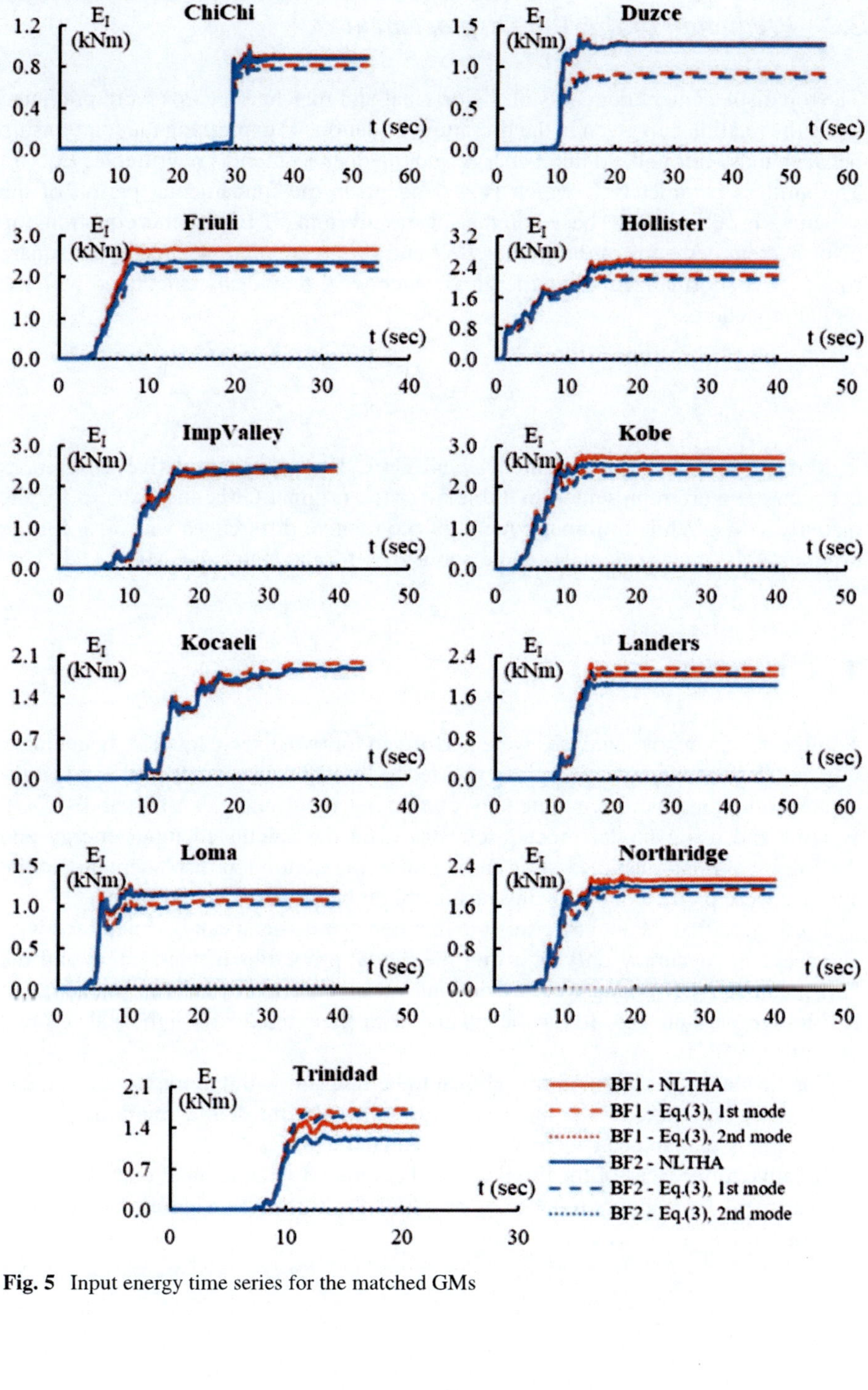

Fig. 5 Input energy time series for the matched GMs

3.2 Prediction of the Top Displacement

The top displacement demands of the original and matched records were predicted using the relationship given in the literature. Equation (4) expressing the relationship between mass-normalized input energy and top displacement is rewritten as Eq. (6). The unitless parameter ζ', which is dependent on the fundamental period of the structure, is taken as 1.10 based on the spectra given in [17]. With this equation, top displacement demands of the frames *BF1* and *BF2* were predicted using the square root of mass-normalized input energies, which were previously calculated with the modal procedures.

$$\delta_{top} = \frac{\sqrt{E_I/m}}{\omega_1 \cdot \zeta'} \tag{6}$$

Maximum δ_{top} values obtained by NLTHA, Eq. (6), and relative differences between them are represented in Table 5 for the original GMs and Table 6 for the matched GMs. While arithmetic mean of the relative differences was 30% for the original GMs, it was calculated to be about 20% for the matched GMs.

4 Conclusions

Nonlinear time history analyses were performed for two diverse low-rise benchmark frames. The seismic input energies imparted to MDOF frames by the selected earthquake excitations were computed by combining input energies of their E-SDOF systems and using modal procedures. Based on the calculated input energy and the empirical relationship existing in literature, maximum top displacements of the systems were predicted. Following results might be driven from the study.

It was seen that the inelastic seismic input energy demand can be calculated with an acceptable accuracy only by using the modal properties of the system and the GM parameters. Less accurate predictions were obtained for Duzce, Kocaeli, and Northridge earthquakes. It can be related with their relatively high peak ground velocities (PGV).

Top displacement demand was shown to be relevant with the maximum seismic input energy imparted into the structure. With a more stable relationship, top displacement demand can accurately be predicted.

Relative differences of the predicted and computed seismic input energy and top displacement demands are reduced by matching the acceleration spectra of the actual records with a target design spectrum.

Table 5 Top displacement estimations for the original GMs

GM #	BF1			BF2		
	$(\delta_{top})_{NLTHA}$ (m)	$(\delta_{top})_{predicted}$ (m)	Relative Diff. (%)	$(\delta_{top})_{NLTHA}$ (m)	$(\delta_{top})_{predicted}$ (m)	Relative Diff. (%)
1	0.033	0.020	41.2	0.036	0.018	51.1
2	0.147	0.054	63.6	0.168	0.052	69.2
3	0.055	0.047	14.0	0.048	0.046	4.6
4	0.034	0.023	33.5	0.034	0.022	36.8
5	0.057	0.046	20.2	0.051	0.045	12.5
6	0.064	0.044	31.4	0.059	0.042	28.9
7	0.042	0.032	23.4	0.043	0.031	27.3
8	0.039	0.035	12.3	0.043	0.032	25.2
9	0.054	0.036	33.1	0.051	0.035	31.8
10	0.096	0.056	42.0	0.098	0.054	45.2
11	0.014	0.011	22.7	0.015	0.011	28.3
Mean			30.7			32.8

Table 6 Top displacement estimations for the matched GMs

GM #	BF1			BF2		
	$(\delta_{top})_{NLTHA}$ (m)	$(\delta_{top})_{predicted}$ (m)	Relative Diff. (%)	$(\delta_{top})_{NLTHA}$ (m)	$(\delta_{top})_{predicted}$ (m)	Relative Diff. (%)
1	0.074	0.038	48.8	0.083	0.037	55.5
2	0.079	0.040	49.6	0.085	0.038	54.9
3	0.057	0.062	8.3	0.051	0.060	17.2
4	0.069	0.060	12.2	0.076	0.058	23.6
5	0.061	0.065	6.4	0.063	0.063	0.1
6	0.055	0.064	18.2	0.057	0.062	9.1
7	0.063	0.057	9.2	0.057	0.056	2.1
8	0.054	0.061	12.1	0.059	0.059	0.3
9	0.059	0.044	26.7	0.061	0.042	31.7
10	0.077	0.057	26.5	0.078	0.055	29.9
11	0.055	0.053	2.7	0.054	0.051	4.2
Mean			20.1			20.8

References

1. Chou, C.C., Uang, C.M.: Establishing absorbed energy spectra—an attenuation approach. Earthquake Eng. Struct. Dyn. **29**(10), 1441–1455 (2000)
2. Uang, C.M., Bertero, V.V.: Evaluation of seismic energy in structures. Earthquake Eng. Struct. Dyn. **19**(1), 77–90 (1990)
3. Fajfar, P., Fischinger, M.A.: Seismic procedure including energy concept. In: Ninth European Conference on Earthquake Engineering. Moscow, Russia (1990)
4. Akbas, B., Shen, J.: Earthquake-resistant design (EQRD) and energy concepts. IMO Teknik Dergi **192**, 2877–2901 (2003)
5. Güllü, A., Yüksel, E., Yalçın, C., Dindar, A.A., Özkaynak, H., Büyüköztürk, O.: An improved input energy spectrum verified by the shake table tests. Earthquake Eng. Struct. Dyn. **48**(1), 27–45 (2019)
6. Decanini, L., Mollaioli, F., Mura, A.: Equivalent SDOF systems for the estimation of seismic response of multistory frame structures. WIT Trans. Built Environ. 57, 101–110 (2001). https://doi.org/10.2495/ERES010101
7. Chou, C.C., Uang, C.M.: Evaluating distribution of seismic energy in multistory frames. In: The 13th world conference on earthquake engineering, Vancouver, B.C., Canada (2004)
8. Lei, C., Xianguo, Y., Kangning, L.: Analysis of seismic energy response and distribution of RC frame structures. In: The 14th World Conference on Earthquake Engineering, Beijing, China (2008)
9. Mezgebo, M.G., Lui, E.M.: A new methodology for energy-based seismic design of steel moment frames. Earthquake Eng. Eng. Vibr. **16**(1), 131–162 (2017). https://doi.org/10.1007/s11803-017-0373-1
10. Dindar, A.A., Yalçın, C., Yüksel, E., Özkaynak, H., Büyüköztürk, O.: Development of earthquake energy demand spectra. Earthquake Spectra **31**(3), 1667–1689 (2015)
11. Akiyama, H.: Earthquake-Resistant Limit-State Design for Buildings. University of Tokyo Press (1985)
12. Kalkan, E., Kunnath, S.K.: Effective cyclic energy as a measure of seismic demand. J. Earthquake Eng. **11**(5), 725–751 (2007). https://doi.org/10.1080/13632460601033827
13. Ucar, T.: Computing input energy response of MDOF systems to actual ground motions based on modal contributions. Earthquakes Struct. **18**(2), 263–273 (2020). https://doi.org/10.12989/eas.2020.18.2.263
14. Sürmeli, M., Yüksel, E.: A variant of modal pushover analyses (VMPA) based a non-incremental procedure. Bull. Earthquake Eng. **13**, 3353–3379 (2015)
15. Sari, A., Manuel, L.: Correlation of structural performance with energy- and strength-based parameters. In: Sixth National Conference on Earthquake Engineering. Istanbul, Turkey (2007)
16. Teran-Gilmore, A.: A parametric approach to performance-based numerical seismic design. Earthquake Spectra **14**, 3 (1998)
17. Mollaioli, F., Bruno, S., Decanini, L., Saragoni, R.: Correlations between energy and displacement demands for performance-based seismic engineering. Pure Appl. Geophys. **168**, 237–259 (2011)
18. Güllü, A.: Determinations of the inelastic displacement demand and response control of steel frame type structures by seismic energy equations. Ph.D. Thesis, Istanbul Technical University, Istanbul (2018)
19. Güllü, A, Yüksel, E, Yalçın, C.: Seismic energy-based design: numerical evaluation of diverse MDOF systems. In: Sixteenth European Conference on Earthquake Engineering. Thessaloniki, Greece (2018)
20. Güllü, A, Yüksel, E.: Piecewise exact solution of the seismic energy balance equation and its verification by shake table tests. J. Eng. Mech., under review
21. SeismoSoft, SeismoMatch—Response spectrum matching software

22. TBEC-2018: Turkish Building Earthquake Code-2018, Specifications for buildings to be built in seismic areas. Ministry of Public Works and Settlement, Ankara, Turkey (2018)
23. CSI Inc., Perform 3D Performance based design software

NDSHA—A Reliable Modern Approach for Alternative Seismic Input Modelling

Mihaela Kouteva-Guentcheva and Giuliano F. Panza

Abstract The energy-based design concept considers earthquake effect as an energy input and how this energy is distributed within the structure. The structural damage indicates that some of this energy was not duly dissipated by the structure. Realistic alternative time histories representation of seismic action is crucial for reliable assessment of the earthquake energy input, transmitted to the structures, as well as for relevant damage capacity of the site-specific seismic actions. The Neo-Deterministic Seismic Hazard Assessment approach, NDSHA formulated at the turn of Millennium provides a reliable base for further cause–failure–impact analyses and can be applied for SHA at local, national, and regional scales. NDSHA, in its standard form, defines the hazard as the largest ground shaking at the site, computed considering a large set of scenario earthquakes, including the maximum credible earthquake (MCE), whose magnitude and focal mechanism is defined from regional seismic history and seismotectonics. This approach allows for a realistic description of the seismic ground motion due to an earthquake of given epicentral distance and magnitude at any point of interest within a given construction site. It relies on the existing acquired expert information—usually available via the technical documentation of the infrastructure projects—e.g., the comprehensive geological reports that provides detailed geological and geophysical data on the site. This contribution provides a brief

M. Kouteva-Guentcheva (✉)
National Earthquake Engineering Center—University of Architecture, Civil Engineering and Geodesy (NEEC-UACEG), Sofia, Bulgaria
e-mail: kouteva_fce@uacg.bg

G. F. Panza
Accademia Nazionale dei Lincei, Rome, Italy

Institute of Geophysics, China Earthquake Administration, Beijing, China

Accademia Nazionale delle Scienze detta dei XL, Rome, Italy

International Seismic Safety Organization (ISSO), Arsita, Italy

Beijing University of Civil Engineering and Architecture (BUCEA), Beijing, China

G. F. Panza
e-mail: giulianofpanza@fastwebnet.it

A. Benavent-Climent and F. Mollaioli (eds.), *Energy-Based Seismic Engineering*, Lecture Notes in Civil Engineering 155,
https://doi.org/10.1007/978-3-030-73932-4_7

description of the worldwide-validated NDSHA tools and illustrates their relevance for the energy-based design concept.

Keywords Seismic input · Earthquake scenario · Neo-deterministic seismic hazard assessment · Realistic seismic loading · Validation

1 Introduction

The need of realistic reliable seismic risk assessment corresponding to the current state of a given structure within the modern earthquake design is raised by the need of providing adequate models of the seismic demand for structural analysis of the large set of various, for age and complexity, buildings, and structures. The breakdown of residential building by construction, available at the official website of the European Union, Fig. 1, shows that in most of the countries the residential buildings, built after 2010 represent less than 10% of the overall stock. The prevailing part of the existing residential buildings was built last century and therefore does not meet the requirement of the modern Eurocodes (e.g., [1]) for design and construction that are mandatory for the specification of European public works, starting from March 2010, and are de facto standard also for the private sector. This new generation modern codes for earthquake resistant design and construction raises more rigorous requirement, starting with the seismic loading modelling. We did not find similar statistics for other buildings or critical infrastructure components, but practitioners know that the picture is not that different. The trend shown by the urbanization rate statistics in Europe (year 2013) indicates that the average distribution of residential

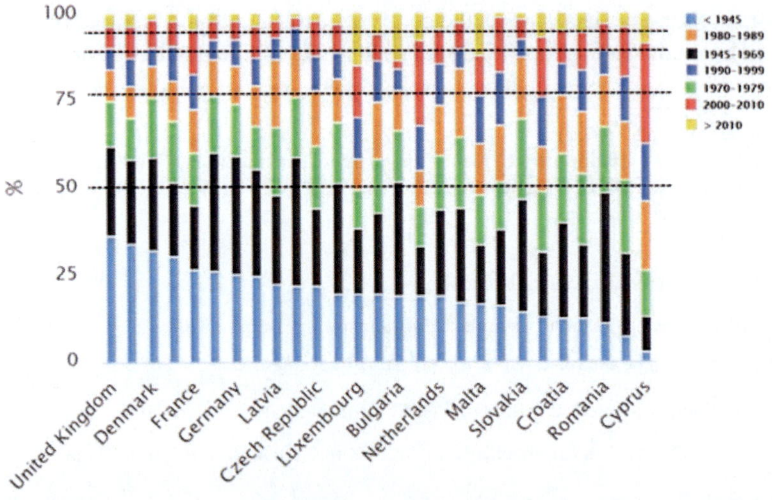

Fig. 1 Breakdown of residential building by construction year (2014), EU Buildings Factsheets | Energy (https://europa.eu)

buildings is subdivided as follows: 42% in the urban centers, 30% in intermediate urban areas, and the remaining 28% in rural areas. In the same analysis it is clearly mentioned that in most EU countries, half of the residential stock was built also before the first thermal regulations, built before 1970 (EU Buildings Factsheets | Energy (https://europa.eu)).

A major challenge in Europe is to perform retrofitting of existing buildings, coupling at the same time both, the new standards for earthquake—resistance [1] and the energy efficiency requirements [2]. Because 80% of the current EU building stock was built before the 90s, while 40% are pre-60s and a considerable amount being even older and classified as cultural patrimony, it is necessary to elaborate specific techniques to maintain this cultural heritage for our future generations. Upgrading the existing EU buildings and the cultural heritage ones is becoming increasingly important due to (1) their poor seismic performance during recent earthquakes (e.g. in Italy, 2012, 2016 or Greece, 2012, 2019) that have resulted in loss of human lives, severe injuries and significant economic losses (2) their low energy performance which increases significantly their energy consumption [3] (https://ec.europa.eu/jrc/en/research-topic/improving-safety-construction/i-resist-plus). Energy retrofit is already part of the EU policy, but in fact, for the purpose of this policy the society needs standing up buildings and structures, during and after future earthquakes. The modern life dynamics including the urbanization rate, the aged building stock, and the evolution of the earthquake-resistant legislation clearly indicate the necessity of reliable control of the seismic risk at any scale—e.g., single structure, municipality, urban area, national, regional.

Reliable risk management relies upon relevant hazard identification and assessment, the latter requiring tools for prognostic qualitative damage assessment. To propose, design and to realize relevant seismic retrofitting, it is necessary to estimate the current stress-strain state of existing structures considering (a) earthquake history load within their live so far, and (b) potential invisible accumulated damages. Structural damage during an earthquake might be the result of excessive deformation or cumulative cyclic damage—in both situations we need reliable methods for estimating displacement demands on structures. Definition of completely satisfactory measure of the capacity of an earthquake to damage a built structure has been a major engineering challenge. Modern energy-based methods are effective tools for a comprehensive interpretation of the behavior observed during destructive events and can provide more insight into the seismic performance. To consider the effect of a particular earthquake or set of seismic events on a given structure, it is necessary to use tools for scenario-based modelling of the relevant earthquake loads.

General aim of the earthquake engineering and of the PBSD procedure is to build earthquake resilient systems aiming to reduce: (i) failure chances and consequences from failures (lives lost, damage, etc.) and (ii) time to recovery of the damaged elements to their normal functions. The necessity of well verified and validated scenario-based tool for seismic input modelling is clearly supported by some recent examples—e.g., L'Aquila–Italy, 2009; Haiti, 2010; Chile, 2010; Japan, 2011; Emilia–Romagna Italy, 2012; Central Italy, 2016; Ischia-Italy, 2017; Gansu–China, 2013. These severe seismic events in urban regions during the last two decades indicated a clear discrepancy between the predicted ground motion levels, estimated via the traditional Probabilistic Seismic Hazard Assessment method, PSHA, and the instrumental observations of the occurred strong ground motion severity. Most modern codes for earthquake resistant design and construction are still based on PSHA maps, despite the fact, that during the last 30 years, with the acquired knowledge and data, an impressive amount of evidence criticizing PSHA have been published expressing various professional points of view: engineers [4–8]; physicists [9, 10]; mathematicians [11], and last, but no least, statisticians [12]. The last generation PSHA model, Global Earthquake Model (GEM-www.globalquakemodel.org/), moves beyond the Global Seismic Hazard Assessment Program (GSHAP-https://www.seismo.ethz.ch/static/GSHAP/index.html) by seeking not only a fully harmonized, transparent and state-of-the-art global seismic hazard but a seismic risk model also. GEM is on the wrong track, if it continues to base seismic risk estimates on the standard method to assess seismic hazard [13].

Not few recent examples that make us to question what we should do towards better, reliable and thrustful seismic input definition for engineering nonlinear structural analysis are published. Some of them, Table 1, have been analyzed by Rugarli et al. [14] who clearly show that those events demonstrated that a PBSD approach based on PSHA is neither reliable nor cost effective.

Table 1 Losses due to failure of the usage PSHA ([14] and references in)

Earthquake	Losses
Christchurch earthquake, New Zealand 22 February 2011 $M_R = 6.2$	• 181 deaths • At least 900 buildings in the business district • Over 10,000 homes had to be demolished • Repairs cost was estimated in about US$ 15–20 billions • The highest cost ever caused by an earthquake in New Zealand *The event exceeded the P = 2500 years acceleration response spectra prescribed by the New Zealand seismic code*
Tohoku earthquake, Japan 11 March 2011 $Mw = 9$	• Devastating tsunami US$ 260 billion
Wenchuan (Sichuan) earthquake, China 12 May 2008 $M = 7.9$	• US $124 billion of direct losses • US$ 100 billion of indirect losses to production and housing sectors
Italy 1944 to 2012	• €181 billion in public funding *The devastating series of earthquakes that struck the country between August and October 2016 indicated recorded spectral accelerations much higher than those with a "return period" of 2475 years given by the Building Code*

2 The Seismic Demand and the Energy-Based Concept

Earthquake response of structures and their foundations is a result of the complex interaction between the seismic input ground motion and the continuously changing dynamic characteristics of the system subjected to the ground motion. To reach a reliable assessment of assets, a complete understanding of both input motion and structural system, and their interaction, is required.

The energy-based design concept considers earthquake effect as an energy input and how this energy is distributed within the structure; the design criterion in the energy-based methods is achieved by the comparison of the structure energy absorption capacity with the earthquake energy input. The structural damage indicates that some of the input energy was not duly dissipated by the structure. In fact, earthquake resistance analyses based on the energy-based design concept provides realistic assessment when the duration of the earthquake is considered and thus, it is consistent with the physics of the system. The seismic energy input must be balanced with the energy dissipated in the structure throughout energy absorption or dissipation mechanisms such as damping or inelastic action (hysteretic energy) while the members remain within the expected performance limits. The seismic performance is directly related to the local and global deformations of the structure. The energy-based parameters are directly related to cycles of response of the structure so they can implicitly capture the effect of ground motion duration, which is practically ignored by conventional spectral parameters [15]. In the Energy Based Design methodology damage is directly related to the input energy, therefore with the damage potential of the input seismic strong ground motion. In Performance-Based Seismic Engineering (PBEE), usually, seismic input to the structure is comprehensively defined by the Intensity Measure (IM) parameter. Another parameter for quantitative seismic hazard assessment is the input energy that considers the contribution of amplitude, frequency content and duration of ground motion, hence, it conveys information on the ground motion duration and the effects of inelasticity and ground motion duration, usually overlooked by the conventional spectral parameters, are implicitly captured by input energy [16].

Modern alternative for modelling the seismic action for the purpose of the energy-based concept is the representation of the ground motion via acceleration time-histories and related quantities (velocity and displacement). Most of the codes for earthquake resistant design recommend the use of a minimum of three to seven different seismic signals. There are two approaches to obtain earthquake-like ground motion time histories for the purposes of the advanced structural analyses in earthquake engineering: (a) natural earthquake recordings and (2) computed seismic time histories, Fig. 2. Depending on the nature of the application and on the information available, the seismic motion can be modelled by recorded or artificial accelerograms, or simulated accelerograms. With the larger availability of high-quality strong motion data various seismic sources in the past decades, natural records of earthquakes have increased significantly, but they remain still intrinsically insufficient for structural analyses. In general, it is necessary to select accelerograms from other regions that have produced significant strong motion recordings. This is an early decision, which, in fact, very often neglects the effect on the ground motion due to the seismic source mechanism (normal, inverse, strike-slip) for inter-plate or intra-plate earthquakes and attributes them incorrectly to the source and path effects. The parameters that characterize the conditions under which strong-motion records queries are generated within the strong motion databases can be grouped into three sets representing: (i) the earthquake source (magnitude, rupture mechanism, directivity, and focal depth); (ii) the path from the source to the recording site (distance and azimuth) and (iii) the nature of the site (surface geology and topography). Matching a design scenario and selected time history is based on reconciling magnitude–focal depth–distance–soil condition. To consider the uncertainties in the evaluation of the parameters that describe the severity (magnitude) and the location of the seismic event (focal depth) it is reasonable to select records within ranges from the design event to increase the possibility of finding a viable record suite. This issue is overcome by the NDSHA thanks to its capability to perform fast parametric sensitivity analyses for the modelled seismic scenario, described by the mentioned set of parameters.

Artificial acceleration records are generated signals that satisfy given engineering criteria but are not related to the physics of earthquake stress wave generation and propagation. Usually, these signals are generated based on three elements: (i) power spectral density, (ii) random phase angle generator and (iii) an envelope function; and the simulated motion can be calculated as the sum of several harmonic excitations [17]. Due to the large number of cycles that are shown by the artificial seismic signals, they may represent unrealistic seismic demand for inelastic structural systems. Uniform hazard and code spectra represent many scenarios characterized by different magnitude, distance, and soil—having a single record representing tens of feasible scenarios may lead to over-conservatism of the ensuing artificial motion [17].

Scenario based generation of realistic earthquake-like signals become possible with the considerable advances in earthquake geophysics, wave propagation modelling and computing facilities. From earthquake engineering perspective, validated models for this generation may be used to compute input motions in areas of the world where natural records do not exist. These approaches are constrained by four specific factors, shown in Fig. 3. The Neo-Deterministic Seismic Hazard Assessment NDSHA method, proposed some twenty years ago, is shown to both reliably and realistically simulate the wide suite of earthquake ground motions that may impact civil populations, as well as their heritage buildings.

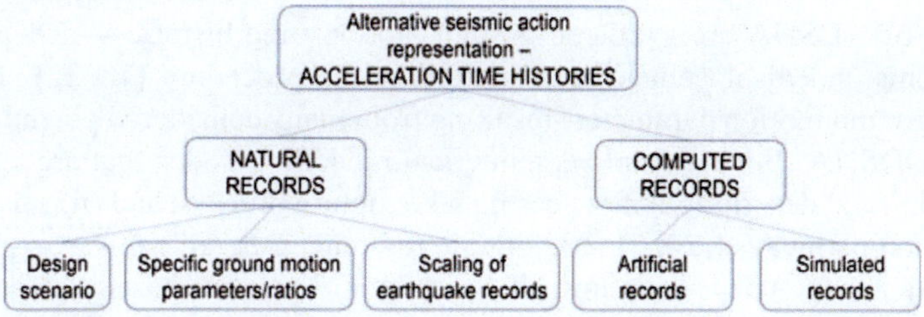

Fig. 2 Alternative seismic action representations

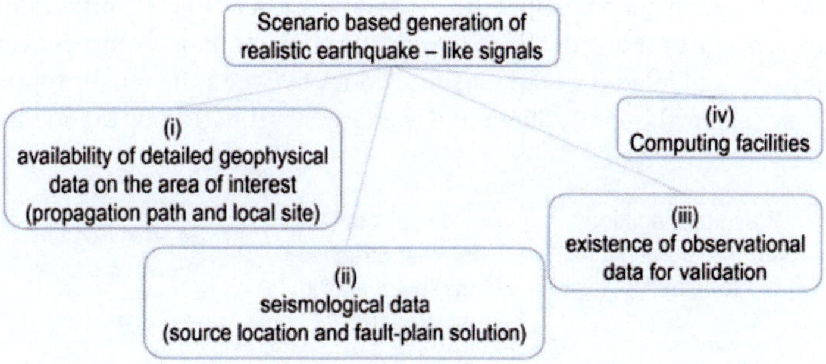

Fig. 3 Factors constraining the scenario-based approaches for generating realistic signals

3 The Neo-deterministic Seismic Hazard Assessment (NDSHA) Approach

Seismic hazard assessment, SHA, aims to estimate the severity of the earthquake ground motion that depends on different factors—earthquake magnitude, site-to-source distance, propagation path, local site conditions, focal depth, source directivity. Neo-Deterministic Seismic Hazard assessment, NDSHA, is a scenario-based approach that provides computational estimation of an envelope of scenario ground-shaking characteristics of both (1) the largest historically observed earthquake within a region; and (2) the Maximum Credible Earthquake MCE, whose magnitude and focal mechanism is defined making use of available geological and geophysical data.

This envelope is calculated by means of physically rooted models based on the available physics-based knowledge on earthquake source and wave propagation processes. Because each scenario is always "a real earthquake", NDSHA does not require considerations of either probabilistic hazard model temporal representations of earthquake "likelihood", or scalar empirical Ground Motion Prediction Equation attenuation models (GMPEs) that cannot account for the tensor nature of the earthquake ground motion [18].

NDSHA provides strong ground motion parameters based on the seismic waves' propagation modelling, accounting for a wide set of possible seismic sources and for the available information about structural models (earthquake scenarios). The end products of NDSHA are synthetic ground motion time histories—accelerograms, velocigrams and displacement time histories—the procedure [18–22]. To predict reliably ground motion parameters resulting from many considered potential seismic sources NDSHA (Fig. 4) employs numerical modelling codes that are based upon the: (i) physical description of the earthquake rupture process; and (ii) seismic wave propagation pathways ([20, 21, 23, 24] and references therein).

A major benefit of adopting NDSHA is the physics-based framework for computing synthetic seismograms that can be performed with different levels of details, depending on the purpose of the analysis—from national scale seismic hazard mapping based on Regional Scale Analysis (RSA) to local seismic hazard assessment mapping based on Site-Specific Analysis (SSA) [14]. In principle, NDSHA is a multi-scenario-based procedure which supplies realistic complete time history ground motions, calculated as the tensor product between the tensor representing in a formal way the earthquake source and the Green's function of the medium.

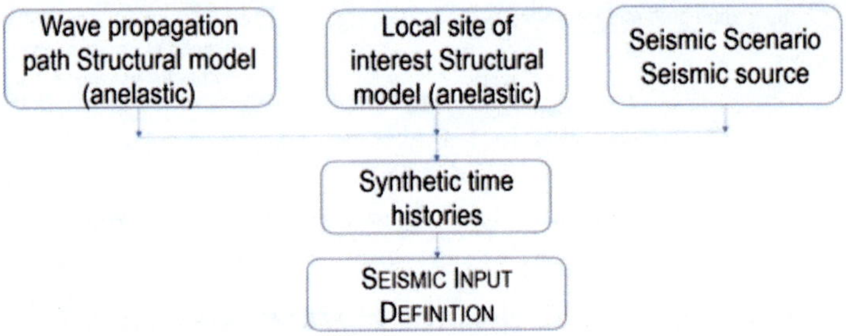

Fig. 4 NDSHA conceptual flowchart

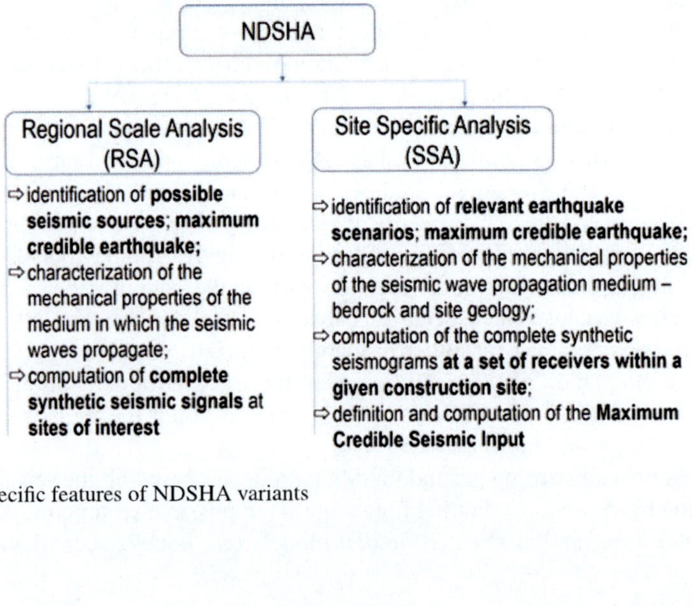

Fig. 5 Specific features of NDSHA variants

Computed seismograms are used to estimate engineering relevant strong ground motion that can be used directly as input for the nonlinear time history analysis of structures parameters—e.g., Peak Ground Acceleration (PGA), Peak Ground Displacement (PGD), Peak Ground Velocity (PGV) and spectral values, Input Energy, etc. The "Regional Scale Analysis" (RSA) is carried out using many possible seismic sources and simplified structural models, representative of bedrock conditions. SSA considers structural and topographical heterogeneities, but also the effect of the source rupture process on the seismic wave yields at a site. So far, the NDSHA method has been applied in many countries, worldwide, at different levels of detail [21, 25]. Recent comprehensive review of the NDSHA achievements in Central and South-eastern Europe has been performed by [26]. Some features of NDSHA can be tested thanks to the development of a web application (https://www.xeris.it/index.html) [27]. A comprehensive illustration of the steps required to perform a RSA and a SSA, starting from NDSHA computations, with a focus on the Italian territory, where technical construction standards [28] explicitly assert that the use of accelerograms generated simulating source mechanism and wave propagation is allowed provided the hypotheses about the seismogenic characteristics of the source and the properties along the pathway are duly justified [20, 21]. Relevant upgrades in the seismograms' computation to meet engineering needs are described by Fasan et al. [29–31] and Magrin et al. [23] and are briefly summarized in Fig. 5.

3.1 Regional Scale Analyses

Multi-performance levels concept, introduced by the Vision 2000 report [32] remains a major contribution within the development of the PBSD philosophy of design [14]. The association of the four defined performance levels (acceptable damage that a building is supposed to achieve during earthquakes of different strength)—Operation-al Limit (OL), Immediate Occupancy (IO), Life Safety (LS) and Collapse Prevention (CP)—with "return periods" of 43, 72, 475 and 970 years that should correspond to a "probability" of exceedance of 69, 50, 10 and 5% in an interval of 50 years is practically unclear and unconvincing [33–35]. Comprehensive analysis of this misleading assumption, for administrators and even dangerous for decision makers, has been recently provided [33].

The NDSHA comes as innovative, already well validated, new paradigm for the definition of the seismic hazard that can be directly used within PBSD solutions. NDSHA estimation is based on the maximum magnitudes expected at a site regardless of their likelihood of occurrence. Ground motion parameters of interest, or so-called Intensity Measures (IM)—e.g., PGA or Spectral Acceleration (SA), computed making use the NDSHA tools [20, 21] are derived with no help of the empirical equations such as Ground Motion Prediction Equations (GMPEs). The basic data uncertainties and limitations—earthquake catalogues and/or lack of satisfactory information on earthquake sources is a common challenge, faced also by NDSHA. The way NDSHA deals with this challenge is to supply the hazard values

Table 2 Example of discrete ranges of hazard values, in geometrical progression close to 2, consistent with real resolving power of the worldwide available data—mostly catalogues of historical earthquakes [36, 40, 41]

Acceleration (g)	0.000 0.001	0.001 0.002	0.002 0.004	0.004 0.008	0.008 0.015	0.015 0.030	0.030 0.060	0.060 0.120
Velocity (cm/s)	0.0 0.5	0.5 1.0	1.0 2.0	2.0 4.0	4.0 8.0	8.0 15.0	15.0 30.0	30.0 60.0
Displacement (cm)	0.0 0.1	0.1 0.5	0.5 1.0	1.0 2.0	2.0 3.5	3.5 7.0	7.0 15.0	15.0 30.0

at national/regional scale as discrete ranges in geometrical progression close to 2 over areas whose extension is consistent with the information content of the basic data, Table 2 [36]. This is a suitable approach also to avoid potential overdetailing of mapping past earthquakes, which might decrease the maps' capacity for prognostic estimation of future shaking (RSA). When necessary, more specific hazard estimates can be obtained at local scale (SSA) [33].

The original NDSHA formulation [20, 21] deals with a first approximation and fundamental problems posed by an adequate description of the physical process of earthquake occurrence. It takes the Maximum Credible Earthquake (MCE)—the largest event physically possible, whose magnitude M at a given site can be tentatively set equal to the *maximum* observed or estimated magnitude *Mdesign* (until proven otherwise) plus some multiple of its accepted global standard deviation (σ_M) [37]. Historical and morphostructural data are used to estimate the maximum in areas without information on faults or sparse data. The predictive capabilities of NDSHA [20, 38] has been systematically verified since 1996, in the framework of different international research projects—UNESCO–IUGS–IGCP 414, NATO SfP IVESB 972266 and 65/219.33/01/12/2000, Marie-Curie Training site grant EVK2-CT-2000-57002, CEI 2007-2009, NATO SfP 980468, INTAS Research Project 05-104-7584. Several valuable collections of papers were published in 2000 [25, 38, 39].

The adequate simulation of far field ground motion is performed computationally very efficiently using the Modal Summation (MS) technique [19, 20, 42]. For the short paths in the near field conditions a Discrete Wave Number technique (DWN) can be applied. The full wave field, including all body waves and near field, is achieved though the DWN implementation [43]. In 2013, Magrin [44] provided analysis of the increase of the computational cost of DWN with epicentral distance-source depth ratio—the greater is the ratio between the epicentral distance and the depth of the source, the more the calculation time grows. The NDSHA evolution encompasses a reasonable optimization between accuracy and CPU time—MS is used in computations at large distances, routinely up to 150 km and DWN—for epicentral distances less than 20 km. The NDSHA RSA synthetic seismograms are then computed over the national territories with a frequency cut off at 10 Hz for each node of a grid of 0.2°×0.2° shifted by 0.1° from the grid of the sources [33]. After the publication of NDSHA maps for various seismic hazard parameters for central and south eastern Europe [38] strong earthquakes occurred in Italy in 2009, 2012, 2016, 2017 and Albania, 2019. These maps of Peak Ground Displacement (PGD), Velocities (PGV), and Design Ground Acceleration (DGA), based on the spectral shape given by European Building Code EC8 for bedrock conditions, indicate that future events may generate seismic load even larger than that observed so far. The NDSHA procedure allows for easy validation and/or update, of the maps, when necessary—e.g., after upgrade of the used data banks with new relevant data. Remarkably, after the mentioned earthquakes, NDSHA maps did not need any updating, since all observed ground shaking parameters have been safely enveloped by the mapped values. For a brief review of numerous NDSHA application in Italy [45], Slovenia [46], Croatia [47], Hungary [48], Romania [49], Republic of Moldova [50], Albania [51], see [26]. As a rule, the seismic source parameters and the local geological features significantly influence the earthquake ground motion [52, 53].

3.2 *Site Specific Analysis*

Traditionally the local site response is evaluated in a simplified manner. A set of coefficients is introduced in the seismic codes to modify the shape of the bedrock response spectra; through these coefficients the shape of the spectral curves is adjusted to the mechanical properties of the superficial ground layer. A widely used technique is also the computation of the local effects using the ratio between the horizontal and the vertical response spectra (H/V ratio) [54]. This method has been demonstrated to be unable, as a rule, to give correct local effects [21] since it is based on the wrong assumption that the vertical ground motion is not affected by the superficial layer(s). Furthermore Nakamura's [55] original assumptions now seem *questionable*. Several later studies (e.g., [55, 56]) have demonstrated that, in presence of thick sediments with poor mechanical properties, the horizontal-to-vertical ratio H/V is more strictly correlated with the *polarization* of Rayleigh waves in the horizontal plane [33, 57]. In nature, "amplifications" of both vertical and horizontal components of motion are strongly dependent not only on the soil and topography characteristics, but also on the incidence angles of the radiated wave field—e.g., it is obvious the variety of the recorded signals at numerous sites due to earthquakes occurring at same seismic sources differing in focal depth and seismic source mechanisms (e.g., [58]). Demonstration that the vertical component of motion can be severely affected by local soil mechanical conditions was provided about 50 years ago [57].

NDSHA SSA is based upon computer simulations exploiting the knowledge about the source process, the path source-to-site, and the local site conditions. The efficiency of the methods, analytical and hybrid, for computing synthetic seismic signals generated by complex seismic sources in laterally heterogeneous media has been documented for many case studies worldwide [18, 20, 36]. SSA can be applied via (a) hybrid technique that combines mode summation for the path to the source and finite difference method for the local profile or (ii) analytical procedure combining mode summation for the path to the source and, when necessary, mode coupling method applied for the specific site of interest [36, 59]. The SSA output consists of a large set of synthetic time histories at predefined "receivers' sites" at the construction site of interest—thus becomes also possible to provide realistic earthquake time histories for long-in-plan structures that might suffer non synchronic foundation displacements during an earthquake [60]. Relevant energy input can be always easily extracted from the available signals' sets. Maximum Credible Seismic Input, MCSI is defined as a Response Spectrum or as a set of accelerograms; MDSI is calculated by means of a large set of NDSHA simulations. Following the NDSHA method, MCSI can be defined at a given site at two levels of detail [31]: (i) "Maximum Credible Seismic Input at bedrock"—RSA outcome, computed without considering the site effects and (ii) "Maximum Credible Seismic Input Site Specific"—result of the SSA considering the local structural heterogeneities. For the later, RSA is used as a reference to choose the most dangerous sources for the site and ground motion parameter of interest and then SSA is carried out for each source-to-site path. The use of source spectra computed by PULSYN06 [61] introduces a stochastic element in NDSHA. MCSI builds upon computed time histories that come as SSA outcome; SSA supplies realistic accelerograms, consistent with the maximum empirically estimated magnitude, possibly incremented by some multiple of its global standard deviation (σ_M), focal mechanism, epicentral distance and soil site conditions.

Examples supplied in numerous publications (see [36] and references therein) prove the capability of the method to realistically model the seismic ground motion in different European metropolitan areas. Very important and useful experience has been gained with the beginning of microzoning actions in large cities in Central Europe—Bucharest, Debrecen, Ljubljana, Naples, Rome, Russe, Sofia and Zagreb (see [36] and references therein). This group is gradually enlarged with Napoli and Catania [62, 63], Thessaloniki [64] and Ohrid [65]. NDSHA SSA is a valuable tool from the engineering point of view, which provides realistic synthetic seismograms representative of the dynamic characteristics of the site of interest. Providing complete site-specific earthquake time histories for the purpose of the dynamic time history structural analysis, this possibility is truly important given that the number of available recorded ground motion is unavoidably very low, particularly for large earthquakes.

3.3 NDSHA and PBSD Concept

Performance-based seismic design (PBSD) aims to couple expected structural performance level with expected levels of seismic ground motion. Structural performance levels are associated with a certain level of structural damage caused by given seismic actions. PBSD, most often, deals with four conventional performance levels defined as: Operational Limit (OL), Immediate Occupancy (IO), Life Safety (LS) and Collapse Prevention (CP) [66]. Recently, reasonable feasible Performance Based Seismic Design (PBSD) procedure has been proposed [31]. It is grounded in a scenario based definition of the seismic input based on the following important considerations, arisen from the analysis of earthquake phenomena: (a) any structure at a given location, regardless of its importance, is subject to the same shaking as a result of a given earthquake; (b) it is impossible to determine with ultimate precision when a future earthquake of a given intensity/magnitude will occur; and (c) insufficient data are available to develop reliable statistics for earthquakes. NDSHA generates efficiently the seismic input at a given site based on the seismic history and the seismogenic zones, seismogenic nodes (for a recent review see [67]), and faults. As it was already mentioned, based on the computed complete earthquake time histories, NDSHA is capable to provide MCSI at two different levels: (i) response spectrum at bedrock as RSA outcome that is like what is proposed by codes and (ii) SSA output—the site-specific seismic input. The variety of uncertainties is mastered through numerous NDSHA simulations and sensitivity analyses rather than quantifying them probabilistically. MCSI occurs to be an excellent quantity that can be associated with the worst structural performance acceptable for a building, called Target Performance Level (TPL)—this TPL can be identified accordingly to the importance of the structure. As indicated by [31] in this way the importance of the structure (risk category) is considered by changing the structural performance level to check, rather than to change the seismic input. The procedure is a generalized enhancement of what has already been applied for the seismic upgrading of several critical and essential buildings—schools of the Trieste Province, Italy [31].

4 Conclusive Remarks

The structural performance levels are associated with predefined quantitative damage assessments, usually defined in terms of storey drift or rotation of plastic hinges that is accepted to occur in the structural elements subjected to given seismic action characterized by certain level of ground motion. Collapse is easily recognizable, so Collapse Prevention Level can be referred to a specific physical phenomenon. Other three performance levels are associated with conventional states of damage. Thus, it is crucial to define relevant ground motion that must be used to check if a performance level has been reached. NDSHA within its two different analyses (RSA and SSA) is a powerful tool to generate a seismic input at a given site based on (i) the seismic history and the seismogenic zones, seismogenic nodes, and (2) the realistic consideration that any structure at a given location, regardless of its importance, is subject to the same shaking because of a given earthquake. This seismic input might be supplied as envelope response spectra or set of complete seismic signals, consistent with the maximum magnitude and focal mechanism of the MCE), epicentral distance and soil conditions of the site of interest and directly usable for all range of the earthquake engineering structural analysis, including the advanced PBDS and energy concepts.

References

1. EN 1998-1 (English): Eurocode 8: design of structures for earthquake resistance—part 1: general rules, seismic actions and rules for buildings (2004)
2. Directive 2012/27/EU of the European parliament and of the council of 25 October 2012 on energy efficiency (2012)
3. Bournas, D.A.: Concurrent seismic and energy retrofitting of RC and masonry building envelopes using inorganic textile-based composites with insulation materials: a new concept. Elsevier Compos. Part B Eng. **148**, 166–179 (2018)
4. Klugel, J.U.: Error inflation in probabilistic seismic hazard analysis. Eng. Geol. **90**(3), 186–192 (2007)
5. Rugarli, P.: Zone Griglie o…Stanze? Ingegneria Sismica 1-2008 (2008)
6. Rugarli, P.: Validazione Strutturale. EPC Libri, Rome, Italy (2014)
7. Rugarli, P.: The role of the standards in the invention of the truth. Acta, Convegno La. Resilienza delle Citta d'Arte ai Terremoti. Accademia Nazionale dei Linei, Rome, 3–4 Nov 2015
8. Rugarli, P.: Primum: non nocere. Acta, Convegno La Resilienza delle Citta d'Arte ai Terremoti. Accademia Nazionale dei Lincei, Rome, 3–4 Nov 2015
9. Bizzarri, A., Crupi, P.: Linking the recurrence time of earthquakes to source parameters: a dream or a real possibility? Pure Appl. Geophys. **171**, 2537–2553 (2013)
10. Peresan, A., Kossobokov, V., Romashkova, L., Magrin, A., Soloviev, A., Panza, G.F.: Time dependent neo-deterministic seismic hazard scenarios: preliminary report on the M6.2 Central Italy earthquake. New Concepts Glob Tecton J 4(3):487–493 (2016)
11. Wang, Z.: Seismic hazard assessment: issues and alternatives. Pure Appl. Geophys. **168**, 11–25 (2010)
12. Stark, P.B., Freedman, D.A.: What is the chance of an earthquake? In: Mulargia, F., Geller, R.J. (eds.). Earthquake Science and Seismic Risk Reduction, NATO Science Series, IV: Earth and Environmental Sciences, vol. 32, pp. 201–213. Kluwer, Dordrecht, The Netherlands (2001)
13. Wyss, M., Nekrasova, A., Kossobokov, V.: Errors in expected human losses due to incorrect seismic hazard estimates. Nat. Haz. **62**, 927–935 (2012)
14. Rugarli, P., Amadio, C., Peresan, A., Fasan, M., Vaccari, F., Magrin, A., Romanelli, F., Panza, G.F.: Neo-deterministic earthquake accelerograms and spectra. A NDSHA approach to seismic analysis. In: Jia, J., Paik, J. (eds.) Engineering Dynamics and Vibrations: Recent Developments (2019)
15. Mollaioli, F., Bruno, S., Decanini, Decanini, L., Saragoni, R.: Correlations between energy and displacement demands for performance-based seismic engineering. Pure Appl. Geophys. **168**, 237–259 (2011). https://doi.org/10.1007/s00024-010-0118-9 (2010 Birkhäuser/Springer Basel AG)
16. Mollaioli, F., Lucchini, A., Donaire-Ávila, J., Benavent-Climent, A.: On the importance of energy-based parameters. In: Conference: COMPDYN 2019 7th ECCOMAS Thematic Conference on Computational Methods in Structural Dynamics and Earthquake Engineering, At: Energy-Based Seismic Design. Crete, Greece (2019)
17. Elnashai, A.S., Di Sarno, L.: Fundamentals of Earthquake Engineering. Wiley, 337p
18. Panza, G.F., Bela, J.: NDSHA: a new paradigm for reliable seismic hazard assessment. Eng. Geol. **275**, 105403 (2020)
19. Panza, G.F.: Synthetic seismograms: the Rayleigh modal summation technique. J. Geophys. **58**, 125–145 (1985)
20. Panza, G.F., Romanelli, F., Vaccari, F.: Seismic wave propagation in laterally heterogeneous anelastic media: Theory and applications to seismic zonation. Adv. Geophys. **43**, 1–95 (2001)
21. Panza, G.F., La Mura, C., Peresan, A., Romanelli, F., Vaccari, F.: Seismic hazard scenarios as preventive tools for a disaster resilient society. Adv. Geophys. **53**, 93–165 (2012)
22. Panza, G.F., Peresan, A., Sanso, F., Crespi, M., Mazzoni, A., Nascetti, A.: How geodesy can contribute to the understanding and prediction of earthquakes. Rend. Lincei. (2017). https://doi.org/10.1007/s12210-017-0626-y

23. Magrin, A., Gusev, A.A., Romanelli, F., Vaccari, F., Panza, G.F.: Broadband NDSHA computations and earthquake ground motion observations for the Italian territory. Int. J. Earthq. Impact Eng. **1**(4), 395–420 (2016)
24. Panza, G., Peresan, A., La Mura, C.: Seismic hazard and strong ground motion: an operational neo-deterministic approach from national to local scale. Geophysics and geochemistry. In: UNESCO-EOLSS Joint Committee, Encyclopedia of Life Support Systems (EOLSS), Developed under the Auspices of the UNESCO. Eolss Publishers, Oxford, UK (2013)
25. Panza, G.F., Irikura, K., Kouteva-Guentcheva, M., Peresan, A., Wang, Z., Saragoni, R. (Eds.): Advanced Seismic Hazard Assessment, Part I: Seismic Hazard Assessment (2011)
26. Kouteva-Guentcheva, M., Cioflan, C.O., Paskaleva, I., Panza, G.F. (2020) NDSHA achievements in Central and South-eastern Europe. In: Panza, G., Kossobokov, V.G., Laor, E., De Vivo, B. (Eds.) Earthquakes and Sustainable Infrastructure: Neodeterministic (NDSHA) Approach Guarantees Prevention Rather than Cure (in print)
27. Vaccari, F.: A web application prototype for the multiscale modelling of seismic input. In: Earthquakes and Their Impact on Society. Springer International Publishing, Cham, pp. 563–584 (2016)
28. NTC2008—Norme tecniche per le costruzioni 2008 (NTC 2008)—D.M. 14 Gennaio 2008 (D.M. 14/1/08) (Chapter 3.2.3.6)
29. Fasan, M., Amadio, C., Noe, S., Panza, G.F., Magrin, A., Romanelli, F., Vaccari, F.: A new design strategy based on a deterministic definition of the seismic input to overcome the limits of design procedures based on probabilistic approaches. In: Convegno ANIDIS 2015. L'Aquila, Italy (2015)
30. Fasan, M., Magrin, A., Amadio, C., Romanelli, F., Vaccari, F., Panza, G.F.: A seismological and engineering perspective on the 2016 Central Italy earthquakes. Int. J. Earthq. Impact Eng. **1**, 395–420 (2016)
31. Fasan, M.: Advanced Seismological and Engineering Analysis for Structural Seismic Design. University of Trieste, Italy (2017)
32. SEAOC: Vision 2000: Performance Based Seismic Engineering of Buildings. Structural Engineers Association of California, Sacramento, California (1995)
33. Rugarli, P., Amadio, C., Peresan, A., Fasan, M., Vaccari, F., Magrin, A., Romanelli, F., Panza, G.F.: Neo-deterministic scenario Earthquake accelerograms and spectra a NDSHA approach to seismic analysis. In: Jia, J., Paik, J.K. (eds.) Engineering Dynamics and Vibrations Recent Developments, pp. 187–241 (2012)
34. Bertero, R.D., Bertero, V.V.: Performance-based seismic engineering: the need for a reliable conceptual comprehensive approach. Earthq. Eng. Struct. Dyn. **31**, 627–652 (2002)
35. Bommer, J.J., Pinho, R.: Adapting earthquake actions in Eurocode 8 for performance-based seismic design. Earthq. Eng. Struct. Dyn. **35**, 39–55 (2006)
36. Panza, G.F., Vaccari, F.: Introduction. In: Panza, G.F., Radulian, M., Trifu, C.I. (eds.) Seismic Hazard of the Circum-Pannonian Region. Pageoph Topical Volumes. Birkhäuser, Basel (2000). https://doi.org/10.1007/978-3-0348-8415-0_1
37. Båth, M.: Introduction to Seismology, p. 395. Wiley, New York (1973)
38. Panza, G., Radulian, M., Trifu, C.I., (eds.) Seismic Hazard of the Circum Panonian Region. Birkhäuser, Pageoph Topical Volumes (2000)
39. Zaicenko, A., Craifaleanu, I., Paskaleva, I. (eds.) Harmonization of Seismic Hazard in Vrancea Zone: With Special Emphasis on Seismic Risk Reduction, 347p (2008)
40. Cancani, A.: Sur l'emploi d'une double échelle sismique des intensités, empirique et absolue. Gerlands Beitr. Geophy. **2**, 281–283 (1904)
41. Lliboutry, L.: Quantitative Geophysics and Geology. Springer, London, UK (2000)
42. Florsch, N., Fäh, D., Suhadolc, P., Panza, G.F.: Complete synthetic seismograms for highfrequency multimode SH-waves. Pure Appl. Geophys. **136**, 529–560 (1991)
43. Pavlov, V.M.: Matrix impedance in the problem of the calculation of synthetic seismograms for a layered-homogeneous isotropic elastic medium. Izv. Phys. Solid Earth **45**, 850–860 (2009)
44. Magrin, A.: Multi-scale seismic hazard scenarios. Ph.D. thesis. University of Trieste, Italy (2013)

45. Peresan, A., Magrin, A., Nekrasova, A., Kossobokov, V.G., Panza, G.F.: Earthquake recurrence and seismic hazard assessment: a comparative analysis over the Italian territory. In: ERES 2013, Transactions of Wessex Institute (2013). https://doi.org/10.2495/ERES130031
46. Zivcic, M., Suhadolc, P., Vaccari, F.: Seismic zoning of slovenia based on deterministic hazard computations. In: Panza, G., Radulian, M., Trifu, C (eds.) Seimic hazard of the circum-Pannonian region, pp. 171–184. Birkhäuser Verlag, Basel (2000)
47. Markusic, S., Suhadolc, P., Herak, M., Vaccari, F.: A contribution to seismic hazard assessment in Croatia from deterministic modeling. In: Panza, G., Radulian, M., Trifu, C. (eds.) Seismic hazard of the Circum-Pannonian Region, Pure and Applied Geophysics, pp. 185–204. Birkhäuser Verlag, Basel (2000)
48. Bus, Z., Szeidovitz, G., Vaccari, F.: Synthetic seismogram based deterministic seismic zoning for the hungarian part of the Pannonian basin. In: Panza, G., Radulian, M., Trifu, C. (eds.) Seismic hazard of the Circum-Pannonian Region, Pure and Applied Geophysics, pp. 205–220. Birkhäuser Verlag, Basel (2000)
49. Radulian, M., Vaccari, F., Mandrescu, N., Panza, G., Moldoveanu, C.: Seismic hazard of Romania: a deterministic approach. In: Panza, G., Radulian, M., Trifu, C. (eds.) Seismic hazard of the circum-Pannonian region. Pure and Applied Geophysics, pp. 221–247 (2000)
50. Sandu, I., La Mura, C.: Vrancea Earthquakes impact on republic of moldova (South Area): a deterministic parametric study. In: Ionescu, C., Vacareanu, R. (eds.) International Conference: 5th National Conference on Earthquake Engineering and 1st National Conference on Earthquake Engineering and Seismology, Bucharest, pp. 173–180 (2014)
51. Muço, B., Vaccari, F., Panza, G., Kuka, N.: Seismic zonation in Albania using a deterministic approach. Tectonophysics **344**, 277–288 (2002)
52. Panza, G., Cioflan, C., Kouteva, M., Paskaleva, I., Romanelli, F.: An innovative assessment of the seismic hazard from the Vrancea intermediate-depth earthquakes: case studies in Romania and Bulgaria. 12ECEE. CD, Ref. 230.32, pp. CD, Ref. 230.32. ECEE, London (2002)
53. Schanz, T., Iankov, R.: Earthquake source and local geology effects on the seismic site response. In: Kouteva, M., Paskaleva, I., Panza, G. (eds.) Coupled Site and Soil-Structure Interaction Effects with Application to Seismic Risk Mitigation. Springer, NATO Security through Science Series C: Environmental Security (2009)
54. Nakamura, Y.: A method for dynamic characteristics estimation of subsurface using microtremor on the ground surface. Railw. Techn. Res. Inst. (RTRI) **30**, 25–33 (1989)
55. Lachet, C., Bard, P.-Y.: Numerical and theoretical investigations on the possibilities and limitations of Nakamura's technique. J. Phys. Earth **42**(5), 377–397 (1994)
56. Konno, K., Ohmachi, T.: Ground-motion characteristics estimated from spectral ratio between horizontal and vertical components of microtremor. Bull. Seism. Soc. Am. **88**(1), 228–241 (1998)
57. Panza, G.F., Schwab, F.A., Knopoff, L.: Channel and crustal Rayleigh waves. Geophys. J. R. Astr. Soc. **30**, 273–280 (1972)
58. Kouteva, M.: Estimates of some ground motion parameters, related to the damage potential of strong intermediate-depth Vrancea earthquakes. Acta Geod. Geophys. Hung. **45**, 356–371 (2010)
59. Panza, G., Peresan, A., Romanelli, F., Vaccari, F.: Earthquake scenarios for seismic isolation design and the protection of cultural heritage. Cultural heritage. In: Istanbul 2011, 5th International Congress "Science and Technology for the Safeguard of Cultural Heritage in the Mediterranean Basin". Istanbul, Turkey (2011)
60. Romanelli, F., Panza, G., Vaccari, F.: Realistic modeling of the effects of asynchronous motion at the base of bridge piers. J. Seismol. Earthq. Eng. **6**(2), 19–28 (2004)
61. Gusev, A.A.: Broadband kinematic stochastic simulation of an earthquake source: a reined procedure for application in seismic hazard studies. Pure Appl. Geophys. **168**, 155–200 (2011)
62. Nunziata, C.: Seismic ground motion in Napoli for the 1980 Irpinia earthquake. In: Panza, G., Paskaleva, I., Ninziata, C.: Seismic Ground Motion in Large Urban Areas, Pageoph, pp. 1239–1265. Birkhäuser Verlag, Basel, Boston, Berlin (2004)

63. Decanini, L.D., Mollaioli, F., Panza, G.F., Romanelli, F.: The realistic definition of the seismic input: an application to the Catania area: earthquake resistant engineering structures, transactions of the built environment, vol. 38, pp. 425–434. WIT Press (1999)
64. Triantafyllidis, P., Suhadolc, P., Hatzidimitriou, M., Anastasiadis, A., Theodulidis, N.: Part I: theoretical site response estimation for microzoning purposes. In: Panza, G., Paskaleva, I., Nunziata, C. (eds.) Seismic Ground Motion in Large Urban Areas, Pageoph, pp. 1185–1205. Birkhäuser Verlag, Basel, Boston, Berlin (2004)
65. Milkova, K., Dumova-Jovanovska, E., Drogreshka, K., Chernih-Anastasovska, D., Pekevski, L., Romanelli, F., Vaccari, F., Panza, G.: Region specific application of neo-deterministic analysis for reliable seismic hazard assessment. In: 16th European Conference on Earthquake Engineering, Thessaloniki (2018)
66. American Society of Civil Engineers—ASCE (2013) ASCE Standard ASCE/SEI 41-13: Seismic Evaluation and Retrofit of Existing Buildings. Reston, Virginia (2013)
67. Gvishiani, A.D., Soloviev, A.A., Dzeboev, B.A.: Problem of recognition of strong-earthquake-prone areas: a state-of-the-art review, Izvestiya. Phys. Solid Earth **56**(1):1–23 (2019) https://doi.org/10.1134/S1069351320010048 (Pleiades Publishing Ltd.)

A Review on Performance-Based Plastic Design Method: Concept and Recent Developments

Sutat Leelataviwat, Piseth Doung, and Nattakarn Naiyana

Abstract The Performance-based Plastic Design (PBPD) Method is a practical seismic design and evaluation approach based on an energy concept. The required frame strength is derived corresponding to a target deformation level and a selected yield mechanism using the energy balance concept. The method directly accounts for inelastic behavior and considers the internal force distribution at the ultimate limit state by ensuring the formation of a pre-selected yield mechanism. Plastic design is specifically used to design the structure to achieve the selected mechanism and ensure uniform damage. Since its inception, the method has been successfully applied to a variety of steel and RC framing systems. The underlying energy equation has also been used for seismic evaluation and retrofitting design of existing structures. This paper provides an overview of the concept, the framework, and the development of the PBPD method. Key aspects of the PBPD method including the energy demand calculation, the force and member strength distribution to achieve uniform damage, and hysteretic consideration are reviewed and discussed. An example that illustrates the PBPD concept is provided. Finally, recent developments of the PBPD method in the literature are also summarized and discussed.

Keywords Performance-based plastic design · Modified energy balance concept · Plastic design

1 Introduction

Current structural design practice around the world is normally carried out by force-based elastic procedures. The inelastic behavior and energy dissipation are accounted for in an implicit and indirect manner. It is well known that, under severe loadings

S. Leelataviwat (✉) · N. Naiyana
King Mongkut's University of Technology Thonburi, Bangkok, Thailand
e-mail: sutat.lee@kmutt.ac.th

P. Doung
Institute of Technology of Cambodia, Phnom Penh, Cambodia

A. Benavent-Climent and F. Mollaioli (eds.), *Energy-Based Seismic Engineering*, Lecture Notes in Civil Engineering 155,
https://doi.org/10.1007/978-3-030-73932-4_8

such as earthquake, blast, impact, etc., the structures may undergo severe inelastic deformations, and the yielding activity and energy dissipation may be unevenly distributed in the structure leading to premature failure of some key members. To achieve more predictable structural performance, design factors such as deformation limit, appropriate design lateral forces, member strength hierarchy, yield mechanism, cumulative energy dissipation must become an integral part of the design process. This is best handled by an energy-based procedure.

This paper presents a review of an energy-based seismic design procedure called the Performance-Based Plastic Design (PBPD) method. The development of this procedure began in the early 2000s [1–6] and culminated in the publication by Goel and Chao [7–9]. The well-known energy balance concept used in deriving inelastic design response spectra for single-degree-of-freedom systems is modified and extended to include the effects of plastic yield mechanism and the distribution of seismic forces along the height of the structure. In this procedure, the inelastic design base shear of the frame is derived corresponding to a maximum target drift and a selected global yield mechanism. Plastic design is then used to ensure the formation of the selected yield mechanism. Since its inception, the method has been successfully applied to a variety of steel and RC framing systems including moment resisting frames, truss moment frames, concentrically and eccentrically braced frames, steel plate shear walls, and RC coupled walls [10–20]. The method has also been applied to new and novel structural systems such as buckling restrained knee braced frames and buckling restrained knee braced truss moment frames among others [21–28]. The method has also been extended to 3D frame structures [29]. The underlying energy equation has also been used for seismic evaluation and retrofitting design of existing structures [30–32].

In this paper. Key aspects of the PBPD method are reviewed and discussed. An example is provided to illustrate the PBPD method. Where appropriate, recent developments of the PBPD method in the literature are also summarized and discussed.

2 PBPD Concept

In the PBPD method, the required design base shear is derived corresponding to a maximum drift level (called target drift) and a selected yield mechanism using the modified energy balance concept [3]. The concept is similar to the energy balance concept introduced by Housner [33]. It is postulated that the input energy which contributes to damage is equal to the sum of the elastic energy and the plastic work done by the structure. In the modified energy balance concept, the energy balance is applied between the fraction of the input energy (E) and the work needed to push the structure monotonically up to the target drift

$$\gamma E = E_e + E_p \tag{1}$$

where E_e and E_p are, respectively, the elastic and plastic components of the energy needed to push the structure up to the target drift, and γ is called the energy factor [3].

Housner [33] showed that the maximum earthquake input energy for a MDOF system, on the average, can be estimated as

$$E = \frac{1}{2} M S_v^2 = \frac{W g C_e^2 T^2}{8\pi^2} \tag{2}$$

where M is the total system mass, S_v is the pseudo-velocity from the elastic response spectrum, C_e is the normalized pseudo-acceleration (S_a/g) with respect to g, W is the weight, and T is the fundamental period. The energy factor can be derived as the ratio of the energy absorbed by the inelastic system to that of the equivalent elastic system [3] and can be given as

$$\gamma = \frac{2\mu - 1}{R_\mu^2} \tag{3}$$

where μ is the ductility ratio and R_μ is the yield force reduction factor. By this definition, the energy factor can be readily computed for a given ductility level using any well-known $R_\mu - \mu - T$ equation. In the PBPD method, the $R_\mu - \mu - T$ relationship developed by Newmark and Hall [34, 35] was used and the energy factor became only a function of ductility and period. The modified energy balance concept is illustrated in Fig. 1. By introducing the energy factor, γ, the ties between the structural strength, the maximum deformation, and input energy are established. The concept can be used to design a structure, given the input energy, by calculating the required strength to limit the deformation within the selected target. The energy balance equation has been shown to be applicable to single-degree-of-freedom systems and approximately so for multi-degree of freedom systems [30]. In Fig. 1, the hysteretic behavior is assumed to

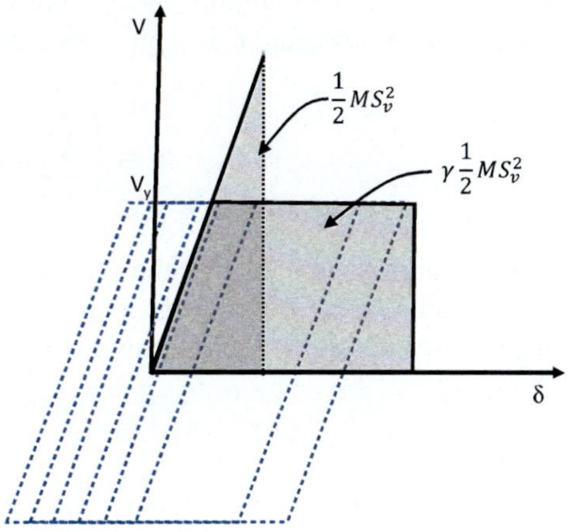

Fig. 1 Modified energy balance concept

be elastic-perfectly-plastic (EPP). However, the method can be extended to non-EPP systems as will be discussed later.

The input energy has been a subject of extensive research. It is well known that the input energy depends on several factors such as ground motion characteristics and the dynamic and hysteretic properties of the system. The energy factor can be derived directly based on a large number of time-history analyses using a specific set of ground motions and systems with varying strength. However, in the PBPD method, the elastic input energy is simply used as a means to derive the energy factor based on the ductility and the strength of the system. The advantage of this approach is that well established $R_\mu-\mu-T$ equations can be readily incorporated into the PBPD framework. Although a consensus has not been reached, given the uncertainty in predicting future earthquake ground motions, the accuracy of Eq. (2) for design purposes is within an acceptable level. However, for special structures, a more complete analysis may be needed. The maximum elastic energy as governed by S_v can also be estimated directly from the total input energy [36].

For the elastic vibrational energy, E_e, Akiyama [37] showed that it can be calculated with a reasonable accuracy by using the elastic input energy corresponding to the yield force

$$E_e = \frac{1}{2}M\left(\frac{T}{2\pi} \cdot \frac{V_y}{W} \cdot g\right)^2 \tag{4}$$

where V_y is the yield base shear. The total plastic energy, E_p, that the structure needs to dissipate can then be found by substituting Eqs. (2), (3), and (4) into Eq. (1). Once a target ductility level is selected and the energy factor is computed, the base shear at yield (V_y) can then be calculated as

$$\frac{V_y}{W} = \frac{-\alpha + \sqrt{\alpha^2 + 4\gamma C_e^2}}{2} \tag{5}$$

where W is the weight of the structure, C_e is the normalized design pseudo acceleration (S_a/g), and α is a parameter given by

$$\alpha = \left(\sum_{j=1}^{n} \lambda_i h_i\right) \frac{\theta_p 8\pi^2}{T^2 g} \tag{6}$$

In the above equation, T is the period, h_i is the height from the ground to the floor level i, and λ_i is the lateral force distribution factor, θ_p is the target plastic drift. The lateral force at level i, F_i, is assumed to be of the form

$$F_i = \lambda_i V_y \tag{7}$$

The force distribution depends on the selected mechanism. In general, the above equation would be derived based on a sway mechanism with the lateral displacements that increase with height above the base. The selection of the yield mechanism [38] is the key step in the design procedure to ensure the desirable inelastic performance of the structure. The sway mechanism is the most efficient yield mechanism that requires the largest amount of external work per unit value of V/W and θ_p [38]. After the design base shear has been determined, a plastic mechanism analysis is used to determine the strength of the Designated Yielding Members (DYMs) due to the lateral design forces. The DYMs depend on the structural system and the strength of these members dictates the amount of plastic energy dissipation available in the system. In the PBPD method, the relative strength of the DYMs at each level is assigned based on a strength distribution pattern, called the shear proportioning factor

$$\beta_i = \left(\frac{V_i}{V_y}\right) \tag{8}$$

where β_i is the shear proportioning factors and V_i is the story shear at level i. The distribution pattern is chosen such that the deformation and the yielding will be reasonably uniform over the height. The PBPD method can be summarized as shown in Fig. 2. It should be noted that the plastic design is an inherent element of the PBPD method as it allows the direct connection between the strength, the displacement, and the energy. Without controlling the mechanism, the tie between the frame target displacement and the energy cannot be ascertained.

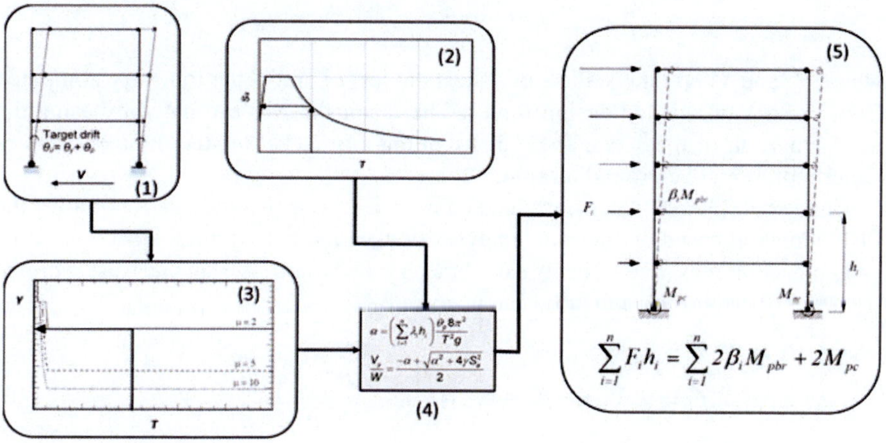

Fig. 2 The PBPD method; (1) Selection of target mechanism and drift; (2) Seismic hazard; (3) Calculation of energy factor; (4) Calculation of base shear; and (5) Plastic design

3 Lateral Force and Story Strength Distribution

The assigned relative story strength and lateral force distribution represent the variation of story strength and stiffness over the height of the structure. As such, the choice of the lateral force distribution and the shear proportioning factor will have a strong influence on the response of the structure. The lateral force distribution in the PBPD method differs from the typical code-based distribution in that the PBPD force distribution is intended for the plastic design of the structure in the inelastic, ultimate, state. To have a relatively uniform maximum story drift along the height, the force distribution, as well as the relative story strength, should be appropriately chosen to follow the applied forces that would be generated during an earthquake. By doing so, the stories with relatively low input story shears would have relatively small strength and stiffness. Similarly, the stories with relatively high input story shears would have relatively large strength and stiffness. This will also ensure that the input energy will be properly dissipated throughout the structure avoiding concentrations of damage in any particular story.

As mentioned earlier, in the PBPD method, the strength of the DYMs in each story level is proportioned based on a parameter called the shear proportioning factor. Leelataviwat et al. [2] first proposed the shear proportioning factor based on the relative distribution of static story shears computed from the assumed linear force distribution. The shear distribution factor is given as

$$\beta_i = (V_i / V_n)^{0.5} \tag{9}$$

where V_i and V_n are the static story shears at level i and at the top story computed from linearly distributed design forces. The factor 0.5 was estimated by using the least-square fit of the actual shear distributions from a set of steel moment frames under representative ground motions.

Chao et al. [39, 40] later introduced a new design later force distribution for the PBPD method based on inelastic analyses of steel moment frames, steel concentrically braced frames, steel eccentrically braced frames, and steel special truss moment frames. The lateral force distribution is given by

$$F_i = \lambda_i V_y = (\beta_i - \beta_{i+1}) \left(\frac{w_n h_n}{\sum_{j=1}^{n} w_j h_j} \right)^{0.50T^{-0.2}} V_y \tag{10}$$

where w_n is the weight of the structure at the top level (level n), h_n is the height from the ground to the top level, and β_i is the ratio of the story shear at level to that of the top story

$$\beta_i = \frac{V_i}{V_n} = \left(\frac{\sum_{j=i}^{n} w_j h_j}{w_n h_n} \right)^{0.50T^{-0.2}} \tag{11}$$

Ideally, the lateral force distribution should also depend on the structural system as the variation of the stiffness and strength along the height are distinct characteristics of each system. For this reason, the lateral force distribution for a specific structural system can be developed based on the results of inelastic structural analyses [41]. The PBPD force distribution does not intentionally aim to have equal energy dissipation in each story level but rather aims at distributing the energy dissipation such that the deformation is approximately uniform over the height following the selected mechanism.

4 Cumulative Energy Dissipation and Damage

It is well known that the structural damage depends not only on the maximum deformation but also on the cumulative energy dissipation (E_p) caused by cyclic load reversals. The PBPD uses maximum displacement (drift) as the design limit state. The modified energy balance equation relies on the monotonic load-displacement response as shown in Fig. 1. However, cyclic energy and low-cycle fatigue damage can be incorporated into the PBPD method following the approach introduced by Fajfar et al. [42, 43]. Since the mechanism is defined in advance in the PBPD method, the deformation of the DYMs can be estimated and the damage due to cyclic load reversals can be assessed using a suitable damage model such as the Park-Ang damage index [44]

$$DI = \frac{\delta}{\delta_u} + \frac{E_{hi}}{Q_y \delta_u} \beta_{di} \tag{12}$$

where DI is the damage index, δ is the maximum displacement, δ_u is the ultimate displacement under monotonic loading, Q_y is the yield strength of the element, E_{hi}, $E_{pi}$is the plastic energy dissipated by the DYM under consideration, and β_{di} is a numerical constant depending on the characteristics of the member being assessed.

The damage index can be computed for the DYMs in the frame. Since the PBPD method uses the displacement as the design parameter, δ of the DYMs can be obtained. By making the plastic design an inherent part of the PBPD method, the relationship between the deformation of the DYMs and the maximum displacement (or target displacement for design purposes) can be established based on the kinematics of the selected yield mechanism.

Similar to the input energy, hysteretic energy component has also been studied by several investigators. The cumulative hysteretic energy is generally believed to be correlated to the input energy [42, 43]. Fajfar et al. [43] showed that the ratio of the cumulative hysteretic energy to the input energy is approximately constant and can vary from 0.5 to 0.8 depending on the ductility. Alternatively, the spectrum for hysteretic energy has also been proposed which allows one to estimate the cumulative hysteric energy based on the PGA and soil type [45]. Assuming that the deformation is relatively uniform following the selected sway mechanism, the hysteretic energy can be assumed to be distributed among the DYMs based on their relative yield strength and the DI can be estimated. The design base shear corresponding to a selected damage index, maximum drift, and seismic hazard can then be obtained using Eqs. (5) and (12). The damage estimation as well as how the hysteretic energy is computed and distributed in PBPD-designed frames are still the subject of on-going research. The research on energy input and the hysteretic energy for MDOF building systems are also relatively limited. Further research is still needed in this area, especially with respect to the PBPD applications.

5 Non-EPP Systems

The PBPD method was first developed for EPP systems and applied to the steel moment frame system that has load-deformation characteristics similar to EPP. However, it was later extended to non-EPP systems. Several approaches have been devised to cope with pinched and degrading hysteretic behavior. In general, there are three approaches that can be applied.

The first approach is by modifying the design base shear equation based on the ratio of dissipated energy by the non-EPP system to that of the EPP system. This approach was first applied to the Concentrically Braced Frame system [7]. The design base shear considering the non-EPP behavior is given by

$$\frac{V_y}{W} = \frac{-\alpha + \sqrt{\alpha^2 + 4(\gamma/\eta)S_a^2}}{2} \tag{13}$$

in which η is defined as the ratio of the energy dissipated by the non-EPP system to that of the EPP system. The concept is illustrated in Fig. 3a.

The second approach is based on calculating an equivalent EPP displacement [32]. The target displacement of a non-EPP can be converted to an equivalent EPP displacement using a modification factor. The factor is generally referred to as the C_2 factor based on FEMA440 [46]. In FEMA440, the C_2 is a modification factor that considers the effects of the pinched hysteresis shape, cyclic stiffness degradation, and strength deterioration on the maximum displacement response. The C_2 coefficient is defined as the maximum lateral displacement of the non-EPP system to that of the EPP system. The equivalent EPP target design drift for the non-EPP system can be found by dividing the intended target drift by the C_2 coefficient. This is expressed as

$$\theta_u^* = \frac{\theta_T}{C_2} \tag{14}$$

where $\theta^*{}_u$ is the equivalent EPP target drift, and θ_T is the target drift for the non-EPP system. The target ductility and the energy factor can also be modified accordingly as follows.

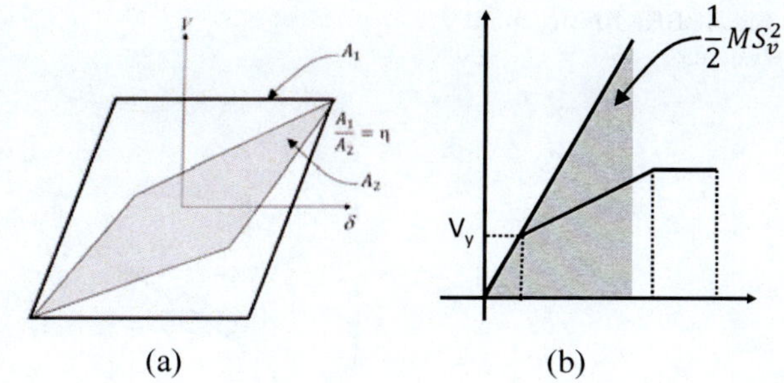

Fig. 3 Modification for Non-EPP systems. **a** Using η factor and **b** Direct Calculation of energy factor for tri-linear system [47]

$$\mu_s^* = \frac{\theta_u^*}{\theta_y} = \frac{\theta_u}{\theta_y C_2} = \frac{\mu_s}{C_2} \tag{15}$$

$$\gamma^* = \frac{2\mu_s^* - 1}{(R_\mu^*)^2} \tag{16}$$

In the last approach, the non-EPP hysteretic behavior is considered directly by modifying the energy factor (γ) [25, 47]. The energy factor for a non-EPP hysteretic behavior can be developed based on a series of time history analyses. The result is the energy factor that is a function of the hysteretic parameters. The modified energy balance equation can then be constructed for the non-EPP system. This approach is the most flexible and, in principle, can be extended to any structural system. The modified energy balance concept for tri-linear systems is illustrated as examples in Fig. 3b.

6 PBPD Example

In this paper, the PBPD method considering damage due to cumulative energy dissipation for a structural system called bucking restrained knee-braced frames (BRKBF) [26] will be presented as an example. BRKBF is an efficient structural steel system based on the applications of buckling-restrained knee braces (BRKB) as illustrated in Fig. 4. The beams are connected to the column using single plate shear connections (SPSCs). SPSCs are simple shear connections that increase the ease of construction and reparability after an earthquake. In this system, the BRKBs are expected to yield while the beams and columns are designed to be elastic. The knee braces are the primary DYMs. Hence, they are expected to deform well into the inelastic range. The design and behavior of this system are discussed in more detail elsewhere [26].

In a BRKBF, the size and brace angle are chosen based on the target deformation and seismic demand. For the mechanism shown in Fig. 4, the strain demand in the BRKBs can be estimated from the frame kinematics. As a first approximation, the equation for the BRKB core plastic strain (ε_b) can be developed assuming a rigid beam and column and assuming that the deformation occurs only in the core [26]

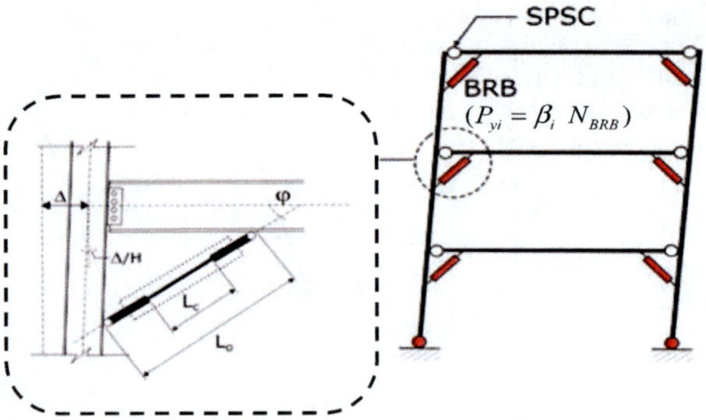

Fig. 4 BRKBF structural system

$$\varepsilon_b = \frac{\Delta \sin(2\varphi)}{2H(L_c/L_o)} \tag{17}$$

where Δ is the sway, φ is the angle of the brace with the beam, L_o is the overall brace length, L_c is the length of the yielding core, and H is the story height or the height of the frame. After the target drift has been chosen, the design base shear and the design lateral forces can be calculated using Eqs. (5), (10), and (11). The required strength of the BRBs can be computed based on the work equation of the selected mechanism

$$\sum_{i=1}^{n}(F_i h_i \theta_p) = \sum_{i=1}^{n}(2\beta_i N_{BRB}\delta_p) + (2M_{pc}\theta_p) \tag{18}$$

where F_i is the lateral force at floor level i, θ_p is the plastic target drift, N_{BRB} is the required axial strength of the BRB at roof level, δ_p is the plastic axial deformation of the BRB computed from Eq. (17), β_i is the shear proportioning factor (Eq. 11) relating the strength of the BRB at floor level i to that of the roof level, and M_{pc} is the plastic moment capacity of the columns at the bases. The above equation is for a one-bay frame but can be easily extended to multi-bay frames. Assuming that the hysteretic energy, E_H, is distributed among the DYMs in proportion to Eq. (18), the DI can be computed by

$$DI = \frac{\delta}{\delta_u} + \frac{\left(\beta_i N_{BRB}\delta_p / \sum_{i=1}^{n}(F_i h_i \theta_p)\right)E_H}{(\beta_i N_{BRB})\delta_u}\beta_{di} = \frac{\delta}{\delta_u} + \frac{\delta_p E_H}{\left(\sum_{i=1}^{n}(F_i h_i \theta_p)\right)\delta_u}\beta_{di} \tag{19}$$

where n is the number of BRBs in each story, N is the number of stories, and δ is the BRB deformation (δ_y+ δ_p) calculating using the target deformation. The plastic energy can be estimated from the input energy based on an equation proposed by Alici and Sucuoğlu [36]. Based on Alici and Sucuoğlu [36], the equivalent velocity for energy calculation can be estimated as follows

$$V_{eq} = [a \cdot e^{-bT} + c]S_v \tag{20}$$

where a and b are numerical factors depending on the period and damping. By assuming the ratio of the cumulative hysteretic energy to the input energy to be 0.8 [43], the hysteretic energy can be computed as

$$E_H = 0.8\frac{1}{2}MV_{eq}^2 \tag{21}$$

The above calculation assumes that the energy characteristics of the PBPD frame are the same as that of an equivalent SDOF system. As mentioned earlier, this still requires further research. However, based on this approximation, the design base shear corresponding to a selected damage index, maximum drift, and seismic hazard can then be obtained using Eqs. (5) and (12).

The above approach has been applied to a 3-story BRKBF shown in Fig. 5 [26]. The design of this system was assumed to be based on a Design Category D with $S_1 = 0.6$ g and $S_s = 1.5$ g at Maximum Considered Earthquake (MCE) level per ASCE [48]. The yield drift of the system was assumed to be 0.7% and the target drift was chosen as 2%. The resulting design base shear for the 2/3MCE level based on Eq. (5) was found to be 0.188. As a first approximation, the factors β_{di} and δ_u are based on a simplified model provided by Andrews et al. [49] with β_{di} of 0.23 and δ_u corresponding to 35% strain. In the calculation, the deformation of the BRB is assumed to be dominated by only the core deformation. The core length was assumed to be 70% of the total BRB length. For this design base shear, the relationships between maximum story drifts, DI values, and seismic intensity are calculated and are illustrated in Fig. 5.

As can be seen from the figure, the maximum story drift increased almost linearly in proportion to the seismic intensity level with increasing damage. Even though, the resulting DI values were relatively low. It should be noted that the damage model [49] used in the calculation was developed primarily for the conventional BRBs. BRKBs are shorter than conventional BRBs and, therefore, may have a different damage characteristic. More accurate DI model for BRKBs should be developed and used. The research is still on-going in this aspect as well as how the input and hysteretic

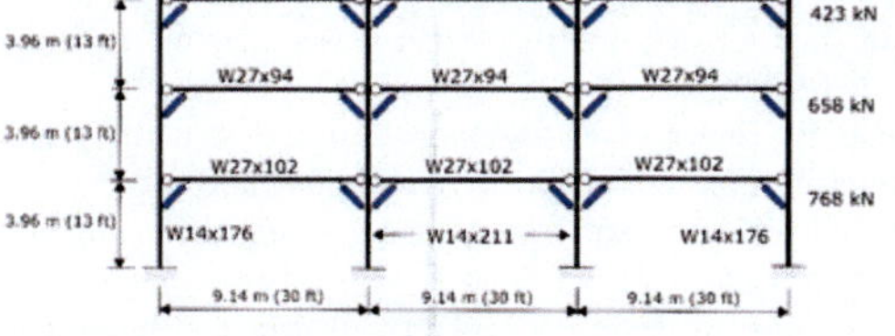

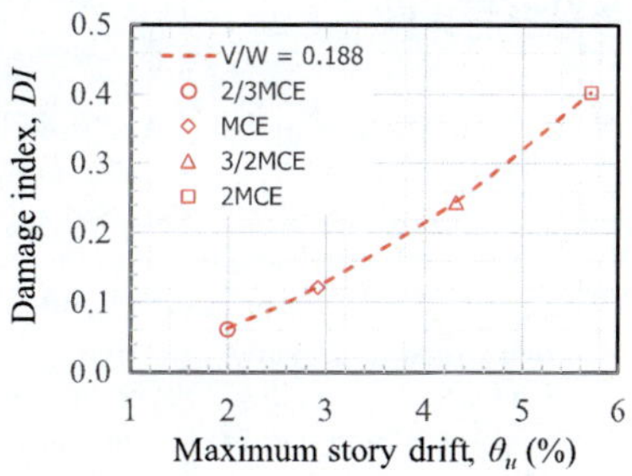

Fig. 5 3-Story BRKBF frame and the expected deformation and damage

energy are estimated and distributed among the DYMs. These are the critical issues that affect the PBPD design calculations. The above example is presented only to illustrate the PBPD framework.

7 Conclusions

The Performance-based Plastic Design (PBPD) Method is a practical seismic design and evaluation approach based on an energy concept. This paper provides an overview of the concept, the framework, and the development of the PBPD method. The key aspects of the PBPD method are the following:

1. The PBPD method relies on the modified energy balance concept in which the energy balance is applied between the fraction of the input energy and the work needed to push the structure monotonically up to the target drift. This energy provides the ties between the energy demand, the required yield strength of the system, and the target deformation. The required base shear strength can be computed for a given ductility level using any well-established $R_\mu - \mu - T$ equation.
2. The lateral force distribution in the PBPD method is derived for the plastic design of the structure in the inelastic, ultimate, state. As such, plastic design is an inherent component of the PBPD method.
3. Cyclic energy and low-cycle fatigue damage can be incorporated into the PBPD method using a suitable damage index.
4. Different load-deformation characteristics of the system can be incorporated in to the PBPD framework.
5. The same energy concept can be applied in seismic evaluation and strengthening design.

Recent developments and an example of the PBPD method are also summarized and discussed in this paper.

References

1. Leelataviwat, S., Goel, S.C., Stojadinović, B.: Toward performance-based seismic design of structures. Earthq. Spectra **15**(3), 435–461 (1999)
2. Leelataviwat, S., Goel, S.C., Stojadinović, B.: Energy-based seismic design of structures using yield mechanism and target drift. J. Struct. Eng. ASCE **128**(8), 1046–1054 (2002)
3. Lee, S.-S.: Performance-based design of steel moment frames using target drift and yield mechanism. University of Michigan (2002)
4. Chao, S.H., Goel, S.C.: Performance-based seismic design of eccentrically braced frames using target drift and yield mechanism as performance criteria. Eng. J. AISC **43**(3), 173–200 (2006)
5. Chao, S.H., Goel, S.C.: Performance-based plastic design of special truss moment frames. Eng. J. AISC **45**(2), 127–150 (2008)

6. Goel, S.C., Chao, S.-H., Leelataviwat, S., Lee, S.-S.: Performance-based plastic design (PBPD) method for earthquake-resistant structures. In: Proceedings of the 14th World Conference on Earthquake Engineering, China (2008)
7. Goel, S.C., Chao, S.-H.: Performance-based plastic design: earthquake-resistant steel structures. International Code Council (2008)
8. Bayat, M.R., Goel, S.C., Chao, S.-H.: Further refinement of performance-based plastic design of structures for earthquake resistance. In: Proceedings of the 14th World Conference on Earthquake Engineering, China (2008)
9. Goel, S.C., Liao, W.C., Bayat, M.R., Chao, S.-H.: Performance-based plastic design (PBPD) method for earthquake-resistant structures: an overview. Struct. Des. Tall Spec. Build. **19**(1–2), 115–137 (2010)
10. Liao, W.C.: Performance-based plastic design of earthquake resistant reinforced concrete moment frames. University of Michigan (2010)
11. Sahoo, D.R., Chao, S.-H.: Performance-based plastic design method for buckling restrained braced frames. Eng. Struct. **32**(9), 2950–2958 (2010)
12. Sahoo, D.R., Chao, S.-H.: Performance-based plastic design (PBPD) of high-rise buckling-restrained braced frames. In: Proceedings of the 15th World Conference on Earthquake Engineering (15WCEE). Lisbon, Portugal (2012)
13. Kharmale, S.B., Ghosh, S.: Performance-based plastic design of steel plate shear walls. J. Constr. Steel Res. **90**, 85–97 (2013)
14. Liao, W.C., Goel, S.C.: Performance-based seismic design of RC SMF using target drift and yield mechanism as performance criteria. Adv. Struct. Eng. **17**(4), 529–542 (2014)
15. Dalal, S.P., Vasanwala, S.A., Desai, A.K.: The performance evaluation of ordinary moment resisting frames designed by performance-based plastic design and limit state design. Int. J. Struct. Eng. **6**(3), 195–211 (2015)
16. Bai, J., Ou, J.: Earthquake-resistant design of buckling-restrained braced RC moment frames using performance-based plastic design method. Eng. Struct. **107**, 66–79 (2016)
17. Chan-Anan, W., Leelataviwat, S., Goel, S.C.: Performance-based plastic design method for tall hybrid coupled walls. Struct. Des. Tall Spec. Build. **25**(14), 681–699 (2016)
18. Karamodin, A., Zanganeh, A.: Seismic design and performance of dual moment and eccentrically braced frame system using PBPD method. Latin Am. J. Solids Struct. **14**(3), 441–463 (2017)
19. Abdollahzadeh, G., Mohammadgholipour, A., Omranian, E.: Seismic evaluation of steel moment frames under mainshock-aftershock sequence designed by elastic design and PBPD methods. J. Earthq. Eng., 1–24 (2018)
20. Abdollahzadeh, G.R., Kuchakzadeh, H., Mirzagoltabar, A.R.: Performance-based plastic design of moment frame-steel plate shear wall as a dual system. Civ. Eng. Infrastruct. J. **50**(1), 21–34 (2017)
21. Shayanfar, M.A., Rezaeian, A.R., Zanganeh, A.: Seismic performance of eccentrically braced frame with vertical link using PBPD method. Struct. Des. Tall Spec. Build. **23**(1), 1–21 (2014)
22. Wongpakdee, N., Leelataviwat, S., Goel, S.C.: Performance-based design and collapse evaluation of buckling restrained knee braced truss moment frames. Eng. Struct. **60**, 23–31 (2014)
23. Yang, T.Y., Li, Y., Leelataviwat, S.: Performance-based design and optimization of buckling restrained knee braced truss moment frame. J. Perform. Constr. Facil. **28**(6), A4014007 (2014)
24. Qiu, C., Li, H., Ji, K., Hou, H., Tian, L.: Performance-based plastic design approach for multi-story self-centering concentrically braced frames using SMA braces. Eng. Struct. **153**, 628–638 (2017)
25. Qiu, C.X., Zhu, S.: Performance-based seismic design of self-centering steel frames with SMA-based braces. Eng. Struct. **130**, 67–82 (2017)
26. Junda, E., Leelataviwat, S., Doung, P.: Cyclic testing and performance evaluation of buckling-restrained knee-braced frames. J. Constr. Steel Res. **148**, 154–164 (2018)
27. Leelataviwat, S., Doung, P., Junda, E., Chan-Anan, W.: Ductile knee-braced frames for seismic applications. Geotech. Geolog. Earthq. Eng. **47**, 149–158 (2019)

28. Qiu, C., Zhao, X., Zhang, Y., Hou, H.: Robustness of performance-based plastic design method for SMABFs. Int. J. Steel Struct. **19**(3), 787–805 (2019)
29. Ghamari, M., Shooshtari, M.: Application of performance-based plastic design (PBPD) method for 3D steel structures. Eng. Struct. (2019). https://doi.org/10.1016/j.engstruct.2019.109649
30. Leelataviwat, S., Saewon, W., Goel, S.C.: Application of energy balance concept in seismic evaluation of structures. J. Struct. Eng. ASCE **135**(2), 113–121 (2009)
31. Jiang, Y., Li, G., Yang, D.: A modified approach of energy balance concept based multimode pushover analysis to estimate seismic demands for buildings. Eng. Struct. **32**(5), 1272–1283 (2010)
32. Khampanit, A., Leelataviwat, S., Kochanin, J., Warnitchai, P.: Energy-based seismic strengthening design of non-ductile reinforced concrete frames using buckling-restrained braces. Eng. Struct. **81**, 110–122 (2014)
33. Housner, G.W.: Limit design of structures to resist earthquake. In: Proceedings of the 1st World Conference on Earthquake Engineering, USA (1956)
34. Newmark, N.M., Hall, W.J.: Seismic Design criteria for nuclear reactor facilities. Report no. 46. Building Practices for Disaster Mitigation, National Bureau of Standards, U.S. Department of Commerce (1973)
35. Newmark, N.M., Hall, W.J.: Earthquake Spectra and Design. Earthquake Engineering Research Institute, California (1982)
36. Alici, F.S., Sucuoğlu, H.: Prediction of input energy spectrum: attenuation models and velocity spectrum scaling. Earthq. Eng. Struct. Dyn. **45**, 2137–2161 (2016)
37. Akiyama, H.: Earthquake-resistant limit-state design for buildings. University of Tokyo Press, Tokyo (1985)
38. Goel, S.C., Furukawa, S.: Role of yield mechanism selection on seismic behavior of steel moment frames designed by performance-based plastic design. In: Proceedings of the 2009 Structures Congress, Austin, Texas (2009)
39. Chao, S., Goel, S.C., Lee, S.S.: A seismic design lateral force distribution based on inelastic state. Proceedings, Proceedings of the 8th US National Conference on Earthquake Engineering, San Francisco, California (2006)
40. Chao, S.H., Goel, S.C., Lee, S.S.: A seismic design lateral force distribution based on inelastic state of structures. Earthq. Spectra **23**(3), 547–569 (2007)
41. Zhang, H., Lian, M., Su, M., Cheng, Q.: Lateral force distribution in the inelastic state for seismic design of high-strength steel framed-tube structures with shear links. Struct. Des. Tall Spec. Build. **29**(17), e1801 (2020)
42. Fajfar, P.: Equivalent ductility factors, taking into account low-cycle fatigue. Earthq. Eng. Struct. Dyn. **21**(10), 837–848 (1992)
43. Fajfar, P., Vidic, T., Fischinger, M.: On energy demand and supply in SDOF systems. In: Fajfar, P., Krawinkler, H. (eds.) Nonlinear Seismic Analysis and Design of Reinforced Concrete Buildings, pp. 48–71. Elsevier (2005)
44. Park, Y.J., Ang, A.H.-S., Wen, Y.K.: Seismic damage analysis and damage limiting design of RC buildings. In: Structural Research Series, vol. 56. University of Illinois, Urbana, Illinois (1984)
45. Dindar, A.A., Yalcin, C., Yüksel, E., Özkaynak, H., Büyüköztürk, O.: Development of earthquake energy demand spectra. Earthq. Spectra **31**, 1667–1689 (2015)
46. FEMA440. Improvement of Nonlinear Static Seismic Analysis Procedure. Federal Emergency Management Agency, Washington, DC (2005)
47. Ke, K., Ke, S., Chuan, G.: The energy factor of systems considering multiple yielding stages during ground motions. Soil Dyn. Earthq. Eng. **71**, 42–48 (2015)
48. ASCE7-10: Minimum Design Loads for Buildings and Other Structures. American Society of Civil Engineers, ASCE/SEI 7-10, Reston, Virginia (2010)
49. Andrews, B.M., Fahnestock, L.A., Song, J.: Performance-based engineering framework and ductility capacity models for buckling-restrained braces. In: The Newmark Structural Engineering Laboratory (NSEL) Report Series Report No. NSEL-012, University of Illinois at Urbana-Champaign (2008)

Effect of Kinematic Hardening and Ductility Ratio on Inelastic Input Energy Spectra of Near-Fault Ground Motions

Taner Ucar and Onur Merter

Abstract In energy-based seismic design of structures, ground motion effect is considered as an energy input to the systems. Consistent development of input energy spectra is of great importance for the energy-based seismic design since the total energy input to structural systems can be practically obtained by means of these graphs. The main purpose of the present study is to investigate the influence of post-yield stiffness ratio and ductility demand on inelastic input energy spectra of near-fault ground motions. A wide range of nonlinear single-degreed-of-freedom (SDOF) systems characterized by their natural periods ranging from 0.02 to 3.0 s and normalized lateral strength are considered. Bilinear elastoplastic (BEP) hysteresis models with six different post-yield stiffness ratios are used to generate the results for constant ductility ratios ranging from 2 to 5. Mean $\pm$ one standard deviation input energy equivalent velocity spectra of a set of 21 near-fault accelerograms exhibiting pulse-like characteristics are computed based on nonlinear time history analyses of SDOF systems with 5% damping. The analytical results have shown that the influence of post-yield stiffness ratio on inelastic input energy spectra of near-fault ground motions can be neglected practically, whereas the influence of ductility ratio is more obvious. Moreover, a transition period of approximately 0.7 s between the increasing and decreasing input energy equivalent velocity spectra based on ductility ratio is identified.

Keywords Input energy spectra · Post-yield stiffness ratio · Ductility demand · Near-fault ground motions · Input energy equivalent velocity

T. Ucar (✉)
Dokuz Eylül University, İzmir 35390, Turkey
e-mail: taner.ucar@deu.edu.tr

O. Merter
İzmir University of Economics, İzmir 35330, Turkey
e-mail: onur.merter@ieu.edu.tr

A. Benavent-Climent and F. Mollaioli (eds.), *Energy-Based Seismic Engineering*, Lecture Notes in Civil Engineering 155,
https://doi.org/10.1007/978-3-030-73932-4_9

1 Introduction

Structures designed for code specific forces are likely expected to experience inelastic response when they are subjected to severe earthquake ground motions (EQGMs). Traditional force-based structural design is usually performed on the basis of the strength concept without considering the direct inelastic response and structural damage caused by EQGMs. The strength capacity of structural members is taken into consideration and GM effects are generally considered as equivalent static lateral loads. These strength-based design procedures are the most common ones in seismic design codes and they generally employ response modification factor R to consider the inelastic behavior indirectly [1–6]. EQ demand is also defined in the form of design acceleration spectra, which are the plots of peak pseudo acceleration responses of several SDOF systems. However, the effect of GM duration is almost lost in response spectra.

One of the major developments in seismic design over the last three decades has been the increased emphasis changing from strength to displacement. Direct displacement-based design philosophy was first proposed by Priestley [7] and it has attracted much attention since then. The main idea of the direct displacement-based design procedure is to design structural systems according to an acceptable level of damage under EQGM effects [8, 9]. The goal of the procedure is to obtain a target displacement profile for structures and to design ductile systems which achieve these target displacement limits. Displacement-based seismic design uses displacement information in the selection of structural systems and it is directly related to the structural damage. However, there are some limitations in displacement-based design procedure with regard to displacement demand estimation, design GM and performance levels [10]. In addition to above, both traditional strength- and displacement-based procedures ignore the effects such as GM duration, frequency content of the motion and hysteretic behavior of structural members [11].

Although not yet implemented in seismic design codes, energy-based procedures have become promising alternative to strength- and displacement-based design methods. Since the first proposal by Housner [12], energy parameters and the use of energy concepts in seismic design have gained great attention especially in the last few decades [11]. It can be referred to as more preferable seismic design method as it includes more GM characteristics compared with strength- or displacement-based seismic designs. Additionally, input energy transmitted to a structure is a more appropriate measure of GM intensity in comparison to commonly used intensity parameters such as peak ground acceleration (PGA), peak ground velocity (PGV) and effective peak acceleration (EPA). Accordingly, energy-based parameters have been considered to be the most rational parameters and energy input to a structure due to an EQ has been referred to as more reliable and stable parameter in EQ resistant design [13]. Many different researchers used energy principles in EQ resistant structural design after Housner [12]. EQ input energy has been researched in detail and several equations have been proposed for estimating the energy input [14–22]. A consistent estimation of input energy is a crucial issue since it is the first step in developing an energy-based seismic design method. However, many questions still need to be solved for rational energy-based design methods in EQ resistant design of structures. As examples, the energy input to a structure subjected to near-fault GMs containing long period velocity pulses should be computed appropriately, and the distribution of this energy throughout structural members should be investigated [13, 23].

The usage of input energy spectra is an effective tool in energy-based seismic design procedures, such as the use of design acceleration spectra in strength-based methods. The computation of input energy spectra offers an important advantage to determine the energy input to structures with the effect of EQGMs. PGA, PGV, the ratio of PGV to PGA, the duration and predominant periods of GMs, fault type and mechanism, source-to-site distance, the magnitude of EQ and soil conditions have been referred in many previous studies as the specific parameters to obtain the seismic input energy spectra [16, 17, 19, 21, 22, 24–29]. Manfredi [18], Benavent-Climent et al. [19], López-Almansa et al. [21], Dindar et al. [22], Fajfar and Vidic [25], Sucuoglu and Nurtug [26], Decanini and Mollaioli [27], Quinde et al. [28], Alıcı and Sucuoglu [29] and Özsaraç et al. [30] made further investigations about the EQ input energy spectra for both elastic and inelastic systems.

Effect of kinematic strain hardening and ductility ratio on inelastic input energy spectra of near-fault GMs has been investigated in this the study. A set of near-fault GM records exhibiting pulse-like characteristics has been compiled from the Pacific Earthquake Research Center (PEER) database [31]. BEP hysteresis model is used in nonlinear modeling and six different post-yield stiffness ratios (α) are considered from 0 to 0.10 with increment of 2%. A wide range of SDOF systems characterized by their natural periods and normalized lateral strength are taken into consideration and nonlinear time history analyses have been conducted for damping ratio of $\xi =$ 5% by means of the software PRISM [32]. Four constant ductility ratios (μ) are selected for SDOF systems, as $\mu = 2, 3, 4$ and 5. Input energy spectra of the selected near-fault GMs have been computed in terms of energy equivalent velocity (V_{eq}) by using Excel program developed by the authors.

2 Input Energy Spectrum

2.1 Basic Concepts

Energy balance equation of a nonlinear SDOF system subjected to EQGM can be derived by integrating the governing equation of motion with respect to the relative displacement of the mass:

$$\int_0^{u(t)} m \cdot \ddot{u}\, du + \int_0^{u(t)} c \cdot \dot{u}\, du + \int_0^{u(t)} f_s(u)\, du = -\int_0^{u(t)} m \cdot \ddot{u}_g(t)\, du \tag{1}$$

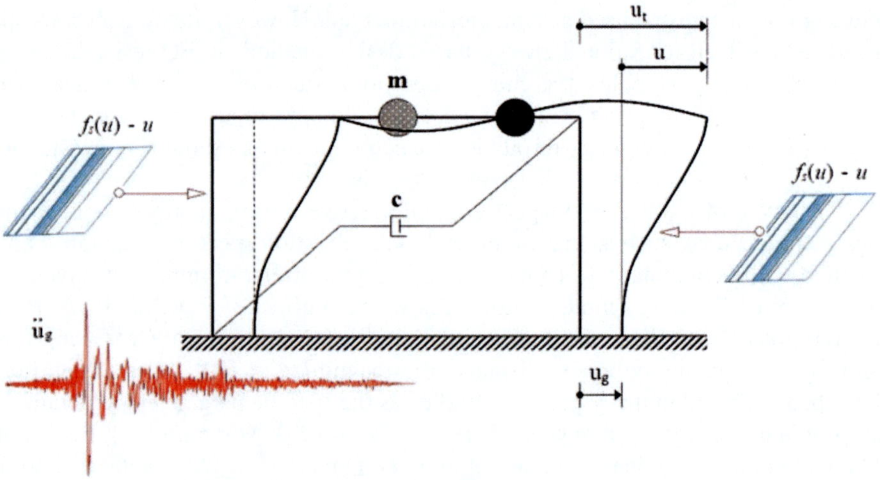

Fig. 1 Fixed-based SDOF system subjected to an EQGM

where u is the relative displacement of the SDOF system with respect to the ground, $\dot{u}$ is the relative velocity, $\ddot{u}$ is the relative acceleration, m is the total mass, c is the damping coefficient, $f_s(u)$ is the resisting force and $\ddot{u}_g(t)$ is the ground acceleration. For convenience, the parameters used in developing Eq. (1) are illustrated in Fig. 1, where u_t is the total displacement of the SDOF system and it is equal to the sum of relative displacement and displacement of the ground ($u_t = u + u_g$).

The energy-balance equation based on the relative motion can be rewritten in terms of the time integral by writing a derivative equality for du ($du = \dot{u}\, dt$) as:

$$\int_0^t m \cdot \ddot{u} \cdot \dot{u}\, dt + \int_0^t c \cdot \dot{u} \cdot \dot{u}\, dt + \int_0^t f_s(u) \cdot \dot{u}\, dt = -\int_0^t m \cdot \ddot{u}_g(t) \cdot \dot{u}\, dt \quad (2)$$

The total energy input to SDOF system can be computed by using the right-hand side term of Eq. (1). In the left-hand side of Eq. (1), the first term represents the kinetic energy (E_K), the second term is the damping energy (E_ξ) and the last term is the total absorbed energy by the system (the sum of elastic strain energy and hysteretic energy). Finally, the input energy can be written by the mass-independent term as:

$$\frac{E_I}{m} = -\int_0^t \ddot{u}_g(t) \cdot \dot{u}\, dt \quad (3)$$

A representative energy-time history graph of an inelastic SDOF structure subjected to EQGM is shown in Fig. 2. The substantial components of input energy

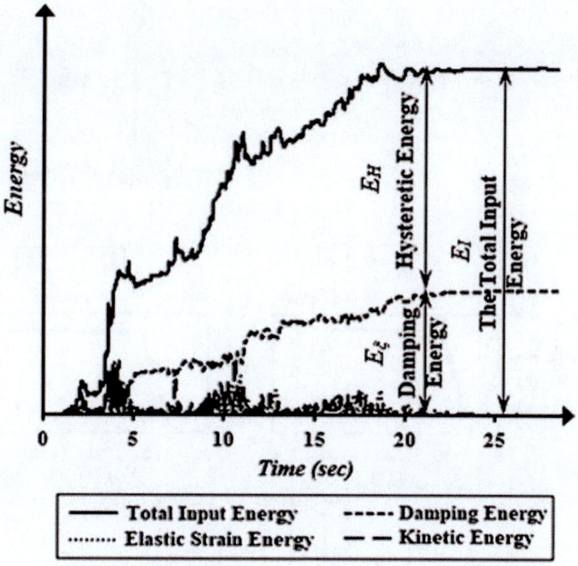

Fig. 2 An energy-time history diagram of an inelastic SDOF system

are the damping energy, and the hysteretic energy which contributes the most to structural damage.

2.2 *Construction of Input Energy Spectrum*

In energy-based seismic design philosophy, the energy demand of an EQ should be less than (or, in limit, should be equal to) the energy dissipation capacities possessed by structure [27]. It is of great importance to accurately compute the seismic input energy demand transmitted to structural systems by the EQGMs in the context of energy-based design. This process involves the detailed and complicated dynamic analyses and so the researchers have sought to obtain practical ways for computing seismic input energy. Many approximate formulas have been proposed for estimation of peak seismic input energy in literature [12, 14–18, 24]. When seismic input energy spectra have been created for a specific EQGM, input energy demands imposed on both elastic and inelastic systems can be obtained via these graphs [18, 19, 21, 22, 25–29, 33–37]. The seismic input energy spectra combine the peak input energy values corresponding to different natural vibration periods (T_n) of SDOF systems (Fig. 3) and there are many factors such as EQ magnitude, distance from the seismic source, soil type, focal mechanism, damping ratio and the type of hysteresis which have an impact on the seismic input energy spectra [22, 27, 28, 34–37].

The input energy can be expressed in terms of input energy equivalent velocity (V_{eq}) in order to facilitate comparison between different GM records:

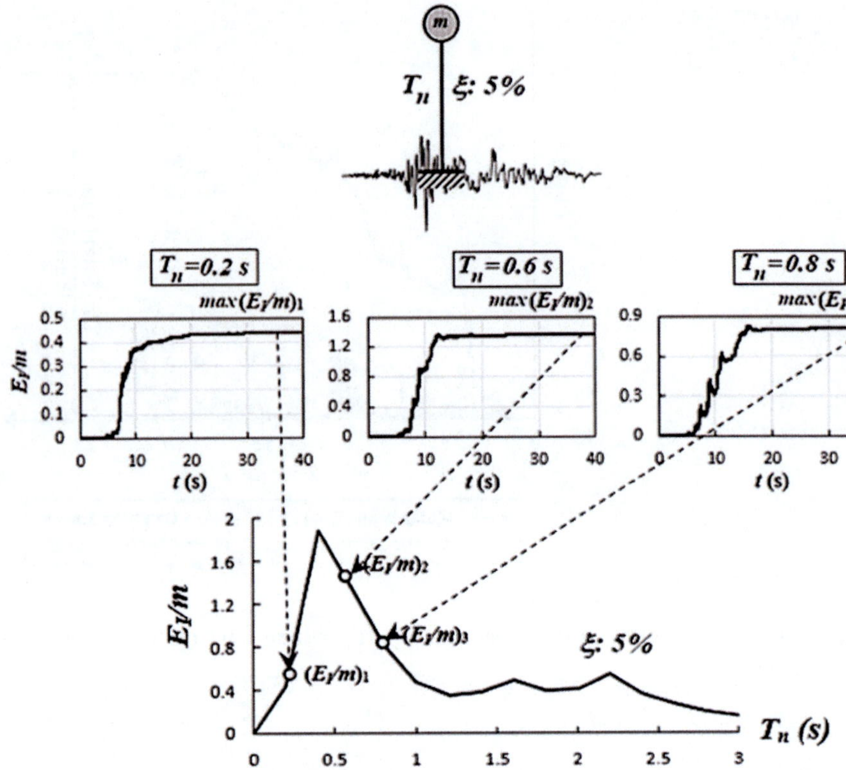

Fig. 3 Construction of input energy spectrum of an individual GM

$$V_{eq} = \sqrt{2E_I/m} \tag{4}$$

For a given GM, the relationship between the input energy expressed in terms of V_{eq} and the natural period of SDOF oscillator can also be plotted as energy input spectrum (i.e., plot of V_{eq} vs. T_n). Henceforth, the energy equivalent velocity will be used as the measure of input energy.

3 Considered SDOF Systems and Hysteresis Models

Analyses of the study are performed on an ensemble of nonlinear SDOF systems with natural vibration periods (T_n) ranging from 0.02 to 3.0 s for four constant ductility ratios as $\mu = 2$, 3, 4 and 5 to cover a wide range of different structures. Velocity response of the systems is obtained by means of PRISM [32], software for seismic response analysis of SDOF systems. Normalized lateral strength index of nonlinear SDOF systems is defined as the ratio of base shear force at which the structures

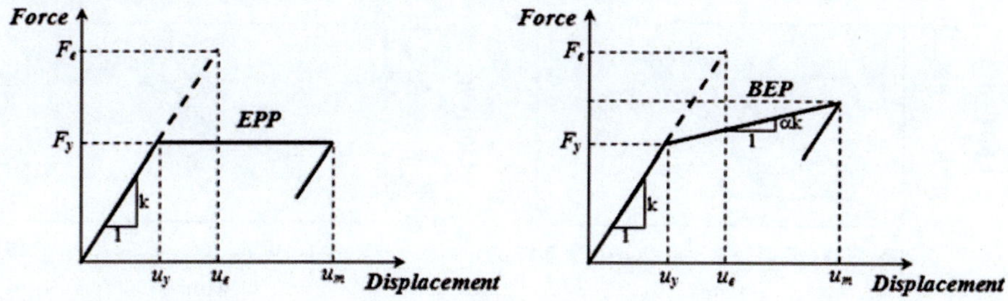

Fig. 4 EPP and BEP hysteresis models

begins its inelastic deformation to seismic weight (i.e., F_y/mg). SDOF oscillators are assumed to have a 5% viscous damping.

A bilinear elastoplastic (BEP) behavior with and without kinematic strain hardening is assumed. Accordingly, six different post-yield stiffness ratios are considered, as $\alpha = 0$, 0.02, 0.04, 0.06, 0.08 and 0.10 to model the nonlinear cyclic force-displacement response in input energy analyses. The first model, considering $\alpha = 0$, corresponds to an elastic-perfectly plastic (EPP) hysteresis model, which is the simplest one and a rough approximation of the real nonlinear behavior of structural systems subjected to cyclic loadings as EQs [38]. It is commonly used to represent the hysteretic behavior of steel structures. EPP and BEP hysteresis models are presented in Fig. 4, where F_y is the yield force, F_e is the elastic force demand, u_y is the yield displacement, u_e is the elastic displacement, u_m is the maximum displacement, k is the elastic stiffness and αk is the post-yield stiffness. Moreover, Fig. 5 shows the cyclic response of nonlinear SDOF oscillator with $T_n = 1.6$ s to GM from 1999 Duzce, Turkey EQ. In this figure, F_y/mg values are the normalized lateral strengths required to achieve constant ductility of $\mu = 5$.

4 Near-Fault GM Dataset

Strong GM data of the study consists of a set of 21 processed near-fault GMs exhibiting pulse-like characteristics. GM records included in the database are compiled from NGA-West2 strong GM database [31]. All far-fault GMs are recorded from EQ events with moment magnitude (M_w) range from 6.0 to 8.0. The faulting mechanism of the near-fault EQs is strike-slip. Site condition of the compiled accelerograms is described in terms of the average shear wave velocity to 30 m depth of subsoil V_{S30} and depicts the features of site class D according to NEHRP site classification (i.e., V_{S30} ranges between 180 and 360 m/s). The impulsive characteristics of near-fault records are characterized with velocity pulses having periods equal to T_p, which vary between 0.5 and 9.5 s and have also been adopted from NGA-West2 strong GM database. Table 1 lists all the records used in the study and

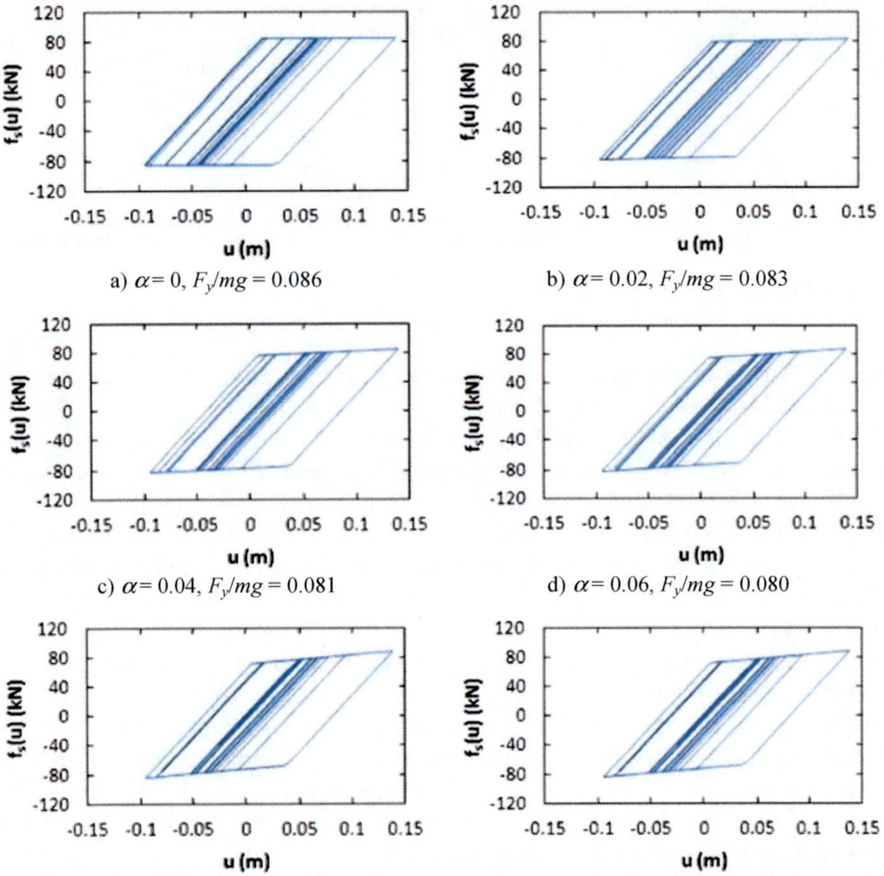

a) $\alpha = 0$, $F_y/mg = 0.086$

b) $\alpha = 0.02$, $F_y/mg = 0.083$

c) $\alpha = 0.04$, $F_y/mg = 0.081$

d) $\alpha = 0.06$, $F_y/mg = 0.080$

e) $\alpha = 0.08$, $F_y/mg = 0.079$

f) $\alpha = 0.10$, $F_y/mg = 0.079$

Fig. 5 Restoring force—deformation response of hysteretic SDOF systems to RSN1602_DUZCE_BOL090 ground motion ($T_n = 1.6$ s and $\mu = 5$)

the distribution of their PGA with respect to M_w and rupture distance (R_{rup}) is demonstrated in Fig. 6. The closest distance to fault rapture surface of the compiled records is chosen to vary from 0.07 to 26.76 km. Near-fault GMs with fling-step effects are not considered in the dataset of the study.

Essential parameters reflecting the intensity, duration and frequency content of the GMs are listed in Table 2, where PGD is the peak ground displacement. The utilized near-fault strong GMs generate relatively high Arias intensity values, whereas the significant duration is relatively small. The ratio of PGV to PGA varies from 0.05 to 0.35 s. PGA/PGV ratio indicates the average duration of acceleration pulse that PGV is reached immediately following the dominant acceleration pulse [26, 39].

Table 1 List of near-fault EQ recordings

Record	T_p[a] (s)	Earthquake name (year)	M_w	R_{JB}[b] (km)
RSN4126_PARK2004_SC1360	0.574	Parkfield-02 (2004)	6	2.85
RSN1602_DUZCE_BOL090	0.882	Duzce (1999)	7.14	12.02
RSN4102_PARK2004_C03360	1.022	Parkfield-02 (2004)	6	2.55
RSN1106_KOBE_KJM000	1.092	Kobe (1995)	6.9	0.94
RSN4098_PARK2004_C01090	1.33	Parkfield-02 (2004)	6	1.66
RSN1119_KOBE_TAZ000	1.806	Kobe (1995)	6.9	0
RSN722_SUPER.B_B-KRN360	2.128	Superstition Hills-02 (1987)	6.54	18.48
RSN159_IMPVALL.H_H-AGR003	2.338	Imperial Valley-06 (1979)	6.53	0
RSN1114_KOBE_PRI000	2.828	Kobe (1995)	6.9	3.31
RSN2114_DENALI_PS10-047	3.157	Denali (2002)	7.9	0.18
RSN171_IMPVALL.H_H-EMO000	3.423	Imperial Valley-06 (1979)	6.53	0.07
RSN181_IMPVALL.H_H-E06140	3.773	Imperial Valley-06 (1979)	6.53	0
RSN182_IMPVALL.H_H-E07230	4.375	Imperial Valley-06 (1979)	6.53	0.56
RSN1176_KOCAELI_YPT150	4.949	Kocaeli (1999)	7.51	1.38
RSN6906_DARFIELD_GDLCS35W	6.23	Darfield (2010)	7	1.22
RSN8606_SIERRA.MEX_CIWESHNE	7.084	El Mayor-Cucapah (2010)	7.2	10.31
RSN900_LANDERS_YER270	7.504	Landers (1992)	7.28	23.62
RSN6897_DARFIELD_DSLCN63E	7.826	Darfield (2010)	7	5.28
RSN6942_DARFIELD_NNBSS13E	8.043	Darfield (2010)	7	26.76
RSN8161_SIERRA.MEX_E12090	8.722	El Mayor-Cucapah (2010)	7.2	9.98
RSN6960_DARFIELD_RHSCS04W	9.394	Darfield (2010)	7	13.64

[a]Velocity pulse period
[b]Joyner-Boore distance

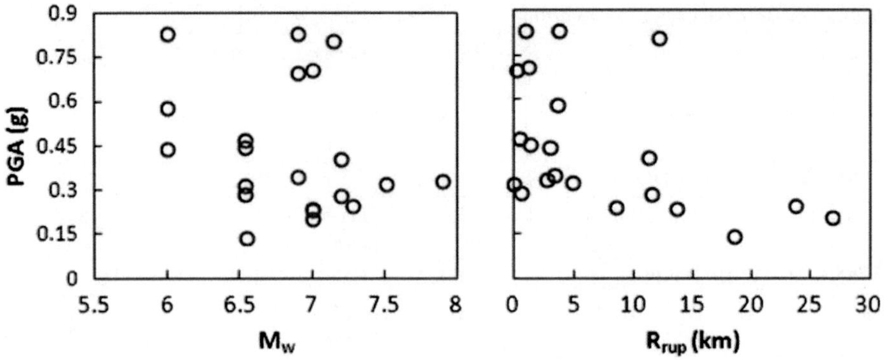

Fig. 6 Distribution of PGA of records with respect to M_w and R_{rup}

Table 2 Characteristics of strong motion records

Record	I_a[a] (m/s)	$D_{a5\text{–}95}$[b] (s)	PGA (g)	PGV (cm/s)	PGD (cm)	V/A[c] (s)
RSN4126_PARK2004_SC1360	1.474	8.10	0.833	39.79	3.34	0.049
RSN1602_DUZCE_BOL090	2.430	9.02	0.806	65.88	13.10	0.083
RSN4102_PARK2004_C03360	0.865	6.54	0.579	37.80	7.37	0.067
RSN1106_KOBE_KJM000	8.393	8.38	0.834	91.11	21.11	0.111
RSN4098_PARK2004_C01090	0.607	7.14	0.440	40.13	9.43	0.093
RSN1119_KOBE_TAZ000	3.070	4.62	0.697	68.41	26.67	0.144
RSN722_SUPER.B_B-KRN360	0.301	12.44	0.139	29.61	7.93	0.217
RSN159_IMPVALL.H_H-AGR003	0.964	13.29	0.287	34.94	9.36	0.124
RSN1114_KOBE_PRI000	1.766	6.76	0.348	90.67	39.31	0.266
RSN2114_DENALI_PS10-047	1.894	22.335	0.333	115.72	53.44	0.354
RSN171_IMPVALL.H_H-EMO000	0.867	8.215	0.317	72.95	34.92	0.235
RSN181_IMPVALL.H_H-E06140	1.541	11.46	0.447	67.02	27.90	0.153
RSN182_IMPVALL.H_H-E07230	1.730	4.795	0.469	113.14	46.94	0.246
RSN1176_KOCAELI_YPT150	1.323	15.08	0.322	71.89	47.33	0.228
RSN6906_DARFIELD_GDLCS35W	4.700	9.965	0.708	100.34	44.57	0.145
RSN8606_SIERRA.MEX_CIWESHNE	1.943	22.89	0.281	52.05	56.38	0.189
RSN900_LANDERS_YER270	0.923	17.58	0.245	51.12	41.72	0.213
RSN6897_DARFIELD_DSLCN63E	1.659	19.565	0.237	67.26	81.29	0.289
RSN6942_DARFIELD_NNBSS13E	0.708	26.895	0.203	56.52	52.32	0.284
RSN8161_SIERRA.MEX_E12090	3.240	31.395	0.406	70.16	62.38	0.176
RSN6960_DARFIELD_RHSCS04W	1.193	23.91	0.234	62.72	59.35	0.273

[a] Arias intensity
[b] Significant duration evaluated as the time intervals between 5 and 95% of Arias intensity
[c] PGV/PGA

5 Results and Discussion

Due to high record-to-record variability typically observed in inelastic time history analyses, mean and mean ± 1 standard deviation (SD) input energy spectra are computed. Thereby, much smoother spectrum than the input energy spectrum of an individual GM record shown in Fig. 3 is constituted and forthcoming observations will therefore based on a set of smooth spectra. Statistical analyses of V_{eq} data computed for individual GM records provide the probability distribution of V_{eq} spectral ordinates at each considered period. The coefficients of variation computed in the study vary from period to period and range from 0.363 to 0.550.

First, the influence of post-yield stiffness ratio on V_{eq} spectra of near-fault GMs is studied. Variation of mean ± 1 SD V_{eq} spectra with post-yield stiffness ratio is presented in Fig. 7 for different constant ductility ratios, which are adopted to be $\mu = 2$, 3, 4 and 5. It can clearly be observed from Fig. 7 that, the influence of post-yield stiffness ratio on V_{eq} spectra is quite small along the considered spectral period range. To be more specific, in short period region, the dependency of V_{eq} spectra on kinematic strain hardening is almost lost, but at intermediate- and long-period spectral regions V_{eq} spectra increase slightly as post-yield stiffness ratio increases. This trend is more obvious for ductility ratio of $\mu = 5$ in Fig. 7d, where the dashed lines represent the considered post-yield stiffness ratios (i.e., $\alpha = 0.02$, 0.04, 0.06, 0.08 and 0.10). Consequently, the results of the study show that the influence of post-yield stiffness ratio can be neglected.

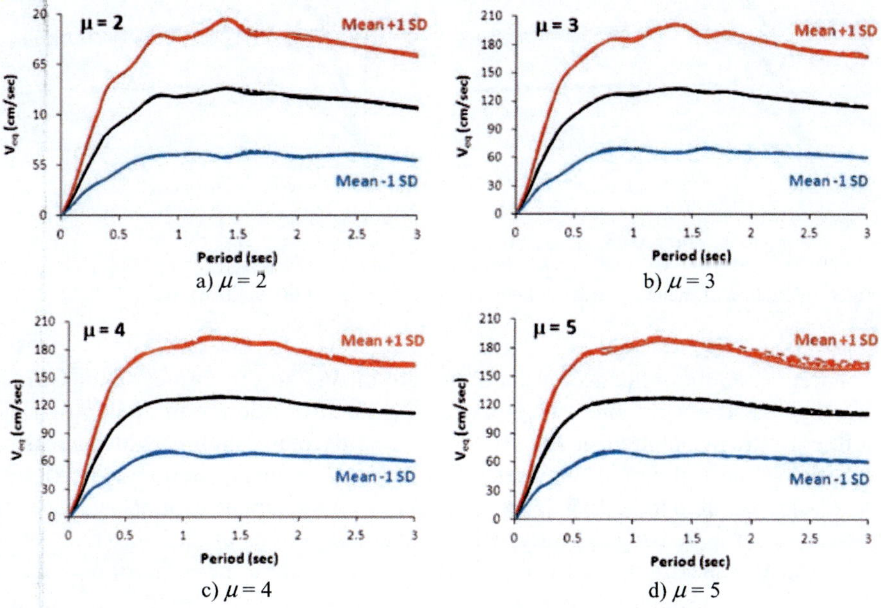

Fig. 7 Mean ± 1 standard deviation V_{eq} spectra for different constant ductility ratios

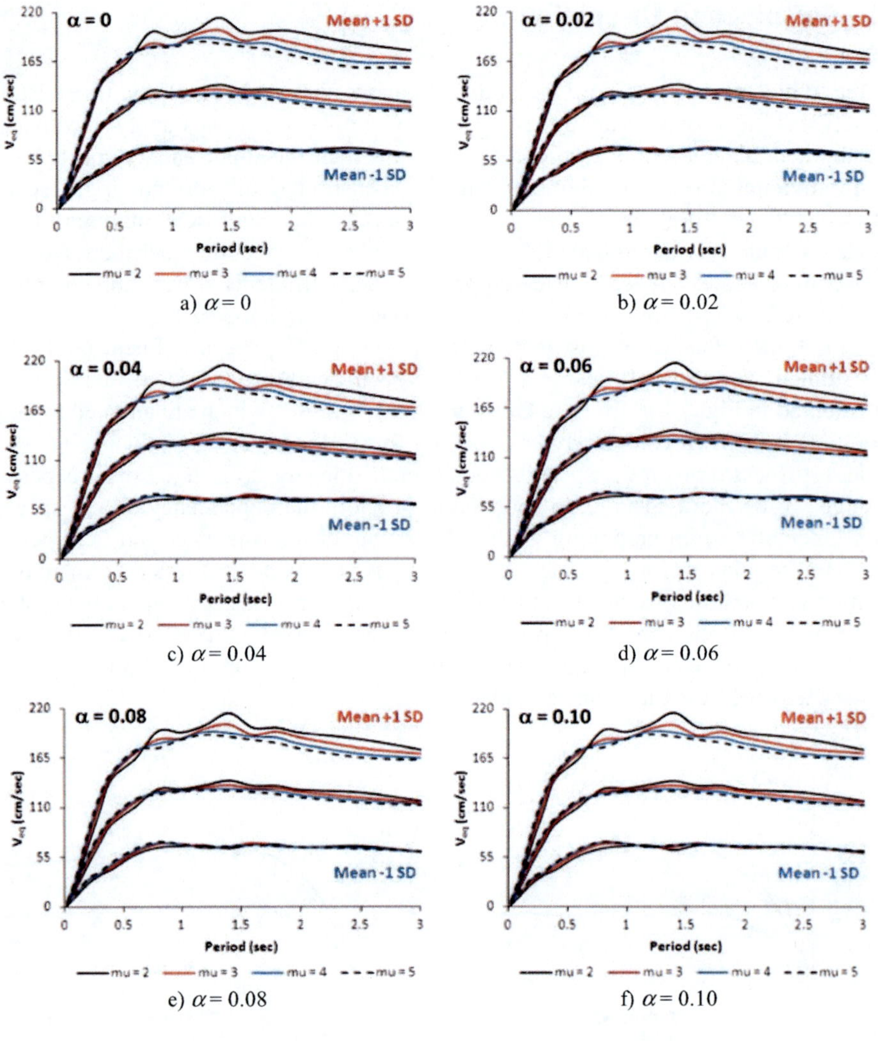

a) $\alpha = 0$ b) $\alpha = 0.02$ c) $\alpha = 0.04$ d) $\alpha = 0.06$ e) $\alpha = 0.08$ f) $\alpha = 0.10$

Fig. 8 Variation of mean ± 1 standard deviation V_{eq} spectra with ductility ratio

Afterwards, the influence of ductility ratio on V_{eq} spectra of near-fault GMs is investigated. The variation of V_{eq} spectra with ductility ratio for different post-yield stiffness ratios is plotted in Fig. 8. V_{eq} spectra decrease slightly with increasing ductility at spectral periods between 0.8 and 3.0 s. For mean spectra, the maximum decreasing is found to be 13% for $\alpha = 0$ and $T_n = 2.5$ s. On the contrary, V_{eq} spectra increase with increasing ductility ratios at spectral periods of 0.02 to 0.6 s. Again for mean V_{eq} spectra, the maximum increasing is found to be 85% for $\alpha = 0$ and $T_n = 0.02$ s, and roughly 55% at $T_n = 0.02$ s regardless of post-yield stiffness ratio. At this spectral region, the increasing rate of V_{eq} spectral ordinates with increasing ductility decreases rapidly as period elongates. For instance, at $T_n = 0.2$ s and $\alpha = 0.10$, the ratio of V_{eq} ($\mu = 5$) to V_{eq} ($\mu = 2$) is 24.6%, which is the highest ratio corresponding to this specific period of mean spectra. Consequently, it can clearly be concluded that $T_n = 0.7$ s is the transition period between the increasing and decreasing V_{eq} spectra, regardless of post-yield stiffness ratio.

6 Conclusions

In order to eliminate high record-to-record variability typically observed in inelastic time history analyses, mean and mean ±1 standard deviation input energy equivalent velocity spectra of near-fault GMs exhibiting pulse-like characteristic are developed and the influence of kinematic hardening and ductility ratio is mainly studied. Basic conclusions derived from the study are as follows:

- In short period region, the dependency of input energy equivalent spectra on post-yield stiffness ratio is almost lost, whereas at intermediate- and long-period spectral regions input energy equivalent spectra increase slightly as kinematic hardening increases. This trend is more obvious for high ductility ratios such as $\mu = 5$.
- Due to weak dependency between input energy equivalent velocity and kinematic hardening, elastic-perfectly plastic hysteresis model, which is the simplest one, can be efficiently adopted.
- Ductility ratio is found to be more influential on inelastic input energy spectra of near-fault ground motions, in comparison to effect of post-yield stiffness ratio.
- Input energy equivalent velocity spectra decrease slightly with increasing ductility ratio at spectral periods between 0.8 and 3.0 s. On the contrary, input energy increase with increasing ductility ratios at periods equal to or shorter than 0.6 s, but the increasing ratio is quite insignificant.
- $T_n = 0.7$ s is observed to be a transition period between the increasing and decreasing V_{eq} spectra, regardless of post-yield stiffness ratio.

References

1. UBC: Uniform Building Code. In: International Conference of Building Officials Officials, Whittier, California, USA (1997)
2. CEN, Eurocode 8: Design of structures for earthquake resistance—part 1: general rules, seismic actions and rules for buildings. European Committee for Standardization, Brussels, Belgium (2004)
3. NBCC: National Building Code of Canada. Associate Committee on the National Building Code, National Research Council of Canada, Ottawa, Ontario (2005)
4. IBC: International Building Code. International Code Council, Country Club Hills, Illinois, USA (2006)

5. ASCE/SEI 7-10: Minimum design loads for buildings and other structures. American Society of Civil Engineers, USA (2010)
6. TBEC: Turkish Building Earthquake Code. The disaster and emergency management presidency of Turkey. In: 18th of March of Official Gazette, Ankara, Turkey (2018)
7. Priestley, M.J.N.: Myths and fallacies in earthquake engineering-conflicts between design and reality. Bull. N. Z. Soc. Earthq. Eng. **26**(3), 329–341 (1993)
8. Priestley, M.J.N., Calvi, G.M., Kowalsky, M.J.: Displacement based seismic design of structures. IUSS Press, Pavia, Italy (2007)
9. Priestley, M.J.N., Calvi, M.G., Kowalsky, M.J.: Displacement-based seismic design of concrete structures. In: Sixth National Conference on Earthquake Engineering, pp. 113–137. İstanbul, Turkey, 16–20 Oct 2007.
10. Moehle, J.P.: Displacement-based seismic design criteria. In: Proceedings of the 11th World Conference on Earthquake Engineering, Mexico (1996)
11. Akbaş, B., Shen, J.: Energy approach in performance-based earthquake resistant design (PB-EQRD). In: 12th European Conference on Earthquake Engineering, Paper Reference 043, London (2002)
12. Housner, G.W.: Limit design of structures to resist earthquakes. In: Proceedings of the 1st World Conference on Earthquake Engineering, Oakland, California, USA (1956)
13. Akbaş, B., Shen, J.: Earthquake resistant design and energy concepts. Techn. J. Dig. **2003**, 865–888 (2003)
14. Akiyama, H.: Earthquake-resistant limit-state design for buildings. University of Tokyo Press, Tokyo, Japan (1985)
15. Kuwamura, H., Galambos, T.V.: Earthquake load for structural reliability. J. Struct. Eng. **115**(6), 1446–1462 (1989)
16. Fajfar, P., Vidic, T., Fischinger, M.: Seismic demand in medium-and long-period structures. Earthq. Eng. Struct. Dyn. **18**(8), 1133–1144 (1989)
17. Uang, C.M., Bertero, V.: Evaluation of seismic energy in structures. Earthq. Eng. Struct. Dyn. **19**(1), 77–90 (1990)
18. Manfredi, G.: Evaluation of seismic energy demand. Earthq. Eng. Struct. Dyn. **30**(4), 485–499 (2001)
19. Benavent-Climent, A., Pujades, L.G., López-Almansa, F.: Design energy input spectra for moderate seismicity regions. Earthq. Eng. Struct. Dyn. **31**, 1151–1172 (2002)
20. Khashaee, P.: Energy-based seismic design and damage assessment for structures. Ph.D. dissertation. Southern Methodist University, Dallas, Texas (2004)
21. López-Almansa, F., Yazgan, A.U., Benavent-Climent, A.: Design energy input spectra for high seismicity regions based on Turkish registers. Bull. Earthq. Eng. **11**(4), 885–912 (2013)
22. Dindar, A.A., Yalçın, C., Yüksel, E., Özkaynak, H., Büyüköztürk, O.: Development of earthquake energy demand spectra. Earthq. Spectra **31**(3), 1667–1689 (2015)
23. Shen, J., Akbaş, B.: Seismic energy demand in steel moment frames. J. Earthq. Eng. **3**(4), 519–559 (1999)
24. Zahrah, T.F., Hall, W.J.: Earthquake energy absorption in SDOF structures. J. Struct. Eng. **110**(8), 1757–1772 (1984)
25. Fajfar, P., Vidic, T.: Consistent inelastic design spectra: hysteretic and input energy. Earthq. Eng. Struct. Dyn. **23**, 523–537 (1994)
26. Sucuoglu, H., Nurtuğ, A.: Earthquake ground motion characteristics and seismic energy dissipation. Earthq. Eng. Struct. Dyn. **24**, 1195–1213 (1995)
27. Decanini, L.D., Mollaioli, F.: Formulation of elastic earthquake input energy spectra. Earthq. Eng. Struct. Dyn. **27**, 1503–1522 (1998)
28. Quinde, P., Reinoso, E., Terán-Gilmore, A.: Inelastic seismic energy spectra for soft soils: application to Mexico City. Soil Dyn. Earthq. Eng. **89**, 198–207 (2016)
29. Alıcı, F.S., Sucuoğlu, H.: Prediction of input energy spectrum: attenuation models and velocity spectrum scaling. In: The 4th International Conference on Earthquake Engineering and Seismology. Anadolu University, Eskisehir, Turkey, 11–13 Oct 2017

30. Özsaraç, V., Karimzadeh, S., Erberik, M.A., Askan, A.: Energy-based response of simple structural systems by using simulated ground motions. Proc. Eng. **199**, 236–241 (2017)
31. PEER: Pacific Earthquake Engineering Research Center Strong Ground Motion Database. https://ngawest2.berkeley.edu/. Last accessed 21 Jan 2021
32. PRISM for Earthquake engineering: a software for seismic response analysis of single-degree-of freedom systems. Department of Architectural Engineering. INHA University, South Korea (2011)
33. Alıcı, F.S., Sucuoğlu, H.: Elastic and inelastic near-fault input energy spectra. Earthq. Spectra **34**(2), 611–637 (2018)
34. Mezgebo, M., Lui, E.M.: Hysteresis and soil site dependent input and hysteretic energy spectra for far-source ground motions. Adv. Civ. Eng., 1–29 (2016)
35. Merter, O.: A study on elastic input energy spectra for actual earthquake ground motions at stiff soil sites. In: COMPDYN 2019, 7th International Conference on Computational Methods in Structural Dynamics and Earthquake Engineering. Crete, Greece, 24–26 June 2019
36. Zhai, C.H., Ji, D.F., Wen, W.P., Lei, W.D., Xie, L.L., Gong, M.S.: The inelastic input energy spectra for main shock-aftershock sequences. Earthq. Spectra **32**(4), 2149–2166 (2016)
37. Karimzadeh, S., Özsaraç, V., Askan, A., Erberik, M.A.: An energy-based seismic response evaluation of simple structural systems with simulated ground motions. In: 4th International Conference on Earthquake Engineering and Seismology, Anadolu University, Eskişehir, Turkey, 11–13 Oct 2017
38. Apostolos, B.: Effects of strength hardening, stiffness degradation, strength deterioration and pinching on the seismic response of SDOF systems. M.Sc. Thesis. Delft University of Technology, Delft (2017)
39. Kalkan, E., Kunnath, S.K.: Effective cyclic energy as a measure of seismic demand. J. Earthq. Eng. **11**(5), 725–751 (2007)

Least-Square Effective Stiffness to be Used for Equivalent Linear Model

Francisco J. Molina and Pierre Pegon

Abstract This work introduces a method for the identification of linear stiffness and viscous damping parameters from a one-degree-of-freedom force-displacement cycle. Using an original approach, the stiffness parameter is derived by a least-square formula from the discrete input force-displacement point coordinates of the loop. The damping ratio is obtained in a classic manner from the quotient of absorbed and elastic energy. The obtained stiffness and damping parameters are proposed to be used, with the known mass, as an equivalent linear mass-spring-damper that should predict the response to a known load for the original nonlinear system associated to the input force-displacement cycle. As an example study, the effectiveness of such prediction is qualitatively shown in the case of the steady-state response to a harmonic load for a particular hysteretic numerical model, by using a range of values for some dimensionless parameters. This kind of study is susceptible to be extended to other kinds of loading and/or numerical and experimental hysteretic models, as well as to other identification procedures available in the literature.

Keywords Linear model · Effective stiffness · Equivalent damping · Least square · Prediction error

1 Introduction

Proposals of equivalent linear mass-spring-damper models that may substitute a non-linear one-degree-of-freedom system are very diffused in the literature and have gained relevance in the last decades especially with the development of displacement based design methods. Assuming that the mass is well known, two parameters (stiffness and viscous damping) need to be determined for the equivalent linear model. For systems with more degrees of freedom, also the choice of the mass represents another parameter, but such systems are out of the scope of the current work.

F. J. Molina (✉) · P. Pegon
Joint Research Centre, Space Security & Migration Directorate, Safety & Security of Buildings E04 Unit, European Commission, Ispra, Italy
e-mail: Francisco-Javier.Molina@ec.europa.eu

A. Benavent-Climent and F. Mollaioli (eds.), *Energy-Based Seismic Engineering*, Lecture Notes in Civil Engineering 155,
https://doi.org/10.1007/978-3-030-73932-4_10

As it is reported in the study produced by Dwairi et al. (2007) for example, the earliest contribution in this field is from Jacobsen (1930), who adopts as equivalent parameters a stiffness equal to the initial one and a damping coefficient based on the equality of the dissipated energy per cycle between the original nonlinear system and the substitute linear system. The same substitute system is adopted by the Direct Displacement-Based Design method for nonlinear structures proposed by Priestley (1993), with the difference of using the secant stiffness at the maximum reached displacement in place of the initial stiffness. The use of the secant stiffness combined with Jacobsen's equivalent damping was proposed by Rosenblueth and Herrera (1964) and is well perceived by the designers because of its simplicity in the computations and its natural integration with the use of linear design spectra. The passage from using the initial stiffness to the secant one and other proposed definitions of stiffness is also referred as period shift in the literature (e.g., Dwairi et al. 2007).

In the current work, for the selection of the stiffness parameter, a different choice is done, which presumably represents an original approach. The stiffness is computed as the slope of the linear regression line of the force values with respect to the displacement values in the loop. This is equivalent to finding a straight line that minimizes the quadratic error of the force values with respect to it. Such slope will be called here least-square stiffness.

The concept of least-square identification is, of course, very diffused and, in the experience of the authors, is also present, although in a different form, in other equivalent linear models such as the Spatial Model, which are able to identify full matrices of stiffness and damping of multi-degree-of-freedom systems from the time response of the system in terms of force and displacement vectors (Ewins 1984). We have proposed and extensively used such model (Molina et al. 1999), particularly for the assessment of the errors in the seismic test setup system (Molina et al. 2011, 2013), but also for the prediction of the response of the structure itself to demand spectra different than the ones associated to the input data (Molina et al. 2016). At the same time, we have been also using the least-square stiffness for obtaining a linear approximation to experimental loops coming from cyclic tests and allowing to do predictions that were confirmed in many cases by the successive seismic experiments on the same specimen (Dal Lago and Molina 2018, for example). The objective of the current work is to present the formulation of the least-square stiffness and the potential use of it.

This article is a shortened version of the report by Molina and Pegon (2017), in which the least-square stiffness identification is combined with Jacobsen's equivalent damping and successively an example study is performed on the prediction error of such model for the steady-state response to harmonic loading. The numerical nonlinear law adopted for such demonstrative study is done by superposing a linear elastic force and a Giuffre and Pinto (1970) dissipative force.

2 Proposed Method for the Identification

The method proposed here allows to identify a linear equivalent stiffness and a viscous equivalent damping from a given restoring force-displacement cycle. The relationship of the force-displacement data with time is ignored so that, in principle, they may come from any viscous or hysteretic model or any experimental measurements.

2.1 Energy Ratio for a Linear Mass-Spring-Damper Oscillator

The equivalence can be based on the assumption that the considered cycle is obtained during a period of oscillation at resonance of a single-degree-of-freedom linear viscous mass-spring-damper. Without knowing the time dependency, for that oscillator, the frequency itself and the mass cannot be identified, but the damping and the stiffness can be estimated from the shape of the cycle if we assume certain conditions to be met. The energy ratio is one of such conditions.

Consider first the motion of a linear mass-spring-damper oscillator:

$$ma + cv + kd = f_{ext}(t) \tag{1}$$

which is written under its canonical form as

$$a + 2\xi\omega v + \omega^2 d = \omega^2 f(t)/k \tag{2}$$

with the natural angular frequency ω and the viscous damping ratio ξ introduced as:

$$\omega^2 = \frac{k}{m} \quad 2\xi\omega = \frac{c}{m} = \omega^2\frac{c}{k} \tag{3}$$

When subjected to an external harmonic force

$$f_{ext}(t) = f_{ext}^0 \sin(\tilde{\omega}t) \tag{4}$$

in steady state, the system oscillates at the same angular frequency $\tilde{\omega}$ of the loading according to

$$\begin{aligned} d(t) &= D\sin(\tilde{\omega}t + \theta) \\ v(t) &= \tilde{\omega}D\cos(\tilde{\omega}t + \theta) \\ a(t) &= -\tilde{\omega}^2 D\sin(\tilde{\omega}t + \theta) \end{aligned} \tag{5}$$

with solution

$$D = \frac{f_{ext}^0/k}{\sqrt{\left(1-\tau^2\right)^2 + (2\xi\tau)^2}} \qquad \theta = \tan^{-1}\left(\frac{2\xi\tau}{\tau^2-1}\right) \qquad \tau = \frac{\tilde{\omega}}{\omega} \tag{6}$$

The maximum amplitude of this oscillation (resonance) is obtained for a frequency close to the natural frequency

$$\frac{\tilde{\omega}_{RESONANCE}}{\omega} = \tau_{RESONANCE} = \sqrt{1-2\xi^2} \tag{7}$$

for which the maximum amplitude of displacement D is reached as

$$D_{RESONANCE} = \frac{f_{ext}^0/k}{2\xi\sqrt{1-\xi^2}} \tag{8}$$

Consider now the total structural (restoring) force

$$f_r = cv + kd \tag{9}$$

as a function of d. The integral of this force over the displacement cycle gives the dissipated energy

$$E_{abs} = \oint f_r(d)dd = \int_0^{2\pi/\tilde{\omega}} f_r(t)vdt = \int_0^{2\pi/\tilde{\omega}} (cv+kd)vdt \tag{10}$$

And, according to Eq. (5), clearly the same results are obtained considering only the viscous force in the integral

$$E_{abs} = \int_0^{2\pi/\tilde{\omega}} cv^2dt = \frac{1}{2}\frac{2\pi}{\tilde{\omega}}cD^2\tilde{\omega}^2 = \pi c D^2\tilde{\omega} \tag{11}$$

Then, the maximum elastic energy stored during the cycle is obtained for $d = D$ as

$$E_{el} = \frac{kD^2}{2} \tag{12}$$

giving an energy ratio of

$$\frac{E_{abs}}{E_{el}} = \frac{2\pi c\tilde{\omega}}{k} = \frac{4\pi\xi\tilde{\omega}}{\omega} \tag{13}$$

which, when exciting at the natural frequency, leads exactly to

$$\frac{E_{abs}}{4\pi E_{el}} = \xi \tag{14}$$

and, at the resonance frequency (7), to approximately the same value

$$\frac{E_{abs}}{4\pi E_{el}} = \xi\sqrt{1-2\xi^2} = \xi + \xi^3 + O\left(\xi^5\right) \approx \xi \tag{15}$$

Formula (14) was proposed by Jacobsen (1930). Note that for applying formula (14), the shape of the loop is used for obtaining the absorbed energy, but an additional condition allowing the estimation of the stiffness is also required for obtaining the elastic energy. In the following section, least-square identification for the estimation of the stiffness is presented.

It must be noted that, in the case of a force-displacement loop coming from a viscous model, the determination of the damping ratio based on the energy quotient (14) may be far from accurate for predictions of the response to harmonic load if the excitation frequency is not confirmed to be close to the natural one (see Eq. (13)). On the other hand, the most interesting (and legitimate) application of such formula should be for loops coming from hysteretic models as it is typical in the literature.

2.2 *Estimation of the Stiffness by a Least-Square (LS) Approach*

Let us define an approximation of the restoring force as a linear function of the displacement

$$\overline{f_r}(t) = f_0 + k_{LS} d(t) \tag{16}$$

where f_0 is the zero ordinate and k_{LS} is the slope of the line. The square "error" or ordinate distance of the restoring force data $f_r(t)$ to the straight line $\overline{f_r}$ (16) is

$$\varepsilon(f_0, k_{LS}) = \int_0^T \left(f_r(t) - \overline{f_r}(t)\right)^2 dt \tag{17}$$

which minimization with respect to the f_0 and k_{LS} parameters yields to the equations

$$\int_0^T f_r(t)dt = f_0 \int_0^T dt + k_{LS} \int_0^T d(t)dt$$

$$\int_0^T f_r(t)d(t)dt = f_0 \int_0^T d(t)dt + k_{LS} \int_0^T d^2(t)dt \tag{18}$$

The solution of Eqs. (18) renders the LS (least-square) identified stiffness

$$k_{LS} = \frac{T\int_0^T f_r(t)d(t)dt - \int_0^T f_r(t)dt\int_0^T d(t)dt}{T\int_0^T d^2(t)dt - \left(\int_0^T d(t)dt\right)^2} \tag{19}$$

and the zero ordinate of the linear model

$$f_0 = \frac{\int_0^T f_r(t)dt - k_{LS}\int_0^T d(t)dt}{T} \tag{20}$$

being

$$T = \int_0^T dt \tag{21}$$

the time duration of the cycle. In fact, Eqs. (19) and (20) represent the linear regression of the force-displacement loop.

Two interesting properties of the LS stiffness estimation (19) for two types of generic problems will be pointed out now in the following two subsections.

First Property: LS stiffness in the case of a linear spring with a viscous damper. In the case of a linear spring with a viscous damper represented by Eq. (9), from the definition (19) we have that

$$\begin{aligned} k_{LS} &= \frac{T\int_0^T (cv(t)+kd(t))d(t)dt - \int_0^T (cv(t)+kd(t))dt\int_0^T d(t)dt}{T\int_0^T d^2(t)dt - \left(\int_0^T d(t)dt\right)^2} \\ &= \frac{T\left(c\int_0^T v(t)d(t)dt + k\int_0^T d^2(t)dt\right) - \left(c\int_0^T v(t)dt + k\int_0^T d(t)dt\right)\int_0^T d(t)dt}{T\int_0^T d^2(t)dt - \left(\int_0^T d(t)dt\right)^2} \end{aligned} \tag{22}$$

Then, assuming a closed displacement loop,

$$\int_0^T v(t)d(t)dt = \int_{d(0)}^{d(T)} ddd = \frac{d^2(T) - d^2(0)}{2} = 0 \tag{23}$$

$$\int_0^T v(t)dt = \int_{d(0)}^{d(T)} dd = d(T) - d(0) = 0 \quad (24)$$

Which results in the property of the LS stiffness of being equal to the stiffness of the original spring independently of the time variation of the displacement

$$k_{LS} = k \quad (25)$$

Second Property: LS Stiffness in the case of a general nonlinear loop of fixed shape. On the other hand, for a general nonlinear hysteretic loop of fixed shape, even though the LS stiffness defined by Eq. (19) may depend on the particular time variation of the displacement and force, its value is not altered when the time variable is multiplied by any constant scale factor λ. Say

$$t = \lambda t_1; T = \lambda T_1 \quad (26)$$

$$d(t) = d_1(t_1); f_r(t) = f_{r1}(t_1) \quad (27)$$

where (27) is a constant loop shape that does not depend on the time scale. Then

$$\begin{aligned}[k_{LS}]_t &= \frac{T\int_0^T f_r(t)d(t)dt - \int_0^T f_r(t)dt\int_0^T d(t)dt}{T\int_0^T d^2(t)dt - \left(\int_0^T d(t)dt\right)^2} \\ &= \frac{\lambda T_1\int_0^{T_1} f_{r1}(t_1)d_1(t_1)\lambda dt_1 - \int_0^{T_1} f_{r1}(t_1)\lambda dt_1\int_0^{T_1} d_1(t_1)\lambda dt_1}{\lambda T_1\int_0^{T_1} d_1^2(t_1)\lambda dt_1 - \left(\int_0^{T_1} d_1(t_1)\lambda dt_1\right)^2} \\ &= \frac{T_1\int_0^{T_1} f_{r1}(t_1)d_1(t_1)dt_1 - \int_0^{T_1} f_{r1}(t_1)dt_1\int_0^{T_1} d_1(t_1)dt_1}{T_1\int_0^{T_1} d_1^2(t_1)dt_1 - \left(\int_0^{T_1} d_1(t_1)dt_1\right)^2} = [k_{LS}]_{t_1} \end{aligned} \quad (28)$$

showing that, assuming a pure hysteretic behaviour, the result does not depend on the adopted time scale, or frequency of oscillation.

2.3 *Proposed Procedure for the Identification Method*

For practical application, the procedure for the estimation of the equivalent linear stiffness and damping starts from the given discrete values of restoring force and displacement of the cycle.

Namely, if the data of the force-displacement loop is given as discrete points

$$f_r(n) = f_r(t_n);\ d(n) = d(t_n) \quad n = 1 \ldots N \tag{29}$$

The minimization of the square "error" of the force values with respect to a straight line

$$\overline{f_r}(n) = f_0 + kd(n) \tag{30}$$

Leads to the LS estimated stiffness given by the linear-regression formula

$$k_{LS} = \frac{N \sum (d(n) f_r(n)) - (\sum d(n)) \sum f_r(n)}{N \sum d(n)^2 - (\sum d(n))^2} \tag{31}$$

which is the discrete version of Eq. (19).

Now, the absorbed energy is computed using the trapezoidal rule as

$$\hat{E}_{abs} = \sum_{n=2}^{N} f_{av}(n)(d(n) - d(n-1)) \tag{32}$$

where $f_{av}(n)$ is the average force

$$f_{av}(n) = \frac{f_r(n) + f_r(n-1)}{2} \tag{33}$$

Then, for the estimation of the elastic energy, the loop amplitude is computed as

$$D = \frac{d_{\max} - d_{\min}}{2} \tag{34}$$

and

$$\hat{E}_{el} = k_{LS} D^2/2 \tag{35}$$

Finally, the damping ratio estimation is derived from expression (14) as

$$\hat{\xi} = \frac{\hat{E}_{abs}}{4\pi \hat{E}_{el}} \tag{36}$$

3 Example of Prediction Error Study for the Proposed Equivalent Stiffness and Damping

In this study, the performance of the proposed method for the identification of equivalent stiffness and damping will be partially assessed. It does not pretend to be a complete study, but rather a working example since it covers only the case of steady-state response to harmonic load for a particular analytical model of restoring force.

3.1 Analytical Hysteretic Model for the Input Force-Displacement Cycle

The model used for the current study comprises two terms. That is to say, the restoring force

$$f_r(d) = f_{el}(d) + g(d) \tag{37}$$

is modelled as the addition of an elastic force f_{el} and a dissipative hysteretic force g.

The elastic force is modelled by a power law on the displacement

$$f_{el}(d) = sign(d) k_1 D_1 \left| \frac{d}{D_1} \right|^q = k_1 d \left| \frac{d}{D_1} \right|^{q-1} \tag{38}$$

where q is a constant exponent, D_1 is a reference displacement point and k_1 is the secant stiffness at D_1.

The dissipative force is modelled as purely hysteretic by a Giuffre and Pinto (1970) law

$$g^* = \frac{d^*}{\left(1 + |d^*|^R\right)^{1/R}} \tag{39}$$

where the normalized force is defined as

$$g^* = \frac{g - g_{RETURN}}{g_{CORNER} - g_{RETURN}} \tag{40}$$

and the normalized displacement as

$$d^* = \frac{d - d_{RETURN}}{d_{CORNER} - d_{RETURN}} \tag{41}$$

as represented in Fig. 1. The asymptote (39) allows to pass smoothly from the initial stiffness k_{DAMP}, to the yield strength g_y. For the first loading branch, the return point $P_{RETURN}(d_{RETURN}, g_{RETURN})$ is taken as the origin of the axis. For successive unloading and reloading curves, the return point is the last point of the previous asymptotic path. The corner point of every loading phase $P_{CORNER}(d_{CORNER}, g_{CORNER})$ is the corner of the asymptote or intersection of its initial stiffness line with the yielding force $+g_y$ or $-g_y$. For example, for the first loading branch, the corner point is $(g_y/k_{DAMP}, g_y)$.

In the case of the exponent q being equal to 1, i.e.,

$$q = 1 \tag{42}$$

the elastic force (38) reduces to the linear case

$$f_{el}(d) = k_1 d \tag{43}$$

and, with the addition of the adopted dissipative law, the resulting restoring force model becomes equivalent to the Menegotto and Pinto (1972) model that allows for linear stress hardening. For values of q larger than 1 or smaller, the model of restoring force defined here allows for representing a wide range of interesting cases with, respectively, increasing or decreasing secant stiffness. Nevertheless, as a first approach for the current study, only the linear elastic force case (42), (43) will be considered.

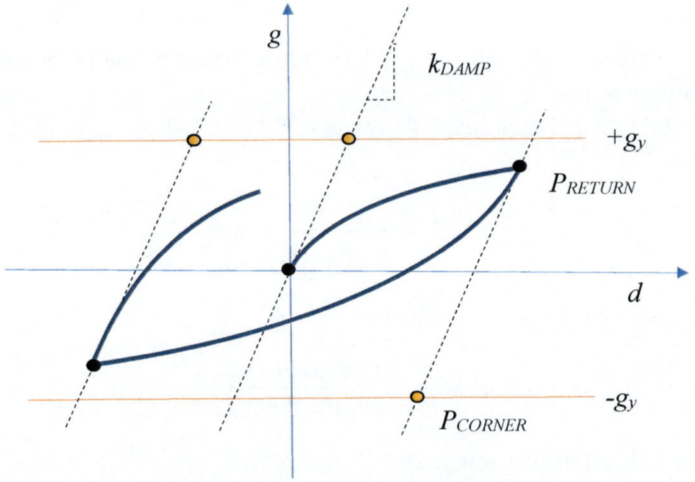

Fig. 1 Dissipative force term defined through a Giuffre and Pinto (1970) model

3.2 Computation of the Steady-State Response for a Harmonic Excitation

This study is limited to the analysis of force-displacement loops that are obtained as the steady-state response of the adopted restoring force model to a harmonic input load. Thus in this case, instead of the linear Eq. (1), the equation of motion is the non linear one

$$ma + f_r(d) = f_{ext}(t) \tag{44}$$

where the restoring force f_r is computed through the model (37) described in the previous section, and the external harmonic load is given by Eq. (4).

In order to obtain the steady-state response, Eq. (44) is solved by using explicit Newmark (1959) method (by taking $\beta = 0$ and $\gamma = 1/2$), which is equivalent to the central difference method. In order to guarantee a high accuracy in the results, the time increment for the integration Δt is obtained by dividing the excitation period

$$\tilde{T} = \frac{2\pi}{\tilde{\omega}} \tag{45}$$

by an integer number which is chosen to be

$$N_{pcyc} = N_{pcyc0} \cdot MAX\left\{1, \text{int}\left(\frac{\tilde{T}}{2\pi\sqrt{m/(k_{DAMP} + k_1)}}\right)\right\} \tag{46}$$

i.e.,

$$\Delta t = \tilde{T}/N_{pcyc} \tag{47}$$

This means that the excitation period is divided in N_{pcyc0} increments at least, depending on the comparison of it with a rough estimation of the minimum natural period for the non-linear system

$$2\pi\sqrt{m/(k_{DAMP} + k_1)} \le T_{natural} \tag{48}$$

For the results that will be presented in the current work, the basic (minimum) number of steps has been taken

$$N_{pcyc0} = 300 \tag{49}$$

However, all the computations have been repeated for a larger number of steps ($N_{pcyc0} = 500$), without significant change in the results for most of the cases, allowing to confirm that the convergence was achieved.

For the selected values of the model and excitation parameters, the integration is performed starting with zero value of displacement, velocity and restoring force. In order to distinguish when the steady state has been reached, at the end of every time interval of duration $\tilde{T}$ (or N_{pcyc} time increments), the displacement amplitude for that excitation period is computed as half of the pick-to-pick displacement within that interval

$$D_{icyc} \leq \left(\max(d)_{icyc} - \min(d)_{icyc}\right)/2 \tag{50}$$

Then, the steady state is considered to be achieved when, between two consecutive cycles, that amplitude does not differ more than a tolerance of, say, 10^{-5} in relative terms, i.e.,

$$\left|D_{icyc} - D_{icyc-1}\right| / D_{icyc-1} \leq 10^{-5} \tag{51}$$

If the steady state is achieved within a limited number excitation periods (say 50), the integration is considered to be finished with a successful result, which is the obtained restoring force-displacement loop during the last excitation period $icyc$, that is to say,

$$[f_r(n) = f_r(t_n); d(n) = d(t_n)]_{n=((icyc-1)N_{pcyc}+1)\ldots icyc \cdot N_{pcyc}} \tag{52}$$

3.3 *Parametric Study of the Prediction Error*

The proposed hysteretic restoring force model (37) and integration method for the numerical generation of the steady-state response of the equation of motion (44) to the harmonic excitation (4) are used for this parametric study. Then the obtained steady-state force-displacement loop (52) is used as input for the determination of the estimated least-square stiffness (31) and the equivalent damping ratio (36). The performance of the estimated stiffness and equivalent damping parameters is afterwards assessed by defining a prediction error of the steady-state displacement amplitude as

$$\varepsilon = (D_{PRED} - D)/D \tag{53}$$

where, as in expression (50), the amplitude D is defined as half of the peak-to-peak displacement within the input interval (52) and, similarly to Eq. (6), the predicted amplitude by the proposed estimated stiffness and damping is

$$D_{PRED} = \frac{f_{ext}^{0}/k_{LS}}{\sqrt{\left(1-\mu^2\right)^2 + \left(2\hat{\xi}\mu\right)^2}} \qquad \mu = \frac{\tilde{\omega}}{\sqrt{k_{LS}/m}} \tag{54}$$

By limiting the current study to the case of forcing the elastic component of the restoring force to be linear, i.e., following Eqs. (42) and (43), the parameters q and D_1 are removed from the model. Now, before presenting results of the error for several ranges of values for some of the parameters of the model, we will transform it from its original variables listed in the implicit equation

$$\overline{h}\left(m, f_{ext}^{0}, \tilde{\omega}, k_1, g_y, k_{DAMP}, R, \varepsilon\right) = 0 \tag{55}$$

to the dimensionless ones used in

$$\overline{H}(\Omega, \Phi, \Lambda, R, \varepsilon) = 0 \tag{56}$$

and defined as it follows

$$\Omega = \tilde{\omega}/\sqrt{k_1/m} \tag{57}$$

$$\Phi = f_{ext}^{0}/g_y \tag{58}$$

$$\Lambda = k_{DAMP}/k_1 \tag{59}$$

with R already introduced by Eq. (39) that was originally defined as a dimensionless constant and with ε defined at Eq. (53) being also dimensionless.

Note that the time instant variable has not been included in (55) since the error (53) is based only on the steady-state loop shape. The basic magnitudes involved in the 8 variables contained in Eq. (55) are 3 (mass, time and space), which, according to Buckingham π theorem, allows to work instead with 5 independent dimensionless variables as it is done in Eq. (56).

Now, by taking as independent parameters the first four ones included in the formulation (56), the following discrete values with all their combinations, have been used for the computation of the error:

$$\Omega = 0.1, 0.2, \ldots, 3.0 \tag{60}$$

$$\Phi = 0.05, 0.10, \ldots, 4.95 \tag{61}$$

$$\Lambda = 10, 100 \tag{62}$$

$$R = 2, 10 \tag{63}$$

Only one page of results is graphically represented in Fig. 2. There, the value of Λ and R are fixed, while Ω and Φ vary for their respective whole assigned range of discrete values specified in Eqs. (60) and (61). However, the axes that have been chosen for the representation in the graphs are μ and $\hat{\xi}$ (see Eqs. (54) and (36)) instead of Ω and Φ in order to try to provide a "physical" idea of the final aspect of the loop (apparent slope and thickness) once the steady state is reached. It must be also commented that for some particular values of the input parameters, no steady state (51) was reached within the integration limit number of time increments, which should be interpreted as a nonlinear resonance with unbounded response. Those points have been removed from these plots. The surface $\varepsilon\left(\mu, \hat{\xi}\right)$ is represented as isolines. The graphs in the upper row use μ and $\hat{\xi}$ as axis (left) or the 3D view with μ, $\hat{\xi}$ and ε as axis (right). The graphs in the lower row are projections of the same isolines, with the same scale of colours, on the $\hat{\xi}$ and ε plane (left) and on the μ and ε plane (right).

For the values of the parameters that have been analysed, some general comments can be made:

- For increasing values of the estimated damping $\hat{\xi}$ (as seen from 0 to 0.8) and decreasing values of frequency ratio μ (from 1 to 0) the error is large and shows a tendency to increase its variability with an upper limit roughly equal to $\hat{\xi}$ (except for the case $\Lambda = 100R = 10$, that shows the largest errors).
- For values of the frequency ratio μ larger than 1.0, the error is always between 0.2 and -0.3, independently of the observed values of estimated damping $\hat{\xi}$.
- Again in that range of the frequency ratio μ larger than 1.0, the error is smaller for $\Lambda = 10$ (between 0.2 and -0.2) than for $\Lambda = 100$ (between 0 and -0.3).
- Combined larger values of Λ and R (which make the loop more similar to a pure friction), increase the variability and amount of error, apart from making more difficult the convergence as it will be shown in the following lines.

After having done all the commented surfaces, the computations were all repeated for a basic number of steps per period increased from the adopted value (49) to

$$N_{pcyc0} = 500 \tag{64}$$

without significant change in the plotted results, except for the case $\Lambda = 100$ $R = 10$. There, for certain combinations of μ and $\hat{\xi}$, the results should be considered not reliable since a convergence was not reached.

4 Conclusions

The proposed method for the estimation of equivalent stiffness and damping from a one-degree-of-freedom force-displacement loop is based on a least-square minimization for the computation of the stiffness, which is assumed to be an original approach,

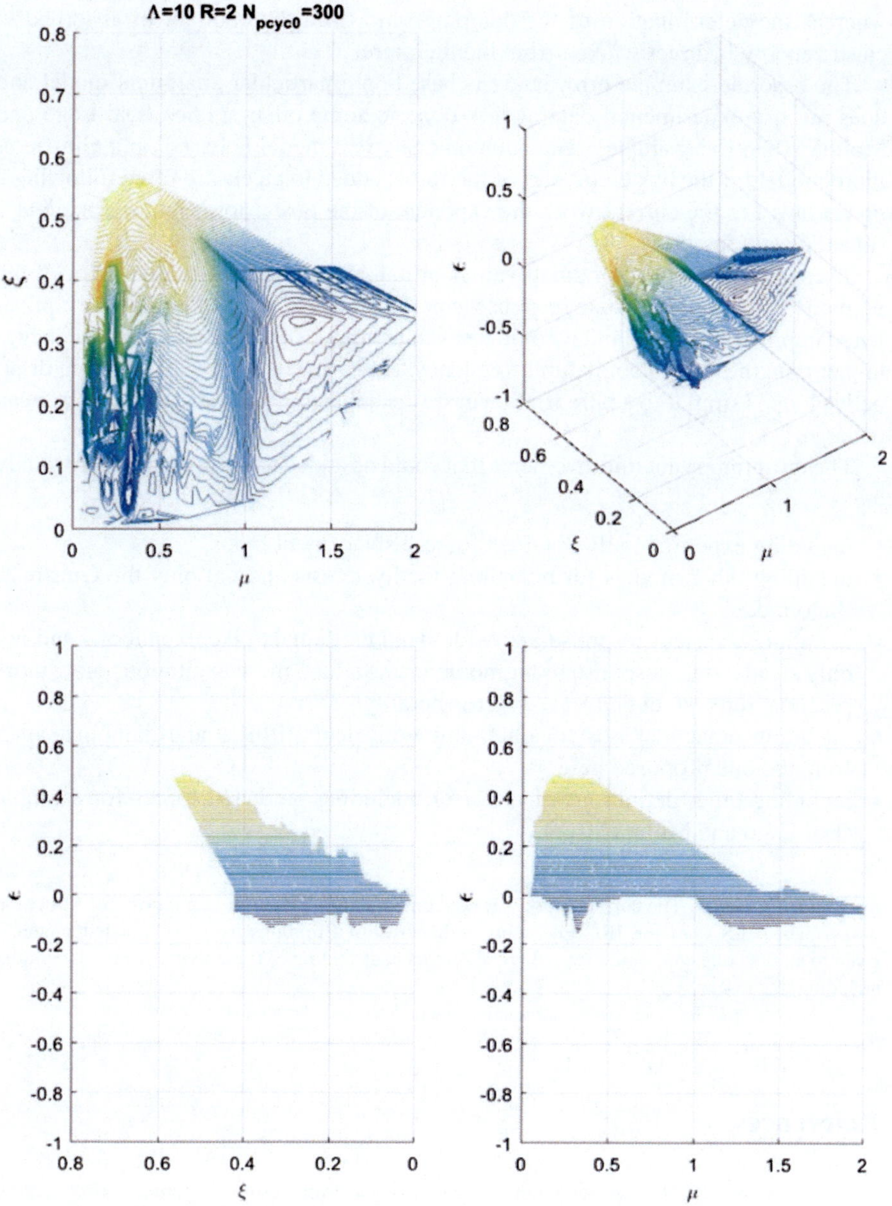

Fig. 2. Tridimensional representation and bidimensional projections of error isolines. Case $\Lambda = 10\ R = 2$.

whereas the determination of the damping ratio from the quotient of absorbed to elastic energy is directly taken from the literature.

The reported example error study is based on a particular analytical model and does not use experimental data as it is done in some other studies (Dal Lago and Molina 2018, for example). The adopted analytical model is based on a Giuffre & Pinto model for the hysteretic part of the force, added to an elastic force following a power law. For the current work the exponent of the power low was kept fixed at 1 (linear elastic force).

The study finds that, for small values of damping ratio, the relative error of the prediction for the steady state response generally tends to be smaller than the equivalent damping ratio value and it could be even better when the excitation frequency is higher than the equivalent natural frequency. However, when the shape of the dissipative loop is similar to a pure friction case, the behaviour is erratic and the error can be large.

The most important improvements that could be added to the presented error study are:

- including experimentally obtained force-displacement loops.
- including other models for modelling the hysteresis and not only the Giuffre & Pinto model.
- including transient response and wide-band earthquake-like excitations and not only steady-state response to harmonic load. In fact, these excitations offer more practical interest, but cover a wider generality.
- including other methods for obtaining equivalent stiffness and damping, apart from the one proposed here.
- including other definitions of the error, including predicted forces for example (Dal Lago and Molina 2018).

Acknowledgements The current work was performed within European Laboratory for Structural Assessment (ELSA) of the JRC E4 Safety & Security of Buildings Unit. Acknowledgement is given to all the colleagues and external collaborators contributing to the activities of the laboratory including this research.

References

Dal Lago, B., Molina, F.J.: Assessment of a capacity spectrum seismic design approach against cyclic and seismic experiments on full-scale precast RC structures. Earthq. Eng. Struct. Dyn. (under review) (2018)

Dwairi, H.M., Kowalsky, M.J., Nau, J.M.: Equivalent damping in support of direct displacement-based design. J. Earthq. Eng. **11**(4), 512–530 (2007)

Ewins, D.J.: Modal testing: theory and practice. Research Studies Press Ltd., Letchworth, Hertfordshire, England; Wiley, New York

Giuffrè, A., Pinto, P.E.: Il comportamento del cemento armato per sollecitazioni cicliche di forte intensità, Giornale del Genio Civile (1970)

Jacobsen, L.S.: Steady forced vibrations as influenced by damping. Trans. ASME **52**, 169–181 (1930)

Menegotto, M., Pinto, P.E.: Method of analysis for cyclically loaded R.C. frames including changes in geometry and non-elastic behavior of elements under combined normal force and bending, Istituto di Scienza e Tecnica delle Costruzioni, University of Rome, Report 32, October (1972)

Molina, F.J., Marazzi, F., Viaccoz, B., Bosi, A.: Stability and accuracy in a hybrid test example. Earthq. Eng. Struct. Dyn. **42**(3), 469–475 (2013)

Molina, F.J., Magonette, G., Pegon, P., Zapico, B.: Monitoring damping in pseudo-dynamic tests. J. Earthq. Eng. **15**(6), 877–900 (2011)

Molina, F.J., Pegon, P., Verzeletti, G.: Time-domain identification from seismic pseudodynamic test results on civil engineering specimens. In: Proceedings of the 2nd International Conference on Identification in Engineering Systems. The Cromwell Press (1999)

Molina, F.J., Pegon, P., Labbé, P.: Stiffness-displacement correlation from the RC shear wall tests of the SAFE program: derivation of a capacity line model. Adv. Mater. Sci. Eng. **2016**, Article ID 5750672, 21 p. (2016). https://doi.org/10.1155/2016/5750672

Molina, F.J., Pegon, P.: Least-square effective stiffness to be used with equivalent viscous damping as equivalent linear model, European Commission, EUR 28818 EN (2017). https://doi.org/10.2760/40703

Newmark, N.M.: A method of computation for structural dynamics. J. Eng. Mech. (ASCE), 67–94 (1959)

Priestley, M.J.N.: Myths and fallacies in earthquake engineering—conflicts between design and reality. Bull. NZ Natl. Soc. Earthq. Eng. **26**(3), 329–341 (1993)

Rosenblueth, E., Herrera, I.: On a kind of hysteretic damping. ASCE J. Eng. Mech. **90**(4), 37–48 (1964)

Key Points and Pending Issues in the Energy-Based Seismic Design Approach

Amadeo Benavent-Climent, Jesús Donaire-Ávila, and Fabrizio Mollaioli

Abstract Since the pioneering works conducted by Housner in the 1960s, energy-based seismic design (EBD) has evolved thanks to many researchers. Notwithstanding, it is overshadowed by the force-based design (FBD) approach, and its implementation in seismic codes is currently limited to structures with energy dissipation systems in Japan. The "Vision 2000" report on future design codes identified EBD as a most promising approach towards the paradigm of performance-based design, however. This paper revises key aspects of EBD, particularly how it differs from FBD and displacement-based design (DBD). Important issues to be addressed in the future so as to implement EBD in the seismic design of conventional structures are underlined. New relations between cumulative energy dissipation and maximum displacement for two different restoring force rules—and for two types of ground motions (with and without pulses)—are presented. Furthermore, a simple means of determining the design-value of the ultimate energy dissipation capacity of non-degrading steel components is put forth.

Keywords Energy based design · Input energy · Energy dissipation demand

A. Benavent-Climent (✉)
Department of Mechanical Engineering, Universidad Politécnica de Madrid, 28006 Madrid, Spain
e-mail: amadeo.benavent@upm.es

J. Donaire-Ávila
Department of Mechanical and Mining Engineering, Universidad de Jaén, 23700 Linares, Jaén, Spain
e-mail: jdonaire@ujaen.es

F. Mollaioli
Department of Structural and Geotechnical Engineering, Sapienza University of Rome, via Gramsci 53, 00197 Rome, Italy
e-mail: fabrizio.mollaioli@uniroma1.it

A. Benavent-Climent and F. Mollaioli (eds.), *Energy-Based Seismic Engineering*, Lecture Notes in Civil Engineering 155,
https://doi.org/10.1007/978-3-030-73932-4_11

1 Introduction

The energy balance of a general single- or multi-degree of freedom (SDOF or MDOF) system subjected to a ground motion can be expressed in terms of relative or absolute input energy, E_{Ir} or E_{Ia}, by Eq. (1).

$$E_{Ir} = E_{kr} + E_{\xi} + E_s + E_H; \quad E_{Ia} = E_{ka} + E_{\xi} + E_s + E_H \tag{1}$$

Here E_{ka} and E_{kr} are the kinetic energies representing the difference between absolute and relative input energy, respectively, E_{ξ} is the damping energy, E_s is the elastic strain energy and E_H is the hysteretic energy. E_{Ir} and E_{Ia} are very close in the practical range of interest, namely 0.3–5s [1]. In the case of forward directivity near fault records that contain a distinguishable acceleration pulse, E_{Ia} may give rise to misleading results, whereas E_{Ir} provides for a more reliable measure of energy [2]. The relative total input energy E_{Ir} at the end of a ground motion serves as the reference for the following discussion. Housner [3] defined the total input energy that contributes to damage E_D as:

$$E_D = E_{Ir} - E_{\xi} \tag{2}$$

Beyond a certain level of plastic deformation on a given structure —i.e. for E_H larger than about four times the product of the yield displacement δ_y and yield strength ${}_sQ_y$ in the case of an SDOF system of mass M and period T—the ratio E_D/E_{Ir} is basically controlled by the fraction of damping of the structure ξ [4]; and the simple Eq. (3) [5] can be used to roughly relate E_D/E_{Ir} with ξ. When the structure undergoes plastic deformation, the elastic vibration energy $E_{kr} + E_{kr}$ at the end of the ground motion is very small in comparison with E_H, and $E_D \approx E_H$.

$$\frac{E_D}{E_{Ir}} = \left[\frac{1}{1 + 3\xi + 1.2\sqrt{\xi}}\right]^2 \tag{3}$$

For design purposes, it is convenient to characterize ground motion directly in terms of E_D ($\approx E_H$) for a givenξ, because the equivalent velocity defined by $V_D = \sqrt{2E_D/M}$ is closely upper-bounded by the pseudo-velocity S_{pv} [3], which can be readily obtained from the pseudo acceleration S_a ($= 2\pi S_{pv}/T$). This is the approach adopted in the energy-based method currently implemented in the Building Standard Law of Japan [6] for structures with displacement-dependent dampers, proposed as well for designing this type of structure in the next generation of EN1998. Energy-based design (EBD) is founded on the solid basis of the energy balance of a structure after Eq. (1); its pivotal virtue (by far, its main advantage) is the fact that the total input energy—either in terms of E_{Ir} or E_{Ia}—depends almost exclusively on mass and period, being approximately the same in an MDOF of total mass M_T and fundamental period T_1 as in an SDOF of the same mass $M = M_T$ and period $T = T_1$. That is, the total

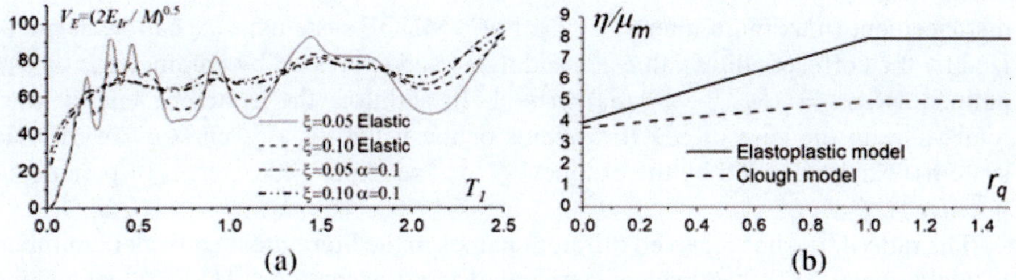

Fig. 1 E_{Ir} spectra for Loma Prieta earthquake (**a**) and η/μ_m ratio (**b**)

input energy is scarcely affected by the viscous damping ratio ξ, the strength, the post-yield stiffness, the hysteretic behavior, the degree of inelastic action (characterized by the ductility factor $\mu = \delta_{max}/\delta_y$, where δ_{max} is the maximum displacement), the number of degrees of freedom of the system or the torsional effects, among other factors. This stability of the total input energy was unveiled by Housner [3] in his pioneering works, then investigated in greater detail by Akiyama [5] for general MDOF systems. It has been corroborated a posteriori by many other researchers [1, 7–12]. For design purposes, the dependence of the total input energy on factors other than M_T and T_I can be neglected, since its variability is minor in comparison with the inherent variability of ground motions. Further, ξ and μ have a smoothing effect on the total energy input spectra that can result in a drastic reduction of the dependency of total input energy upon small variations of T, around a certain value of T_I as ξ and/or μ increases. This is exemplified in Fig. 1a for the Loma Prieta earthquake (10/18/1989, Saratoga-Aloha Ave, 90), where E_{Ir} is expressed in terms of equivalent velocity $V_E = \sqrt{2E_{Ir}/M}$ and $\alpha(={}_sQ_y/Mg)$.

The stability of E_{Ir} is extended to the hysteretic energy E_H by Eq. (3), and this allows one to assert that the total amount of hysteretic energy demand E_H imposed by a given earthquake on a structure with a fraction of damping ξ would depend almost exclusively on M_T and T_I. In addition to the stability of the total input energy, other reasons make energy a better ground for seismic design than forces or displacements, as summarized by Fardis [13]. Beyond the elastic range, forces on the structure are controlled by its strength, not by the earthquake. Thus, the inadequacy of using forces to characterize the loading effect of an earthquake is obvious.

The focus can therefore be placed on the relationship between energy and maximum displacement. In contrast to the total input energy, depending basically on M_T and T_I, maximum displacement is strongly influenced by the characteristics of structure, and particularly by two key features [14]: (i) hysteretic behavior and (ii) the presence of an elastic component working in parallel with the elastoplastic component. As for the second feature, the "weight" of the elastic component with respect to the elastoplastic one can be characterized by the ratio $r_q = {}_f\overline{Q}_m/{}_sQ_y$, where ${}_f\overline{Q}_m$ is the average of the maximum force sustained by the elastic component in the positive and negative domains, and ${}_sQ_y$ the yielding force of the elastoplastic element, both referred to a given story in the case of MDOF systems. The maximum displacement (maximum inter-story drift of a MDOF systems) $\delta_{\max}$ can be normalized by the corresponding value at yielding δ_y and expressed by parameter μ or μ_m defined as $\mu_m = (\delta_{\max} - \delta_y)/\delta_y)=\mu - 1$. In addition, the hysteretic energy (the total value in the case of SDOF systems or the value for a given story in MDOF systems) can be divided by the product ${}_sQ_y\delta_y$ and expressed in terms of parameter $\eta = E_H/({}_sQ_y\delta_y)$.

The ratio η/μ_m has received different names in the literature: "equivalent number of yield cycles" [7], "equivalent number of yield excursions" [15] or "equivalent number of plastic cycles" [16]. In the first and last the term "cycle" is confusing, because in an ideal elastic-perfectly-plastic system, the plastic deformation accumulated in one cycle (with equal positive and negative displacements) is $4\mu_m$. Hence it is preferable to refer to $n_{eq} = \eta/\mu_m = \eta/(\mu - 1)$ as the "equivalent number of yield excursions" following Khashaee et al. [15], and use the name "equivalent number of cycles" for $n_{eq}/4$, that is, for $\eta/4\mu_m$. Akiyama [14] evaluated the η/μ_m ratio in MDOF systems having the same M_T and T_1, different hysteretic behaviors (elastoplastic with and without Bauschinger effect and Clough model) and different r_q, subjected to a given earthquake. He found that η/μ_m varies with the hysteretic model, and particularly with r_q. Figure 1b shows the η/μ_m proposed by Akiyama for design purposes. Accordingly, the maximum displacement in terms of μ_m for a given hysteretic energy dissipation demand expressed by η can be significantly reduced by enhancing the role of the "elastic component" (i.e. by increasing r_q). This dependence of the maximum displacement on the characteristics of the structure emphasizes the importance of understanding the loading effect of the earthquake in terms of total input energy (the stable incumbent), without mixing it with other parameters such as displacement. What the earthquake really imparts to the structure is energy [13], not force or displacement. The displacements are a by-product of the input energy. Once the hysteretic energy demand (at the story level for a multistory building) is evaluated in view of the total input energy, the maximum displacement $\delta_{\max}$ (i.e. inter-story drifts in multi-story buildings) can be estimated for a particular structural type from $n_{eq} = \eta/\mu_m$.

Equations (1), (2) can be applied to an SDOF system of mass M and period T subjected to a given ground motion characterized by E_{Ir}. The EBD approach assumes that E_{Ir} is the same regardless of whether the system responds in the elastic or in the inelastic range. If the system remains elastic, attaining a maximum force ${}_sQ_{\max,e}$ and a maximum displacement $\delta_{\max,e}$, the elastic vibrational energy ($E_{kr}+ E_s$) is $E_{kr} + E_s = {}_sQ_{\max,e}\delta_{\max,e}/2 = {}_sQ^2_{\max,e}T^2/(8\pi^2M)$, and Eqs. (1), (2) give $E_D = {}_sQ^2_{\max,e}T^2/(8\pi^2M)$. If the system responds in the inelastic range with ${}_sQ_y$ and δ_y denoting the force and displacement at yielding, the vibrational elastic energy can be approximated [5] through $E_{kr} + E_s = {}_sQ_y\delta_y/2 = {}_sQ_y^2T^2/(8\pi^2M)$; E_H can be expressed by $E_H = \eta_s Q_y\delta_y = \eta_s Q_y^2T^2/(4\pi^2M)$; and Eqs. (1), (2) give $E_D = {}_sQ_y^2T^2/(8\pi^2M) + \eta_s Q_y^2T^2/(4\pi^2M)$. Equating the second member of the energy-balance equations for elastic and inelastic systems yields Eq. (4), where the forces are expressed in non-dimensional form by $\alpha_p = {}_sQ_y/Mg$ and $\alpha_e = {}_sQ_{max,e}/Mg$.

$$\frac{\alpha_p}{\alpha_e} = \frac{1}{\sqrt{2+1}} \tag{4}$$

Formally, Eq. (4) is similar to the Eq. (5) proposed by Veletsos and Newmark [17] to relate elastic and inelastic SDOF systems.

$$\frac{\alpha_p}{\alpha_e} = \frac{1}{\sqrt{2\mu_m + 1}} \tag{5}$$

However, there is a key difference between Eqs. (4) and (5). The former uses the (cumulative) hysteretic energy dissipated by the system during cyclic reversals imposed by the earthquake, while the latter is obtained equating the area below the force-displacement curve up to the maximum displacement (i.e. "equal area" rule herein). This area represents the energy that is stored and dissipated by the system under monotonic loading up to the maximum displacement. Veletsos and Newmark adopted this energy as an aid to phenomenally relate the elastic and the inelastic responses of the system, but they did not attempt to evaluate the energy input by the earthquake on the structure. The misunderstanding that Veletsos and Newmark used Housner's energy concept hindered the development of the EBD [5].

Substituting $\eta = n_{eq}\mu_m$ in Eq. (4) and solving gives $\mu_m = (\alpha_e^2/\alpha_p^2 - 1)/(2n_{eq})$. Recalling that $\delta_{\max,p}/\delta_{\max,e} = \mu\alpha_p/\alpha_e = (1+\mu_m)\alpha_p/\alpha_e$ and after some algebra, the ratio $(\delta_{\max,p}/\delta_{\max,e})_{Akiyama}$ obtained by applying the EBD approach is given by Eq. (6):

$$\left(\frac{\delta_{\max,p}}{\delta_{\max,e}}\right)_{Akiyama} = \frac{\alpha_p(2n_{eq} - 1)}{2n_{eq}\alpha_e} + \frac{\alpha_e}{2n_{eq}\alpha_p} \tag{6}$$

The counterpart expression of the ratio $\delta_{\max,p}/\delta_{\max,e}$ by applying the Newmark and Veletsos approach given by Eq. (5), $(\delta_{\max,p}/\delta_{\max,e})_{Newmark}$, is obtained by substituting Eq. (5) in $(\delta_{\max,p}/\delta_{\max,e})_{Newmark} = \mu(\alpha_p/\alpha_e) = (1+\mu_m)(\alpha_p/\alpha_e)$, and this gives $(\delta_{\max,p}/\delta_{\max,e})_{Newmark,1} = \{[1/(\alpha_p/\alpha_e)] + (\alpha_p/\alpha_e)\}/2$. Newmark and Veletsos proposed an alternative way to relate the displacements of the elastic, $\delta_{\max,e}$, and inelastic, $\delta_{\max,p}$, systems that assumes they are equal, that is $(\delta_{\max,p}/\delta_{\max,e})_{Newmark,2} = 1$ (i.e. the well-known "equal displacement" rule implemented in most seismic codes in the framework of FBD), and gives $\alpha_p/\alpha_e = 1/\mu = 1/(\mu_m + 1)$.

Figure 2a shows $\delta_{\max,p}/\delta_{\max,e}$ against (α_p/α_e) obtained by applying the EBD approach and the Newmak and Veletsos approach. Obviously, the curve corresponding to Newmark and Veletsos's "equal area" rule coincides with the curve obtained by applying the EBD for $n_{eq} = 1$, but for the rest of values of n_{eq} the differences are significant. It is observed that under the EBD approach, the lowest n_{eq} leads to maximum displacements in inelastic systems that are generally higher than in the counterpart elastic system. Lower values of n_{eq} are associated with pulse-like ground motions, while higher n_{eq} values correspond to far-field earthquakes. If, for

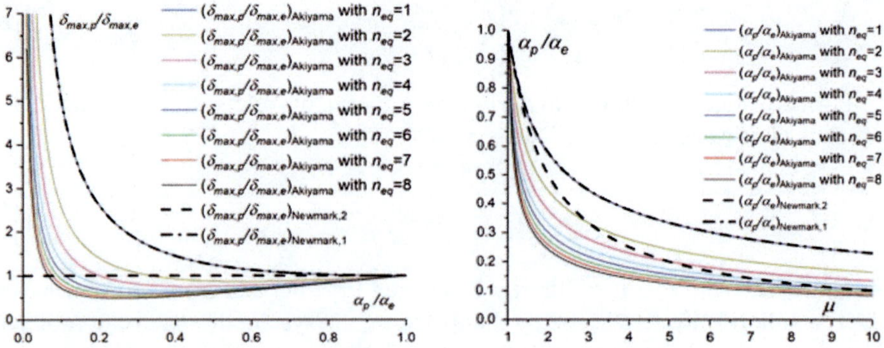

Fig. 2 Comparison between EBD approach and Newmark and Veletsos's equations

the sake of simplicity, a single n_{eq} is adopted for design purposes, past studies [7, 14–16] suggest that $n_{eq} = 4{,}0$ is a reasonable value. For $n_{eq} = 4.0$, the EBD approach gives $\delta_{\max,p}/\delta_{\max,e}$ ratios up to 35% smaller than the "equal displacement" rule for $\alpha_p/\alpha_e > 0.14$, and up to 3 times larger for $0.04 < \alpha_p/\alpha_e < 0.14$. Figure 2b illustrates μ against α_p/α_e. It is seen that the reduction of the required strength in the case of an inelastic response as opposed to the elastic one for a given μ is also notably different depending on the approach used.

Finally, past studies show that replacing a non-linear system with an equivalent pseudo-linear system and representing hysteretic energy as an equivalent linear viscous damping energy is acceptable only if the level of non-linearity is below a certain limit [18, 19]. For general systems that combine —in parallel— an elastic component with an elastoplastic component, Akiyama [14] quantified this limit as $r_q \geq 1$. If $r_q < 1$, the possibility of damage concentration (in terms of hysteretic energy demand) in one or a few stories becomes a key aspect of the seismic design. The occurrence of such damage concentration in highly non-linear systems cannot be captured with displacement-based methods, since they are based on the use of pseudo-linear equivalent systems. Under the energy-based approach, it is possible to grasp the distribution of hysteretic energy dissipation demand among stories and take into account the damage concentration effects in highly non-linear systems, as briefly discussed in Sect. 2. This constitutes an additional and important advantage of the EBD approach over the displacement-based approach.

2 Input Energy and Maximum Displacement Demands

Characterizing seismic action in terms of total input energy is the essential element of the energy-based approach, but attention must also be paid to the maximum displacement $\delta_{\max}$—it is one of the components of the damage (in addition to hysteretic energy), it governs the second order effects, and it controls the damage in non-structural elements. The relation between input energy and maximum displacement was briefly addressed in the Introduction, and is discussed in greater depth here, on the basis of extensive nonlinear time history analyses. Firstly, we can look at some differences observed on n_{eq} depending on the signal characteristics. Figure 3 offers a sketch of the fault responsible for the 1979 Imperial Valley earthquake, together with the El Centro array. The acceleration and velocity time histories of both records are included. The Houston Rd record, obtained in a forward-directivity position, is representative of a pulse-like record, while the Bonds Corner record was recorded in a neutral or backward-directivity position. Consequently, the two records are characterized by different amplitude, frequency content, and duration feature. Both are fault-normal components. Constant ductility spectra were computed for the Houston Rd and Bonds records by carrying out iterative nonlinear dynamic analyses on SDOF systems to identify the normalized strength α ($={}_sQ_y/Mg$) required to attain the target μ. OpenSees [20] was used, the restoring forces following the bilinear Clough model [21], representative of RC structures. Figure 4 shows the spectra of the hysteretic energy normalized by M, $\overline{E}_H = E_H/M$, α and n_{eq} for $\mu = 4$. Even though the peak E_H is similar for the two records, the E_H-T spectra are notably different. In the Bonds spectrum, the peak is around the characteristic period of the soil, while in the Houston record it is on the period of the pulse. Also different are the n_{eq}-T and the α-T spectra. Focusing on the range of periods around the characteristic period of the motion (i.e. 0.02–1 s for Bond Corner, and 1.0–6 s for Houston Rd), n_{eq} varies between 4 and 10 for the non-pulse record (Bonds), while it is basically constant at 2 for the pulse-record (Houston). It is worth recalling that n_{eq}, α, μ, $\overline{E}_H$ and T are not independent; from their very definition they are related by $n_{eq} = 4\overline{E}_H\pi^2/[(\mu - 1)\alpha^2T^2g^2]$. It is also worth noting that n_{eq} presents the largest values as well as higher variability with T in the short period range (representative of masonry structures, for example), while it tends to a roughly constant value as T increases.

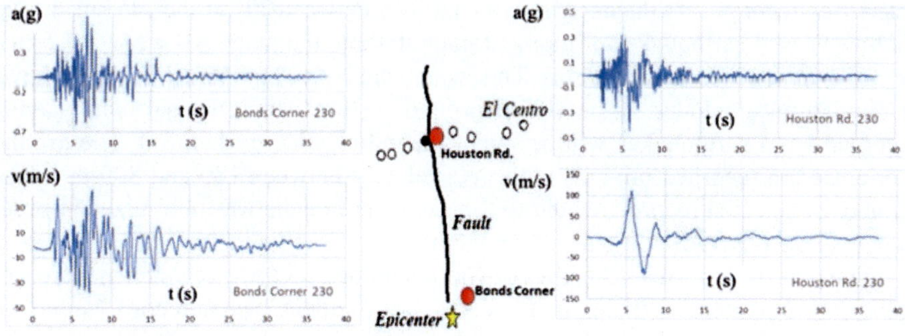

Fig. 3 A 1979 Imperial Valley Earthquake, Houston Rd. (pulse-like record) and Bonds Corner records. Strike-slip fault representation with the location of the station plus the recorded acceleration and velocity time-histories

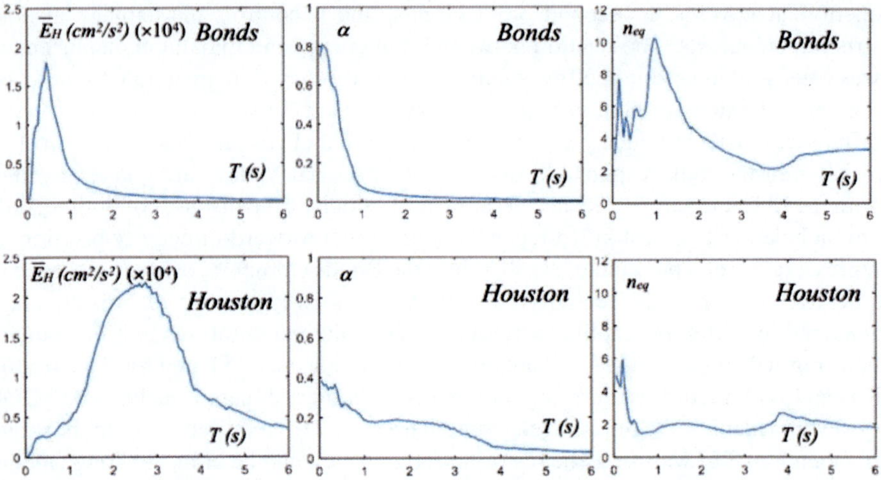

Fig. 4 $\overline{E}_H$, α and n_{eq} spectra of Bonds and Houston ground motions

The study of the η-μ_m (or η-μ) was extended to a greater number of ground motions and fixed values of μ ranging from $\mu = 2$ to $\mu = 8$. To this end, SDOF systems representing RC structures with fundamental periods $T = 0.3$ s and $T = 1.0$ s were subjected to two record sets (near-field and far-field). Furthermore, two different constitutive models, the bilinear Clough model [21] and the bilinear Modified Clough model, were considered. Both incorporate a reduction in the reloading stiffness. In the first, the unloading stiffness remains constant, whereas in the second the unloading stiffness k_u varies following the law $k_u = k_y(\delta_m/\delta_y)^a$—$k_y$ being the lateral initial stiffness, δ_m the previous maximum displacement in the current domain, and a a constant set as $a = -0.50$. The first set includes 50 ordinary ground motions selected from the NGA database (https://peer.berkeley.edu/nga/) [22] without impulsive characteristics, recorded at a distance from the fault D_f between 20 km and 145 km, during shallow crustal earthquakes of moment magnitude in the range $5.0 < M_w < 7.6$ that occurred in active tectonic regions. The second dataset was obtained from Al Shawa et al. [23] and includes 50 near-fault ground motions (fault-normal components) recorded from earthquakes with moment magnitudes larger than 5.0. The ground motions were recorded on soils with v_{S30} velocity between 180 and 800 m/s. The number of selected records related to the same earthquake was minimized in order to reduce the possibility of bias in the results.

Figure 5 shows the results using the Clough model. It can be seen that η increases along with μ, regardless of the fundamental period and the type of record. Moreover, although the variability of η for a given μ grows with μ, the dispersion in results remains nearly constant, with an average coefficient of variation COV between 0.42 and 0.56, where the lowest values correspond to far-field records and the highest to the near-field ones. The figure also shows, with a red line, a linear least-squares fitting regression analysis conducted with the results, $\eta_{lsf} = a_{lsf}\mu + b_{lsf}$. The coefficients of the regression model are reported in Table 1. A growing dispersion of the data around the regression model is seen as μ increases, and it is higher for $T = 0.3$ s than for $T = 1.0$s. More precisely, for $T = 0.3$s the residual variance increases continuously from 7.82 ($\mu = 2$) to 461.12 ($\mu = 8$) for near-fault records, and from 11.16 ($\mu = 2$) to 673.43 ($\mu = 8$) for far- field records. In the case of $T = 1.0$s, the continuous increase is from 3.37 ($\mu = 2$) to 220 ($\mu = 8$) for near-fault records, and from 4.66 ($\mu = 2$) to 271.52 ($\mu = 8$) for far-field records.

A lower envelope was moreover calculated from a linear least-squares fitting of the points with lesser η, as is represented by a dash-red line in Fig. 5. For comparison purposes, the design, $\eta_{Aki\_des} = a_{Aki\_des}\mu + b_{Aki\_des}$, and lower limit, $\eta_{Aki\_lowlim} = a_{Aki\_lowlim}\mu + b_{Aki\_lowlim}$, relations proposed by Akiyama [14] for the Clough constitutive model, are also shown with solid and dash green lines. The values of parameters a_{Aki_des}, b_{Aki_des}, a_{Aki_lowlim}, $b_{Aki\_lowlim.}$ are given in Table 1. As can be seen, the lower envelope obtained in this study is slightly beneath the lower bound proposed by Akiyama; but still, the design value is close (yet below) the linear

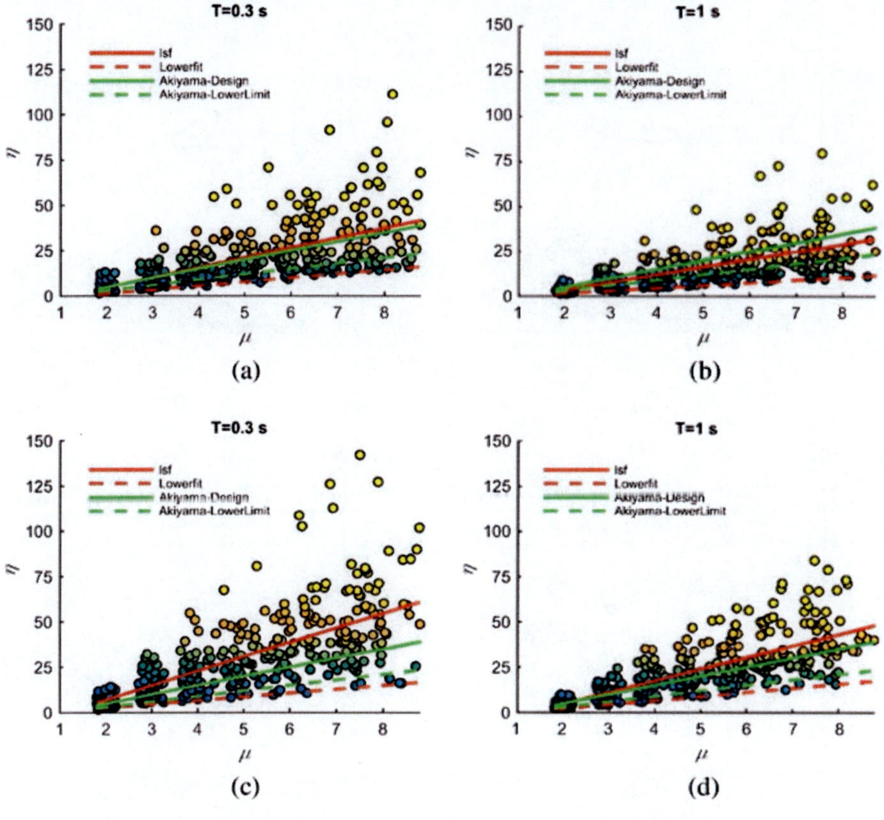

Fig. 5 η-μ relationship for SDOF systems with the Clough model: Near-field records for $T = 0.3$ s (**a**) and $T = 1$ s (**b**); Far-field records for $T = 0.3$ s (**c**) and $T = 1$ s (**d**)

Table 1 Linear regression coefficients and Akiyama's η-μ relations (Clough model)

Record	T(s)	a_{lsf}	b_{lsf}	a_{Aki_des}	b_{Aki_des}	$a_{lowerfit}$	$b_{lowerfit}$	a_{Aki_lowlim}	b_{Aki_lowlim}
Near-Field	0.3	5.41	−5.89	5.00	−5.00	2.17	−3.02	3.00	−3.00
	1.0	4.27	−5.12	5.00	−5.00	1.47	−1.36	3.00	−3.00
Far-Field	0.3	7.93	−8.92	5.00	−5.00	2.06	−1.62	3.00	−3.00
	1.0	6.37	−7.94	5.00	−5.00	2.24	−2.36	3.00	−3.00

regression conducted in this study. Figure 6 and Table 2 show the counterpart results, using the Modified Clough model. The COVs obtained in this case (between 0.40 and 0.55) are very similar to those obtained with the Clough model. As expected, the regression models show values lower than those pertaining to the Clough Model. It is also evident that the dispersion of the η-μ with the Modified Clough Model is higher than for the Clough model, but without significant differences between near- and far-field records. A noteworthy, growing dispersion of the data is seen around the regression model as μ increases, and it is higher for $T = 0.3$ s than for $T = 1.0$s,

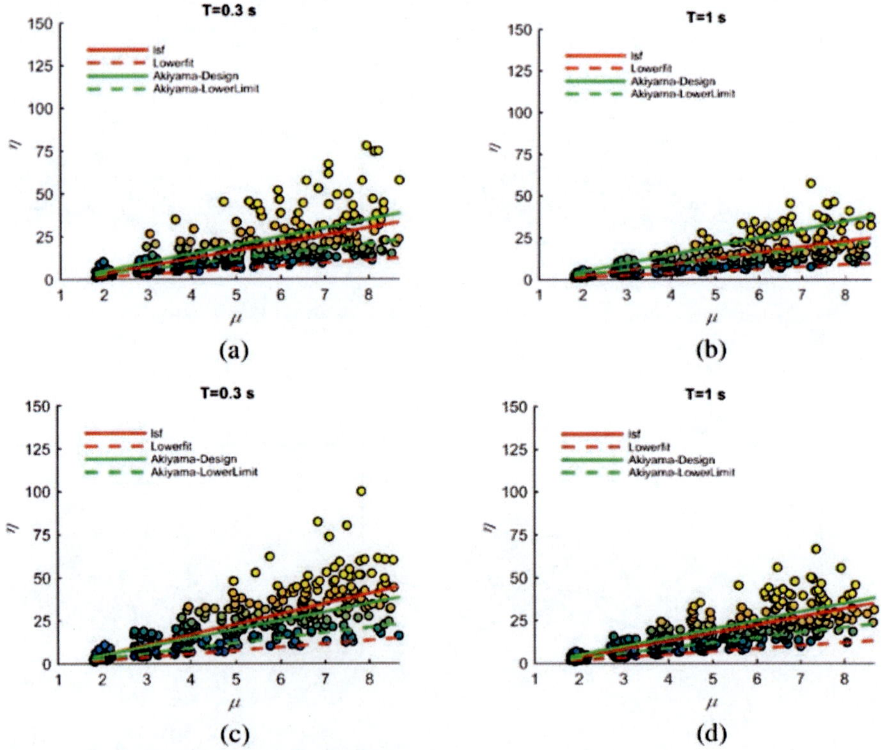

Fig. 6 η-μ relationship for SDOF systems with the Modified Clough model: Near-field records for $T = 0.3$ s (**a**) and $T = 1$ s (**b**); Far-field records for $T = 0.3$ s (**c**) and $T = 1$ s (**d**)

Table 2 Linear regression coefficients and Akiyama's relations η-μ (Modified Clough model)

Record	T(s)	a_{lsf}	b_{lsf}	a_{Aki_des}	b_{Aki_des}	$a_{lowerfit}$	$b_{lowerfit}$	a_{Aki_lowlim}	b_{Aki_lowlim}
Near-Field	0.3	4.45	−5.62	5.00	−5.00	1.65	−1.97	3.00	−3.00
	1.0	3.37	−4.22	5.00	−5.00	1.27	−1.25	3.00	−3.00
Far-Field	0.3	5.97	−7.00	5.00	−5.00	1.94	−2.02	3.00	−3.00
	1.0	4.72	−5.92	5.00	−5.00	1.70	−1.55	3.00	−3.00

but the residual variances are about one-half of those obtained for the Clough model. More precisely, for $T = 0.3$s the residual variance increases from 3.97 ($\mu = 2$) to 261.10 ($\mu = 8$) for near-fault records, and from 4.39 ($\mu = 2$) to 285.24 ($\mu = 8$) for far-field records. In the case of $T = 1.0$s, the continuous increase is from 1.14 ($\mu = 2$) to 124.83 ($\mu = 8$) for near-fault records, and from 2.22 ($\mu = 2$) to 128.69 ($\mu = 8$) for far-field records.

While current knowledge of the total input energy or the total hysteretic energy demand imposed by an earthquake is quite satisfactory, predicting its distribution among the stories is a very tricky problem. It is strongly influenced by the features of the structure, the characteristics of the ground motion, or the higher mode effects, among other factors. Past studies attempted to predict the height-wise distribution of hysteretic energy demand using equivalent SDOF systems within the context of modal pushover analyses [24, 25]. This approach works reasonably well for low levels of non-linearity and for particular structural types such as moment resisting frames that develop an ideal, full strong column-weak beam mechanism (without pinching or strength/stiffness degradation effects), but not for general structures. It should be emphasized that most real structures, particularly RC frames designed according to current seismic codes, are not expected to develop an ideal full strong column-weak beam mechanism because the column-to-beam strength ratios at beam-column connections, $\sum M_c / \sum M_b$, that would guarantee such a mechanism—$\sum M_c / \sum M_b$ on the order of 4—[26] are economically unfeasible and far above the values required in codes ($\sum M_c / \sum M_b \approx 1,2 - 1,3$). Thus, column yielding associated with the formation of intermediate mechanisms must be expected [27]. Moreover, the unavoidable deviations of actual material properties with respect to their nominal values or the rationalization of the size of structural members can cause concentration of inelastic energy dissipation demands in a given story that must be taken into account when assessing its energy dissipation capacity against the demand. A straightforward approach to estimating the distribution of inelastic energy dissipation demand among stories—taking into account the possibility of such damage concentration applicable to high levels of non-linearity—was proposed by Akiyama [5] and has been adopted by the current Building Standard Law of Japan [6] for designing structures with displacement-dependent energy dissipation devices. According to Akiyama's approach, Eq. (7), the distribution of hysteretic energy demand E_{Hs} among the stories, $E_{Hs} / \sum_{k=1}^{N} E_{Hk}$, depends on: (i) the extent to which the yield shear force coefficient of the s-th story, $\alpha_s = Q_{ys} / (\overline{m}_s Mg)$, deviates from an "optimum" value $\alpha_{s,opt}$ that would make η approximately equal in all stories; (ii) the eccentricity between the center of stiffness and the center of mass of the story, e_{ox}, with respect to its torsional radius, r_x, and (iii) a factor n that weights the proneness of the structural type to concentrate damage. Akiyama recommended the adoption of $n = 4$ for structures that are hardly prone to concentrate damage (e.g. strong column-weak beam systems) and $n = 12$ for structures prone to damage concentration (e.g. strong beam-weak column systems). Several distributions have been proposed in the literature for the optimum yield shear coefficient distribution $\overline{\alpha}_{s,opt} = \alpha_{s,opt}/\alpha_1$ [5, 28–30]; Eq. (8) shows the one adopted by the Building Standard Law of Japan [6].

$$\frac{E_{Hs}}{E_H} = \frac{s_s(p_s p_{ts})^{-n}}{\sum_{k=1}^{N} s_k(p_k p_{tk})^{-n}} \tag{7}$$

$$\overline{\alpha}_{s,opt} = 1 + \left[\frac{1}{\sqrt{\overline{m}_s}} - \overline{m}_s\right]\frac{2T_1}{1 + 3T_1} \tag{8}$$

In the above equations, $p_s = \alpha_s/\alpha_{s,opt}$; $\overline{m}_s = \sum_{k=s}^{N} m_k/M_T$; m_k is the mass of the k-th level, N the total number of stories; p_{ts} is a parameter that ranges from $p_{ts} = 0.85$ for $(e_{ox}/r_x) \geq 0.30$ to $p_{ts} = 1$ for $(e_{ox}/r_x) \leq 0.15$; and $s_s = (\overline{\alpha}_{s,opt}\overline{m}_s)^2 k_1/k_s$ where k_s is the initial elastic lateral stiffness of the s-th storey.

3 Energy Dissipation Capacity

Evaluating the capacity of structural members under seismic loadings in terms of energy is by far the most challenging issue and remains the largest gap in our knowledge [13]. The difficulties in characterizing energy dissipation capacity stem, among other factors, from its dependence on the loading path [31, 32] and from the random nature of ground motions. However, the gap of knowledge is different for non-deteriorating structural components (e.g. steel members subjected to bending/shear deformations and axial forces that are exempt of local/global buckling), and for those that present strength/stiffness deterioration under the cyclic displacements imposed by earthquakes (e.g. RC elements). Non-deteriorating structural components exhibit restoring force curves, Q-δ, of the type shown in Fig. 7a. The main feature is that the increment of restoring force and the increment of the imposed displacement have the same sign until the component approaches failure. This allows one to decompose the Q-δ curve into the so-called "skeleton part" and "Bauschinger part", as shown in Figs. 7a, b [33]. This decomposition can be applied to any arbitrary Q-δ curve, and makes it possible to characterize the energy dissipated until failure with the following non-dimensional parameters:

$$_S\eta = \frac{_SW_u^+}{Q_y\delta_y} + \frac{_SW_u^-}{Q_y\delta_y};\ _B\eta = \frac{_BW_u^+}{Q_y\delta_y} + \frac{_BW_u^-}{Q_y\delta_y};\ _{ep}\eta = \frac{_S\delta_u^+}{\delta_y} + \frac{_S\delta_u^-}{\delta_y} : \eta = {_S\eta} + {_B\eta} \tag{9}$$

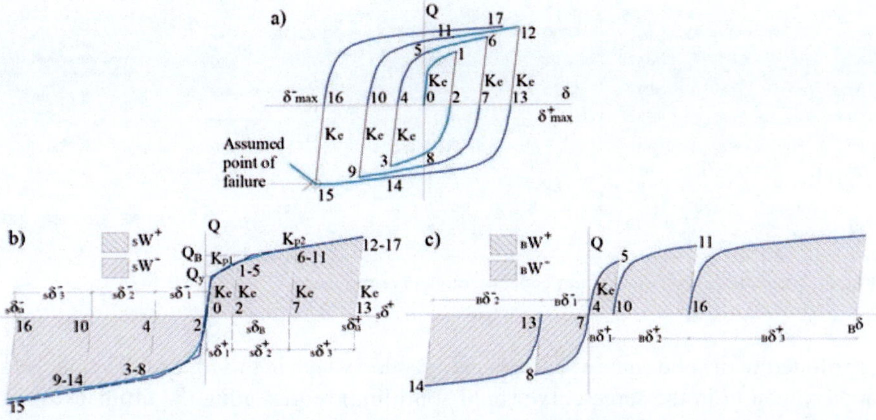

Fig. 7 Decomposition of the Q-δ curve of a non-deteriorating structural component: **a** Q-δ curve; **b** skeleton part; **c** Bauschinger part

where Q_y and δ_y are the force and displacement at yielding, the meaning of the rest of the parameters being shown in Figs. 7b, c. A large number of different types of steel components (beams, columns, rods, plates with slits, etc.) having stable Q-δ curves were tested until failure in past studies [34–36] and the values of $_S\eta$, $_B\eta$ and $_{ep}\eta$ have been calculated. Deserving special mention is the extensive experimental campaign (full-scale shake table tests) conducted in Japan shortly after the 1995 Kobe earthquake to characterize the seismic behavior and ultimate energy dissipation capacity of steel beams, columns, and beam-column subassemblages [37–45]. These tests showed that $_B\eta$ can be simply related with $_{ep}\eta$ by $_B\eta = a_{ep}\eta + b$, where the constants a and b depend on the geometry of the member (specifically, details in the region subjected to plastic deformations) and on the properties of the steel; hence a and b must be determined experimentally. In turn, $_S\eta$ can be readily expressed in terms of $_{ep}\eta$ once the shape of the skeleton curve (Fig. 7b) is known. The skeleton curve can be obtained either experimentally or numerically, in view of the geometry and mechanical properties of the material. The skeleton curve can be idealized with a polygonal defined by Q_y, δ_y, Q_B, K_{p1}, K_{p2}, as shown with dash lines in Fig. 7b. From the equations that relate $_{ep}\eta$ with η, and $_{ep}\eta$ with $_S\eta$, the relation betwen $_s\eta$ and η can be readily obtained. Here, $\eta(= {_S\eta} + {_B\eta})$ represents the normalized ultimate energy dissipation capacity of the structural member. As an example, Figs. 8a–c show with symbols (circle, triangle, diamond) the pairs $({_{ep}\eta}, {_B\eta})$, $({_{ep}\eta}, {_S\eta})$, $({_{ep}\eta}, \eta)$ obtained for displacement-dependent metallic dampers subjected to different patterns of static cyclic loadings until failure [46]. The source of energy dissipation of these dampers was the out-of-plain bending deformation of the web of I-shaped steel sections. Figure 8d shows with diamond-shaped symbols the results of the static cyclic tests in the $_s\eta - \eta$ plane. Two similar metallic dampers were installed in a structure and subjected to dynamic loadings on a shake table that reproduced the 1985 Campano-Lucano earthquake recorded at Calitri [46]. The results in terms of $({_s\eta}, \eta)$ at failure

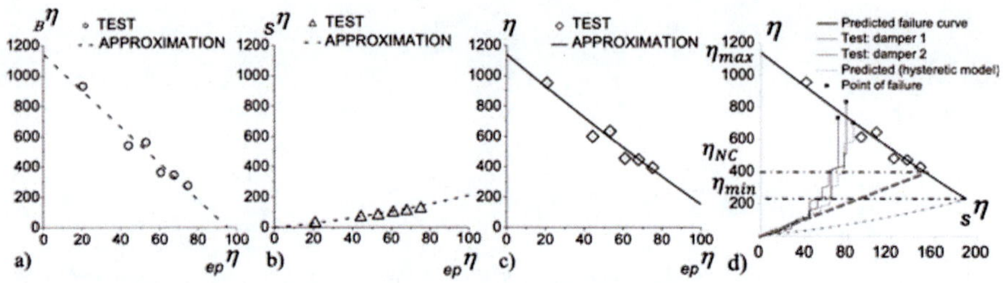

Fig. 8 Ultimate energy dissipation capacity of steel components

are plotted with solid square symbols in Fig. 8d. As seen in this figure, all symbols lay approximately in the same curve (bold solid line) representing the ultimate energy dissipation capacity of the steel component.

As seen in Fig. 8d, the normalized ultimate energy dissipation energy of the steel component η depends on the amount of energy consumed on the skeleton part, ${}_s\eta$, whose value is influenced by the load path followed by the steel component in the ${}_s\eta - \eta$ plane. Figure 8d shows with thin solid lines the load path traced by the two metallic dampers tested on the shake table with the Calitri record. This path is different for each ground motion and is also influenced by structural characteristics. Adopting η_{min} in Fig. 8d for design purposes would be overconservative, since it implies neglecting the energy dissipated by Bauschinger effects. Akiyama [47] investigated the relation between η and ${}_s\eta$ in steel beams and columns with stable restoring force characteristics subjected to axial forces and bending along the strong axis, and concluded that the relation between η and ${}_s\eta$ can be approximated by $\eta = [(1 + p)/(0.39 + 1,09p)]{}_s\eta$, where p is the axial force ratio (positive in compression) $p = N/(f_y A)$, f_y is the yield stress of the steel, and A the member cross area. This equation is plotted with a red dash line in Fig. 8d, and the ordinate η_{NC} of the intersection point with the bold solid curve may serve as the design value of the ultimate energy dissipation capacity of the steel member. This η_{NC} would be associated with the "near collapse" limit state. In light of η_{NC}, values for other limit states could be established, e.g. (3/4)η_{NC} for the limit state of damage limitation. The simple load path represented by the red line of Fig. 8d was obtained for p in the range 0–0.6, and was derived from a limited number of ground motions. Further research is needed to establish more refined design-oriented load paths that take into account the characteristics of the ground motion (e.g. impulsive vs. Non-impulsive), a larger range of p, and different structural systems. Additional studies are likewise needed to extend the approach explained above to more complex loading situations (e.g. columns subjected to biaxial moments).

In contrast to the steel components, the hysteretic behavior of RC structural members typically exhibits pinching and strength/stiffness degradation that render decomposition of the force-deformation curves impossible, as reflected in Fig. 7. Characterizing the ultimate energy dissipation capacity of RC structural components while taking into account the influence of the loading path is, indeed, a very challenging problem that has not yet been resolved.

4 Conclusions

This paper revised important aspects of the EBD approach with emphasis on the differences regarding FBD as opposed to DBD, pointing out key issues that need to be addressed in future research.

First, characterizing the loading effect of the earthquake in terms of total input energy (not mixing it with other parameters such as displacements) is a key aspect of EBD that must be kept in mind when developing this methodology. The main advantages of EBD resides in the stability of the total energy input and the fact that it is uncoupled from other aspects such as strength, level of damping, hysteretic behavior, etc., that can be neglected for design purposes.

Second, displacements are a by-product of the input energy, and the maximum values can largely differ depending on the characteristics of the structure and the ground motion. Once the hysteretic energy demand (at the story level in the case of multi-story buildings) is evaluated from the total input energy, the maximum displacements (i.e. inter-story drifts in multi-story structures) should be estimated from the relation between η and μ. Extensive nonlinear time history analyses with two different hysteretic models (Clough model and Modified Clough model) and far-field and near-fault records were conducted in this study. Although the variability of η for a given μ was found to grow with μ, the dispersion in results remains nearly constant: the average coefficient of variation (COV) is slightly lower (0.42) for far-field records than for near-field ones (0.56). The relations between η and μ obtained from linear least-squares fitting regression analyses are close to the design values recommended by Akiyama for the two hysteretic models investigated, yet slightly above them for far-field records (i.e. Akiyama's equation is conservative), while slightly below for near-fault records (i.e. Akiyama's equation is not conservative).

Third, evaluating the energy dissipation capacity of structural members under seismic loadings is cumbersome due to its dependence on the loading path. This would be one the most challenging issue at hand, or the largest gap in knowledge; still, a clear difference should be drawn between non-deteriorating structural components (e.g. steel members exempt of buckling effects), and deteriorating structural members (e.g. RC elements). For the non-deteriorating steel members, decomposition of the force-displacement curve into the skeleton and Bauschinger parts constitutes a very convenient way to capture the dependence of the ultimate energy dissipation capacity on the loading path. Once the ultimate energy dissipation of the member is characterized in this way, design values of the energy dissipation capacity associated with different limit states can be set, as long as the "design path" followed by the structural element during the "design earthquake" has been established. Further research should aim to define appropriate energy dissipation demand paths for design purposes.

Acknowledgements This research was funded by Spanish Ministry of Economy and Competitivity, research project reference MEC BIA2017 88814 R and received funds from the European Union (Fonds Européen de Dévelopment Régional). This work has been also supported by the Italian Ministry of Education, University and Research (MIUR), Sapienza University of Rome (Grant n.

RG11916B4C241B32) and the Italian Network of University Laboratories of Seismic Engineering (ReLUIS).

References

1. Uang, C., Bertero, V.: Evaluation of seismic energy in structures. Earthq. Eng. Struct. Dynam. **19**, 77–90 (1990)
2. Kalkan, E., Kunnath, S.: Relevance of absolute and relative energy content in seismic evaluation of structures. Adv. Struct. Eng. **11**(1), 1–18 (2008)
3. Housner, G.W.: Limit design of structures to resist earthquakes. In: Proceedings of the 1st World Conference on Earthquake Engineering, 1–20, Berkeley, California (1956)
4. Benavent-Climent, A., Pujades, L.G., Lopez-Almansa, F.: Design energy input spectra for moderate-seismicity regions. Earthquake Eng. Struct. Dyn. **31**, 1151–1172 (2002)
5. Akiyama H.: Earthquake-resistant limit-state design for buildings. University of Tokyo Press (1980). English version 1985
6. Building Standard Law of Japan. Building Research Institute. Earthquake-resistant structural calculation based on energy balance. The Building Center of Japan, Tokyo, Not. 461 (2009)
7. Zahrah, T., Hall, J.: Earthquake energy absorption in SDOF structures. J. Struct. Div. ASCE **110**(8), 1757–1772 (1984)
8. Fajfar, P., Vidic, T., Fischinger, M.: On energy demand and supply in SDOF systems. In: Fajfar, P., Krawinkler, H. (eds.) Nonlinear seismic analysis and design of reinforced concrete buildings, pp. 41–62. Elsevier (1992)
9. Fajfar, P., Vidic, T.: Consistent inelastic design spectra: hysteretic and input energy. Earthq. Eng. Struct. Dyn. **23**(5), 523–537 (1994)
10. Sucuoglu, H., Nurtug, A.: Earthquake ground motion characteristics and seismic energy dissipation. Earthq. Eng. Struct. Dyn. **24**(9), 1195–1213 (1995)
11. Surahman, A.: Earthquake-resistant structural design through energy demand and capacity. Earthq. Eng. Struct. Dyn. **36**(14), 2099–2117 (2007)
12. Cheng, Y., Lucchini, A., Mollaioli, F.: Proposal of new ground-motion prediction equations for elastic input energy spectra. Earthq. Struct. **7**(4), 485–510 (2014)
13. Fardis, M.N.: From force- to displacement-based seismic design of concrete structures and beyond. In: Pitilakis, K. (ed.) Recent Advances in Earthquake Engineering in Europe. ECEE 2018. Geotechnical, Geological and Earthquake Engineering, vol. 46. Springer, Cham (2018)
14. Akiyama, H.: Earthquake-resistant design method for buildings based on energy balance. Gihodo Shuppan Co. Ltd, Tokyo (in Japanese) (1999). Spanish version Ed. Reverté (2002)
15. Khashaee, P., Mohraz, B., Sadek, F., Lew, H.S., Gross, J.L.: Distribution of Earthquake Input Energy in Structures. NISTIR Report 6903 (2003)
16. Manfredi, G., Polese, M., Cosenza, E.: Cumulative demand of the earthquake ground motions in the near source. Earthq. Eng. Struct. Dyn. **32**, 1853–1865 (2003)
17. Veletsos, A.S., Newmark, N.M.: Effect of inelastic behaviour on the response of simple systems to earthquake motions. In: Proceedings of the 2nd World Conference on Earthquake Engineering (II), pp. 895–912 (1960)
18. Shibata, A.: Structural Analysis for Seismic Loading. Moriquita Shuppan Ltd. (1981)
19. Kasai, K., Fu, Y.: Seismic analysis and design using VE dampers. In: Proceedings of a New Direction in Seismic Design, Tokyo (1995)
20. OpendSees. University of California, Berkeley
21. Clough, R.W., Johnson, S.B.: Effect of stiffness degradation on earthquake ductility requirements. In: Proceedings of Japan Earthquake Engineering Symposium (1966)
22. Quaranta, G., Mollaioli, F.: Analysis of near-fault pulse-like seismic signals through Variational Mode Decomposition technique. Eng. Struct. **193**, 121–135 (2019)

23. AlShawa, O., Angelucci, G., Mollaioli, F., Quaranta, G.: Quantification of energy-related parameters for near-fault pulse-like seismic ground motions. Appl. Sci. **10**, 7578 (2020)
24. Chou, C.C., Uang, C.M.: A procedure for evaluating seismic energy demand of framed structures. Earthq. Eng. Struct. Dyn. **32**, 229–244 (2003)
25. Prasanth, T., Ghosh, S., Collins, K.R.: Estimation of hysteretic energy demand using concepts of modal pushover analysis Earthq. Eng. Struct. Dyn. **37**, 975–990 (2008)
26. Kuntz, G.L., Browning, J.: Reduction of column yielding during earthquakes for reinforced concrete frames. ACI Struct. J. **100**(5), 573–580 (2003)
27. Moehle, J.P., Hooper, J.D., Lubke, C.D.: Seismic design of reinforced concrete special moment frames, A Guide for Practicing Engineers, NIST GCR 8-917-1 (2008)
28. Benavent-Climent, A.: An energy-based method for seismic retrofit of existing frames using hysteretic dampers. Soil Dyn. Earthq. Eng. **31**, 1385–1396 (2011)
29. Donaire-Avila, J., Benavent-Climent, A.: Optimum strength distribution for structures with metallic dampers subjected to seismic loading. Metals **10**(127), 1–27 (2020)
30. Donaire-Ávila, J., Lucchini, A., Benavent-Climent, A., Mollaioli, F.: New optimum distribution of lateral strength of shear-type buildings for uniform damage. J. Struct. Eng. Mech. **76**(3), 279–291 (2020)
31. Erberik, A., Sucuoglu, H.: Seismic energy dissipation in deteriorating systems through low-cycle fatigue. Earthq. Eng. Struct. Dyn. **33**, 49–67 (2004)
32. Benavent-Climent, A.: An energy-based damage model for seismic response of steel structures. Earthq. Eng. Struct. Dynam. **36**(8), 1049–1064 (2007)
33. Kato, B., Akiyama, H., Yamanouchi, H.: Predictable properties of structural steels subjected to incremental cyclic loading. In: IABSE Symposium on Resistance and Ultimate Deformability of Structures Acted on by Well Defined Loads, Lisbon (1973)
34. Akiyama, H., Takahashi, M., Shi, Z.: Ultimate energy absorption capacity of round-shape steel rods subjected to bending. Trans. AIJ **475**, 101–112 (1995)
35. Benavent-Climent, A., Akiyama, H.: Development of a brace-type hysteretic damper by using round-shape steel rods. J. Constr. Eng., AIJ **501**, 125–133 (1997)
36. Benavent-Climent, A., Sang- Hoon, O., Akiyama, H.: Ultimate energy absorption capacity of slit-type steel plates subjected to shear deformations. Transactions AIJ **503**, 139–147 (1998)
37. Hasegawa, H., Takeuchi, T., Iwata, M., Yamada, S., Akiyama, H.: Experimental study on dynamic behaviour of unbonded braces. Architect. Inst. Japan J. Technol. Des. **9**, 103–106 (1999)
38. Akiyama, H., Yamada, S., Minowa, C., Teramoto, T., Otake, F., Yabe, Y.: Experimental method of the full scale shaking table tests using the inertial loading equipment. Architect. Inst. Japan J. Struct. Constr. Eng. **505**, 139–146 (1998)
39. Akiyama, H., Yamada, S., Matsumoto, Y., Matsuoka, S., Ogura, K., Kitamura, H.: Study on fracture of beam-to-column connections by means of full scale shaking table tests. Architect. Inst. Japan J. Struct. Constr. Eng. **512**, 165–172 (1998)
40. Akiyama H., Yamada S., Takahashi M., Katsura D., Kimura K., Yahata S. Full scale shaking table tests of the exposed-type coluymn bases. Architectural Institute of Japan J. Struct. Constr. Eng. **514**, 185–192 (1998).
41. Akiyama, H., Yamada, S., Matsumoto, Y., Tanaka, H., Miyama, T., Terada, T.: Full scale shaking table tests of cold-formed RHS columns. Architect. Inst. Japan J. Struct. Constr. Eng. **521**, 133–140 (1999)
42. Akiyama, H., Yamada, S., Matsumoto, Y., Matsuoka, S., Ohtake, F., Sugimoto, H.: Transition from ductile to brittle fracture of full scale beam-to-column connections caused by temperature. Architect. Inst. Japan J. Struct. Constr. Eng. **522**, 105–112 (1999)
43. Akiyama, H., Yamada, S., Benavent, A., Terada, T., Kitamura, H., Ogura, K.: Full scale shaking table tests of flexible-stiff mixed structure with brace-type hysteretic dampers. Architect. Inst. Japan J. Struct. Constr. Eng. **541**, 171–178 (2001)
44. Akiyama, H., Yamada, S., Matsumoto, Takeuchi, T., Sugimoto, H.: Full scale shaking table tests on ultima seismic resistance of advanced type of beam-to-column connections. Architect. Inst. Japan J. Struct. Constr. Eng. **551**, 141–148 (2002)

45. Akiyama, H., Yamada, S., Takahashi, M., Matsumoto, Y., Kumagai, H., Fukuda, K.: Full scale shaking table test of RC beam-to-column connections. Architect. Inst. Japan J. Struct. Constr. Eng. **565**, 111–118 (2003)
46. Benavent-Climent, A., Morillas, L., Vico, J.M.: A study on using wide-flange section web under out-of-plane flexure for passive energy dissipation. Earthq. Eng. Struct. Dyn. **40**, 473–490 (2011)
47. Akiyama, H., Takahashi, M.: Influence of Bauschinger effect on seismic resistance of steel structures. J. Struct. Constr. Eng. Trans. AIJ **418**, 49–57 (1990)

A Damage Index for the Seismic Evaluation of Buckling-Restrained Braces

J. A. Oviedo-Amezquita, N. Jaramillo-Santana, C. A. Blandon-Uribe, and A. M. Bernal-Zuluaga

Abstract This paper focuses on evaluating whether a buckling-restrained brace (BRB) element should be replaced after an earthquake event. For this purpose, a damage index is proposed based on the experimental data obtained from a series of tests conducted on different BRB specimens. In total 19 full-scale BRB specimens were manufactured with local industry and workforce, and tested. 14 BRB specimens were tested using a low-cycle loading protocol, and five BRB specimens were tested using a high-cycle fatigue loading protocol. For the low-cycle protocol, the axial strain in the BRB core was continuously increased until failure, while for the fatigue protocol, the axial strain in the core was increased from zero to 1.5%, and continued at 1.5% strain until core failure occurred. The proposed damage index has been calibrated based on the experimental results, and is capable of considering the effect of the maximum core strain attained as well as the cumulative deformation effect. A qualification scale has been assigned to the proposed damage index as a tool for evaluating whether the BRB element should be replaced or left on site. Furthermore, a series of nonlinear dynamic analyses were carried out on a sample building in order to validate the proposed damage index when subjected to ground motions. Finally, the results suggest that the proposed damaged index can be useful in structural design practice.

Keywords Buckling-restrained brace · Hysteretic damper · Damage index · Fatigue · Experimental test

J. A. Oviedo-Amezquita (✉) · N. Jaramillo-Santana · A. M. Bernal-Zuluaga
F'C Control & Design of Structures SAS, Medellín, Colombia
e-mail: joviedo@efeprimace.co

C. A. Blandon-Uribe
EIA University, Envigado, Colombia

A. Benavent-Climent and F. Mollaioli (eds.), *Energy-Based Seismic Engineering*, Lecture Notes in Civil Engineering 155,
https://doi.org/10.1007/978-3-030-73932-4_12

1 Introduction

Among modern design philosophies, the Performance-Based Seismic Design methodology (PBSD), a design methodology based on the behavior of the structure considering the concept of damage tolerance in structural, non-structural and building equipment components, has become very popular. This design philosophy allows for the evaluation of the seismic reliability of building structures against earthquakes, and has been widely used for the design and retrofit of building structures based on different guidelines and design codes [1–5]. In general, building structures designed under the PBSD concept have a higher level of reliability than those in which this concept has not been applied. One key aspect when evaluating the seismic performance of a building structure under the PBSD methodology has to do with the definition of the acceptance criteria for structural elements, for a particular performance level; for instance, the ASCE 41-17 [5] document presents the acceptance criteria for several structural elements and materials. A complementary procedure for assessing the seismic performance of a structural element or a whole system is carried out through the use of a damage index qualification. To date, several methodologies for damage index determination are available, particularly for reinforced-concrete and steel members [6–9]; some of them correspond to energy-based damage indexes.

The use of Buckling-Restrained Braces (BRBs) in building structures is still gaining popularity as an interesting alternative for the seismic design and retrofit of building structures. Several studies are still being published, reporting not only advances in the development of BRB elements, but also in the design methodologies (e.g., [10–20]). Recently, Speicher and Harris [21] presented the seismic performance assessment for a suite of buckling-restrained braced frames (BRBFs) in which they recommended that the acceptance criteria for BRBs listed in the ASCE 41-17 be re-examined. Therefore, the authors proposed an adjustment factor to account better for the cumulative deformation effect (obtained from nonlinear dynamic analysis) and the maximum deformation capacity of BRBs. In their work, however, no influence of BRB properties was considered, such as the core plastic length.

The present work is part of a larger investigation aimed at introducing two methodologies, one for defining the acceptance criteria for BRBs and another for determining a damage index (DI) when BRBs are subjected to seismic actions. Both methodologies have been established based on a set of experimental results [22–24] obtained from a series of tests carried out on a BRB prototype developed in Colombia. Thus, this paper introduces the methodology for the DI determination, targeted at the post-earthquake evaluation of BRB elements, and as a tool to determine whether the BRB should be replaced or not. The validity of the proposed methodology is confirmed through a series of nonlinear dynamic analyses on a sample building that incorporates BRBs. The results herein presented are expected to contribute to ongoing efforts on improving PBSD methodologies.

2 Experimental Program

2.1 Tested BRB Specimens

In total 19 full-scale buckling-restrained braces (BRBs) were designed and manufactured with local industry and workforce. All BRBs were subjected to cyclic loading in tension and compression. Two types of loading protocol were considered: the low-cycle protocol (LC), which gradually increased the axial strain in the BRB core until failure, corresponded to an even more demanding loading protocol than that required by the Colombian code (or AISC 341-10), and the high-cycle protocol (HC) was used to induce fatigue failure in the BRB core. The 19 specimens were divided and tested into three research groups (RG). Table 1 shows the characteristics of each RG, the research goals and the structural characteristics of all 19 specimens; details on each group can be found elsewhere [22–24]. Figure 1, on the right, illustrates the brace length (L_B), plastic length (L_P), buckling-restrained length (L_R) and the brace deformation between measuring points (δ_B). Figure 1, on the left, illustrates the three different working zones into which the BRB is divided: Zone 1 corresponds to the elastic connection zone, Zone 2 corresponds to the elastic transition zone, and Zone 3 corresponds to the plastic zone (all axial inelastic deformation occurs within L_P). It is worth mentioning that the brace deformation δ_B is considered to be adequate for practical use now that it can be readily measured on site through any displacement sensor or similar; brace deformation of Zone 1 is generally very small. δ_B includes the axial deformation of Zone 3 and both Zones 2.

All tested prototypes had the same length ($L_B = 2{,}351$ mm) and steel type (ASTM A-36), but they differed from each other in: (1) global buckling safety factor, (2) plastic length, and (3) unbonding gap. Here, it is important to note that according to coupon tests on the steel core, the yield stress for the RGI and RGII was 290 MPa, while 310 MPa for the RGIII. The tests carried out for RGs I and II aimed at understanding mainly the influence of the global buckling safety factor and the plastic length on the performance of the BRB prototype. The tests performed for the RGIII were carried out to understand the performance of prototypes under a fatigue-type load. For the latter purpose, two prototypes of BRBs were designed and built; specifically, three specimens of PR1 and two of PR2, for a total of five BRB specimens. Figure 2 shows the experimental set-up.

2.2 Loading Protocols

As mentioned, two types of loading protocols were used for the experimental program: the first protocol referred to as LC was used for the RGs I and II, and the second protocol referred to as HC was used for the RGIII. Thus, all specimens were subjected to cyclic tensile and compressive loading in which the axial deformation in the plastic length of the brace was gradually increased from zero up to a

Table 1 Nominal mechanical properties of tested BRBs

Research goal	Protocol	RG	Prot	#S	Steel core					Buckling restrainer		
					w/t	A_{sc}	L_P (% L_B)	P_y	δ_{By}	GB-SF	L_R	Gap
Global buckling	LC	RGI	PR1	3	10.5	855	1,511 (64)	248	–	2.10	1,797	0.8
			PR2	3	10.5	855	1,511 (64)	248	–	3.30	1,797	0.8
Plastic length, global buckling, unbonding gap	LC	RGII	PR1	2	10.5	855	1,511 (64)	248	2.53	3.28	1,797	0.5
			PR2	2	10.5	855	705 (30)	248	2.06	3.28	1,797	0.5
			PR3	2	10.5	855	1,511 (64)	248	2.53	4.72	1,797	0.5
			PR4	2	10.5	855	1,511 (64)	248	2.53	4.82	1,797	1.0
Fatigue	HC	RGIII	PR1	3	10.5	855	1,261 (54)	265	2.05	3.10	1,797	0.5
			PR2	2	10.5	855	705 (30)	265	1.76	3.10	1,797	0.5

RG: research group (I, II, III); Prot.: prototype identification (PR); #S: number of specimens per prototype; w/t: steel core width-to-thickness ratio; A_{sc}: steel core area, [mm^2]; L_P: plastic length, L_{P64}, L_{P54} or L_{P30}, [mm]; P_y: nominal yield strength, [kN]; δ_{By}: deformation between measuring points at core yield, [mm]; GB-SF: global buckling safety factor; L_R: buckling-restrained length, [mm]; Gap: unbonding gap, [mm]

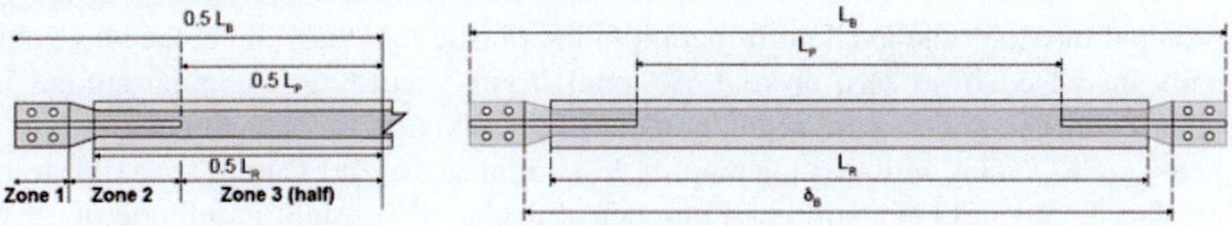

Fig. 1 BRB specimen definitions

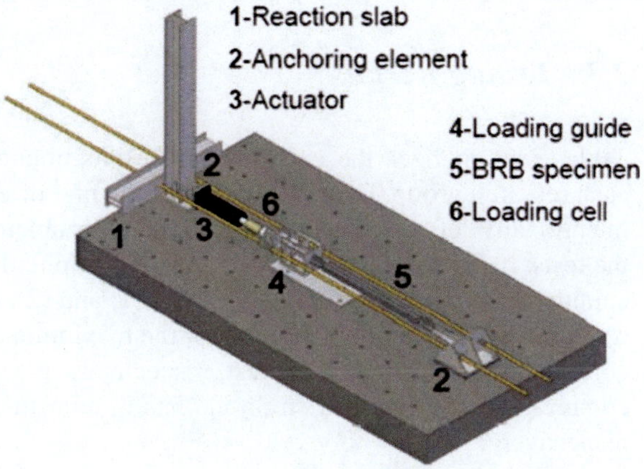

Fig. 2 Experimental set-up for BRB testing

specific value of axial strain according to the protocol used. The LC loading protocol, shown in Fig. 3a, was chosen based on previous research programs carried out in Japan [12] in order to obtain comparable data with past studies. The axial strain of the steel core ε_p was increased from zero up to 3.0% strain, which is 50% larger than the maximum strain required by the current Colombian design regulations (NSR-10 [25]). It is worth mentioning the NSR-10 specifications for BRBs are based on the loading protocol given by the document AISC 341-05 [26]. Thus, LC protocol meets the conditions described in NSR-10 and the strain sequence is more demanding in terms of cumulative inelastic deformation. Figure 3b shows the loading protocol HC, in which Nc_f stands for the cycle number at which failure occurred. All specimens of the RGIII were subjected to the HC cyclic tensile and compressive loading

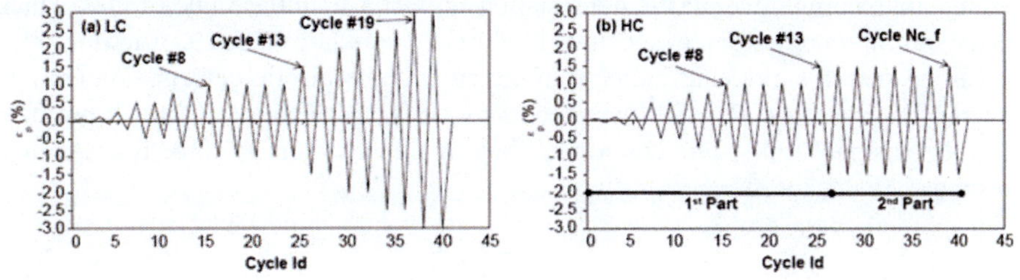

Fig. 3 Loading protocols: **a** LC, and **b** HC

protocol in which the axial deformation in the plastic length of the brace was gradually increased from zero up to 1.5% axial strain, according to similar studies in other countries and current regulations [27, 28]. This protocol is divided into two parts: (1) 12 cycles with varying amplitude, similar to that of LC until the axial strain reaches 1.5%, and (2) a sufficient number of cycles of constant amplitude of 1.5% axial strain, until failure is obtained. Additional details of each loading protocol can be found elsewhere [22–24].

2.3 Testing Results

Table 2 summarizes the experimental results obtained in the testing program for each research group. Here, the results of the first three specimens of RGI have been intentionally left out the table due to some problems in the set-up at the time of the tests. In Table 2, η and ϖ stand for the cumulative plastic deformation and the cumulative strain energy, defined by Eq. (1) and (2), respectively [12] (see Figs. 4a, b). Moreover, ε_{max} and μ_{max} stand for the maximum axial core strain and maximum core ductility obtained in the test, respectively. β and ω stand for the well-known compression strength and strain hardening adjustment factors for BRB elements, respectively.

$$\varpi = E_t / W_y = E_t / \left(P_y \delta_y\right) \tag{1}$$

$$\eta = \frac{\delta_1 + \delta_2 + \delta_3 + \cdots}{\delta_y} \tag{2}$$

Here, it is important to mention that parameters μ, η and ϖ were calculated over the length δ_B; thus, the ductility μ is given by Eq. (3). Where, δ_B is the brace deformation between measuring points (see Fig. 1). Based on the results of all test conducted to the RGII, a limit for ε_p was set to 2.5% as a conservative deformation capacity for the BRB prototype, just before presenting local compression failure. Therefore, $\mu_{2.5}$ stands for the maximum ductility obtained at an axial strain equals to 2.5% ($\varepsilon_p = 0.025$). Additional details on the response parameters obtained in all tests can be found elsewhere [22–24]. In Table 2, it is important to note that all specimens withstood a cumulative plastic deformation η greater than the value of 200, which is the minimum requirement specified by different standards [25, 29, 30]. Moreover, it can be seen a significant increase in the cumulative plastic deformation (η) and energy dissipation (ϖ) when the specimens are subjected to the fatigue protocol HC; for instance, all specimens were able to increase in almost three times their η response.

$$\mu = \frac{\delta_B}{\delta_{By}} \tag{3}$$

Table 2 Test results

RG/Prot	Spec	η	ϖ	β	ω	ε_{max} (%)	μ_{max}	$\mu_{2.5}$	Nc-f
RGI /PR2	2–1	259	259	1.09	1.26	1.5	9.40	N/A	N/A
RGI/PR2	2–2	357	304	1.14	1.29	2.0	12.52	N/A	N/A
RGI/PR2	2–3	324	301	1.09	1.26	2.0	12.39	N/A	N/A
RGII/PR1	1–1	594	518	1.27	1.20	3.0	17.64	15.12	N/A
RGII/PR1	1–2	431	527	1.19	1.37	2.5	15.51	15.51	N/A
RGII/PR2	2–1	287	291	1.02	1.19	3.0	10.93	9.18	N/A
RGII/PR2	2–2	345	366	1.07	1.34	3.0	11.29	9.55	N/A
RGII/PR3	3–1	359	367	1.01	1.35	2.5	15.22	15.22	N/A
RGII/PR3	3–2	486	554	1.24	1.40	3.0	18.16	15.18	N/A
RGII/PR4	4–1	544	643	1.12	1.50	3.0	18.21	15.17	N/A
RGII/PR4	4–2	597	746	1.25	1.41	3.0	18.72	15.60	N/A
RGIII/PR1	1–1	1,693	2,356	1.07	1.38	1.5	11.09	N/A	58
RGIII/PR1	1–2	1,868	2,356	1.07	1.26	1.5	10.93	N/A	64
RGIII/PR1	1–3	1,848	2,306	1.07	1.41	1.5	10.36	N/A	63
RGIII/PR2	2–1	1,354	1,508	1.04	1.15	1.5	7.65	N/A	70
RGIII/PR2	2–2	1,600	2,058	1.24	1.29	1.5	7.53	N/A	83

As for the failure mode of the specimens, experimental results showed that the specimens subjected to the LC protocol failed mainly in a local-buckling failure mode in the steel core due to the capacity loss of the buckling restrainer. On the contrary, the specimens subjected to the HC protocol failed in a tensile-fracture failure mode in the steel core due to fatigue. Here, it is important to mention that these two failure modes are accepted in case of BRB elements. Details on different failure modes can be found elsewhere [31, 32].

From Fig. 4c, it is clear that the maximum ductility in the response history somewhat restrains the cumulative plastic deformation and energy dissipation. In other words, the larger the maximum axial strain (or ductility) in the response history, the lower the value of η and ϖ. This is to be expected since a large ductility demand imposes large forces on the buckling-restraining mechanism, leading it to failure. It is also important to note that BRBs of RGII were able to withstand larger ductility demands than those of the RGI since some prototype design aspects were changed to improve the deformation capacity in the specimens. On the other hand, [33] mentioned minimum limits for η required on BRBs installed into a building structure: which are $\eta = 200$ in case of the design earthquake (DE) intensity, and $\eta = 400$ in case of the maximum considered earthquake (MCE). In [33], Life Safety (LS) performance level was considered for the DE earthquake and Collapse Prevention (CP) for the MCE earthquake. They also mentioned other experimental studies which have reported large values of cumulative plastic deformation ($\eta = 1700$); the results of η obtained in the RGIII are comparable with those large values. In Fig. 4c it can also be observed that all RGIII specimens reached higher values of η than the limits proposed

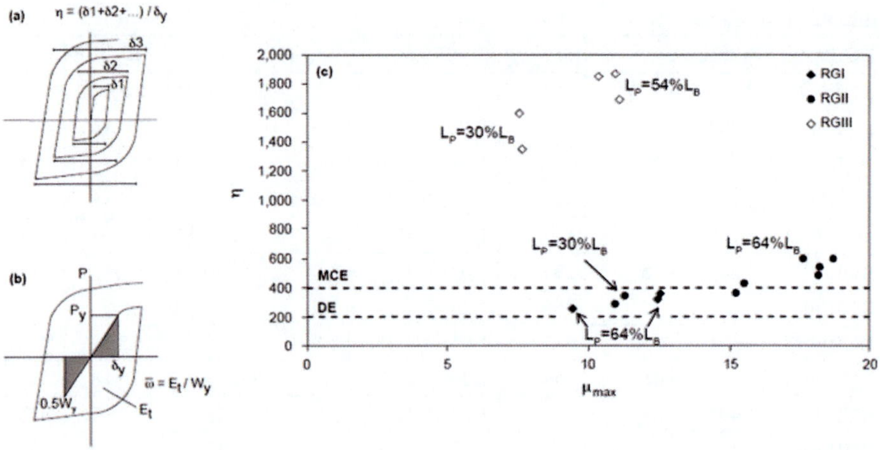

Fig. 4 Response parameter definition and test results: **a** η [12], **b** ϖ [12], and **c** RGs test results

by [33] for the two levels of seismic intensity. In case of the specimens of the RGII, all reached values of η higher than the limit for the DE level, and a few specimens reached values higher than the limit for the MCE level. Here, the latter corresponds to specimens having the larger plastic length (L_{P64}). This clearly indicates the great influence of L_P on the energy dissipation capacity of BRB elements.

3 Proposed Damage Index Concept

3.1 Tested BRB Specimens

Based on the test results, Eq. (4) represents the proposed damage index (DI), which takes into account the two failure modes described in Sect. 2.3.

$$DI = F_1^{\alpha} F_2^{(1-\alpha)} \tag{4}$$

In Eq. (4), F_1 and F_2 are to represent the contribution of the Type-A compressive core failure mode and Type-B tensile core failure mode, respectively. The α coefficient stands for a weighting factor, depending on the maximum axial deformation withstood (between measuring points) δ_{Bmax} and the plastic length L_P. In Eq. (5), a_0 and b_0 coefficients are set to: $a_0 = 0.5$ (for equal participation of factors F_1 and F_2) and $b_0 = -15$ (for better matching with the experimental results). It is important to note that the experimental results showed that the contribution to the calculation of DI is an inversely-proportional type process between both failure modes (Type A and B). Thus, factors F_1 and F_2 can be determined by Eqs. (6) and (7), respectively.

$$\alpha = a_0 + b_0 \frac{\delta_{B\,\max}}{L_P} \tag{5}$$

$$F_1 = \frac{\delta_{B\,\max}}{\delta_c} \tag{6}$$

$$F_2 = \frac{\eta_{\max}}{\eta_c} \tag{7}$$

At the maximum deformation demand, F_1 represents the effect of the maximum deformation, and F_2 represents the effect of cumulative plastic deformation. δ_c and η_c stand for the characteristic capacities observed under qualification tests, obtained from Eqs. (8) and (9), respectively.

$$\delta_c = \delta_u - \sigma_\delta \tag{8}$$

$$\eta_c = \eta_u - \sigma_\eta \tag{9}$$

In Eqs. (8) and (9), δ_u and η_u are the ultimate (maximum) deformation and ultimate cumulative plastic deformation measured in qualification tests, or at a particular "ultimate" state defined by the manufacturer, respectively. Here, it is worth mentioning that although specimens of the RGII were able to withstand axial core strains up to 0.030, δ_u was set to the brace deformation δ_B when ε_p reached 0.025, and η_u is the corresponding cumulative deformation obtained from Eq. (2). Moreover, σ_δ and σ_η are the standard deviations for δ_u and η_u, respectively, among the qualification tests. It should be noted that a different set of δ_c and η_c is required for each plastic length L_P, and that η_c also depends on the failure type (or loading history type).

Table 3 shows the resulting characteristic capacities and the damage index (DI) calculation, while Fig. 5 depicts the evolution of damage index until failure. The resulting value of η_c shown in Table 3 varies between 80 and 90% of the ultimate average capacity η_u recorded in the tests, for both RGs II and III; and the resulting value of δ_c varies between 95 and 100% of the ultimate average capacity δ_u recorded in the tests. The small variation of δ_c is because tests were performed under a displacement-controlled scheme. As seen in the last column of Table 3, DIs obtained from Eq. (4) can have values larger than unity; this is because characteristic values are taken conservatively smaller than the actual capacities measured in the tests. A mean characteristic DI value for L_{P30} and L_{P64} specimens can be set as 1.03 and 1.09, respectively. Another aspect to highlight has to do with the contribution of factors F_1 and F_2 to the DI calculation. As expected, the cumulative effect represented by the factor F_2 contributes more to the DI calculation in case of RGIII. As seen in Fig. 5, the evolution of DI varies with the loading protocol. In the RGII, the slope is larger and tends to remain constant until failure occurs, while the slope tries to follow a bilinear behavior in the RGIII, slightly changing at a DI of about 0.2–0.3. The value of η equals to 200 corresponds to the minimum deformation capacity required for qualification testing [30]. From Fig. 5, the following DIs can then be assigned for η

Table 3 Characteristic properties and damage index calculation

RG/Prot	Spec	η_c	δ_c (mm)	α	δ_{Bmax}/L_P	F1	F2	DI
RGII/PR1	1–1	398	38.24	0.118	0.025	1.01	1.29	1.25
RGII/PR1	1–2	398	38.24	0.110	0.026	1.02	1.08	1.08
RGII/PR2	2–1	261	18.73	0.100	0.027	1.05	1.01	1.01
RGII/PR2	2–2	261	18.73	0.085	0.028	1.04	1.05	1.05
RGII/PR3[a]	3–1	–	–	–	–	–	–	–
RGII/PR3	3–2	398	38.24	0.117	0.026	1.00	1.06	1.05
RGII/PR4	4–1	398	38.24	0.117	0.026	1.00	1.08	1.07
RGII/PR4	4–2	398	38.24	0.108	0.026	1.03	1.02	1.02
RGIII/PR1[a]	1–1	–	–	–	–	–	–	–
RGIII/PR1	1–2	1,833	32.50[b]	0.235	0.018	0.63	1.00	0.90
RGIII/PR1	1–3	1,833	32.50[b]	0.249	0.017	0.64	1.02	0.91
RGIII/PR2	2–1	1,321	18.73	0.205	0.020	0.74	1.03	0.96
RGIII/PR2	2–2	1,321	18.73	0.218	0.019	0.64	1.00	0.91

[a]Specimens were left out of calculation due to a minor premature failure observed in the tests (see [22, 23] for details)
[b]Extrapolated values

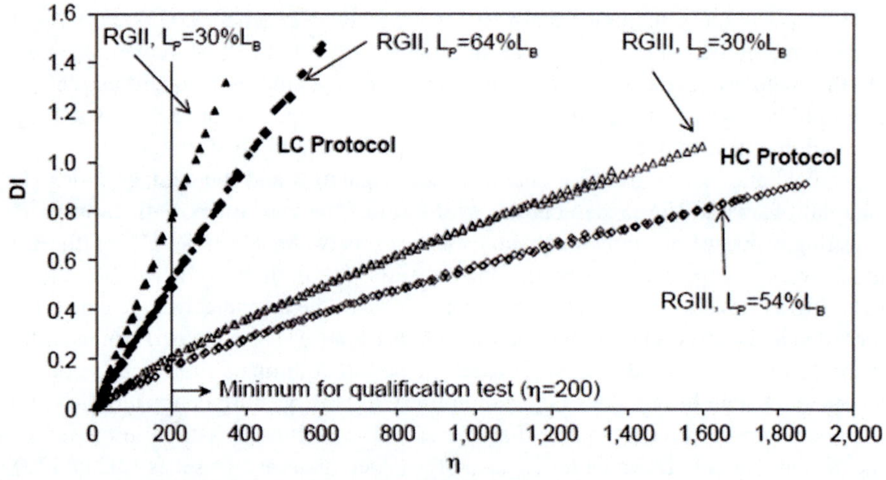

Fig. 5 Damage index (DI) evolution during tests from Eq. (4)

= 200: for RGII: DI = 0.5 for L_{P64} and 0.75 for L_{P30}; and for RGIII: DI = 0.2 for L_{P54} and L_{P30}. It is again clear that the maximum deformation in the response history limits the cumulative plastic deformation; that is, the larger the maximum brace deformation (or ductility), the lower the cumulative plastic deformation capacity.

Table 4 Proposed DI levels

Damage level	Range	Description
Severe	DI $\geq$ 0.7	BRB should be replaced immediately
Moderate	0.3 < DI < 0.7	BRB performance should be further investigated. Project owner's participation is required for deciding if the BRB is left on site or not
Slight	DI $\leq$ 0.3	BRB needs not be replaced unless physical damage is observed

As most damage indexes, the proposed DI requires a qualitative evaluation for design purposes. In case of BRBs, the evaluation should aim at defining whether a BRB specimen should be replaced or not after withstanding a particular loading history (for instance, an earthquake). From the practical viewpoint, a specific methodology for determining the parameter η from the BRB response history is then needed; an example of such a method can be found elsewhere [31]. Thus, and based on the test results, Table 4 shows a proposal for such an evaluation. Here, the limits DI = 0.7 and DI = 0.3 were also chosen to provide safety factors of about 1.5 (1/0.7) and 3 (1/0.3), respectively. Both safety factors are considered appropriate for accounting for capacity variations in the BRB due to fabrication process.

4 Discussion on Earthquake Response

4.1 Analyzed Braced-Frame Building Structure

A series of earthquake nonlinear dynamic analyses (NLDA) was carried out on a sample building in order to investigate the proposed methodology. Figure 6 shows the three-dimension six-story reinforced concrete (R/C) structure used as a sample building for dynamic analyses. The symmetric plan consists of 3 by 4 bays each of 7 m with a typical height of 3.5 m. The gravitational loads (dead and live) per unit area are assumed to be the same for all stories, with a typical superimposed dead load of 3.23 kN/m^2 and a typical live load of 4.0 kN/m^2 (for hospital use). Two pairs of BRBs in each direction were installed into the R/C main frame, as shown in Fig. 6; BRBs of each floor share the same structural properties. Thus, a total of 48 BRBs were installed into the R/C main frame, grouped into six different BRB types. The structural design was established based on the Colombian seismic code (NSR-10) for R/C moment-resisting frames, and BRBs were sized and designed according to the technical advice of the BRB manufacturer in Colombia. Table 5 summarizes the structural properties of the sample building. In Table 5, P_{ysc}, K_{Bwp} and K_{Bp} correspond to the yield strength (for a nominal yield stress of 250 MPa), workpoint-to-workpoint axial stiffness and post-yield axial stiffness of BRB. L_{Bwp} is the workpoint-to-workpoint length of the brace. The total seismic weight of the three-dimension sample building is 29,742 kN, and the fundamental period is 0.78 s (with cracked concrete cross-sections).

4.2 Input Earthquake Motions

Three different acceleration records were selected for the NLDAs: Armenia (Colombia, 1999), Imperial Valley (USA, 1979) and Loma Prieta (USA, 1989). Two seismic levels were considered for the NLDAs: the DE and MCE earthquake levels. All records components were modified to match the DE design spectrum, according to the requirements given in [25] and [34], and then were cut to retain the 95% of the Arias intensity index [35]. Adjusted input motions are then multiplied by a factor of 1.50 for the MCE level. Table 6 summarizes the input motions used for the NLDAs, and Fig. 7 shows the response spectra of the spectrum-matched records. In Table 6, PGA-mod and T_d correspond to the PGA and the time duration of modified records, respectively.

The ETABS computer program [36] was used for all analyses, where BRBs were modeled trough nonlinear Link elements using the structural properties in Table 5 and a Wen type hysteresis model with a shape factor of 20. Here, only the degree of freedom for axial behavior of braces was activated, and Link both ends were considered as pin-end connections. Moreover, nonlinear hinges were assigned to each end of columns and beams; characteristics of the R/C plastic hinges were defined based on [5], through the feature of automatic hinges of the ETABS software. For all nonlinear analyses, the inherent viscous damping ratio is 0.02.

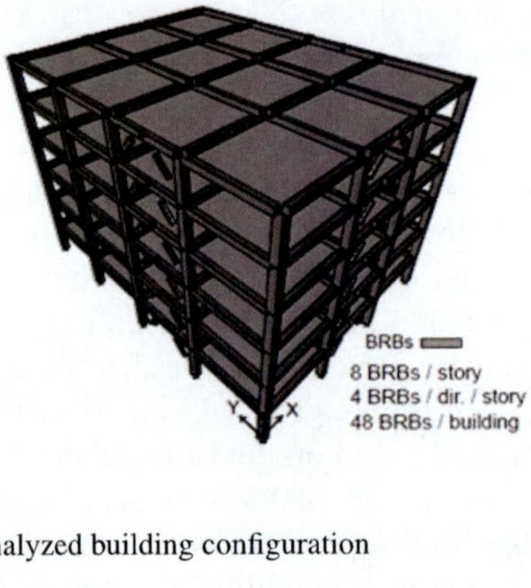

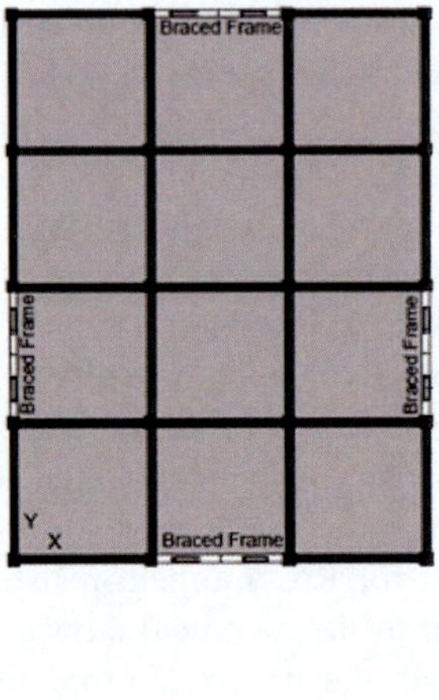

Fig. 6 Analyzed building configuration

Table 5 Structural properties of the analyzed building

Element	Section geometry (m)	Concrete strength (MPa)	P_{ysc} (kN)	K_{Bwp}; K_{Bp} (kN/mm)
1st–6th story cols	0.60 × 0.60	28	–	–
1st story braced frame columns	0.70 × 0.70	28	–	–
1st–6th floor beams	0.40 × 0.55	21	–	–
1st story BRBs	$L_{Bwp} = 4.95$ $L_B = 4.0$ $L_P = 2.4$	–	1,125	300.5; 8.32
2nd story BRBs		–	1,575	420.6; 11.65
3rd story BRBs		–	1,260	336.5; 9.32
4th story BRBs		–	938	250.4; 6.94
5th story BRBs		–	540	144.2; 3.99
6th story BRBs		–	225	60.1; 1.66

Table 6 Input earthquake motion for DE level

EQ. Source	Station	Input motion	PGA (cm/s^2)	PGV (cm/s)	PGA-mod (cm/s2)	Td (s)
Armenia	U.Quindio	A-1	580	251	358	17
		A-2	518	264	311	17
Loma Prieta	Corralito	LP-1	632	549	308	33
		LP-2	473	467	297	33
Imperial Valley	Cerro Prieto	IV-1	154	189	282	52
		IV-2	165	113	286	52

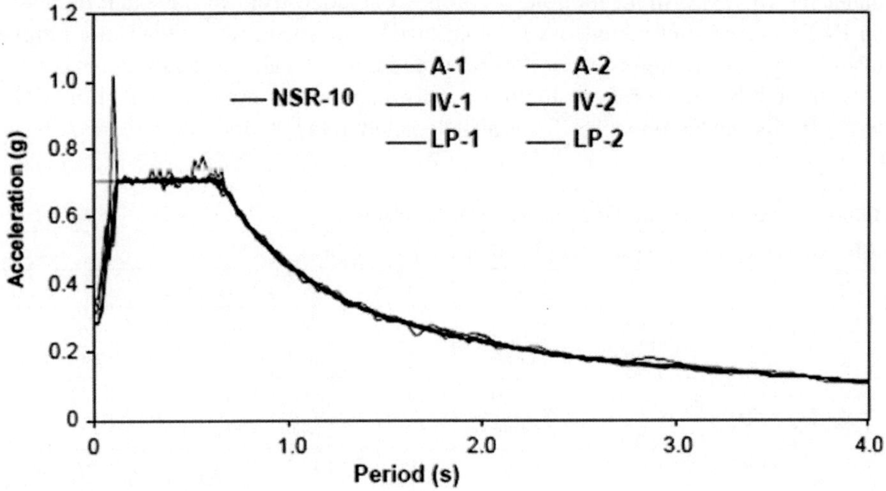

Fig. 7 Response spectra of modified input motions (DE level)

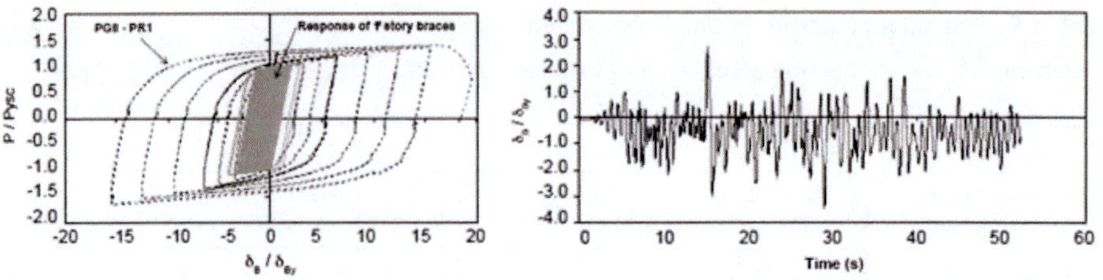

Fig. 8 Response of 1st-story BRBs under the MCE-IV motion: **a** normalized force-deformation, and **b** time-deformation time-history

4.3 Damage Index Calculation

Table 7 shows the maximum earthquake demands and damage index obtained from Eq. (4) for the 1st-story BRBs for the MCE-IV input motion. Figure 8a shows the normalized force-deformation response of all braces of the first story under the MCE-IV input motion. Analysis results here are focused on the BRBs of the first story since they presented the higher earthquake demands. Values for δ_c and η_c were defined based on test result (see Sect. 3). Thus, δ_c was set to $0.9\delta_u$, where δ_u is the brace deformation (δ_B, see Fig. 1) when ε_p reaches 2.5%. Here, it is important to note that since brace lengths of the sample building are larger than those of the test, the value of δ_B increases up to 60.3 mm, leading to $\delta_c = 54.6$ mm. As for η_c, it was set through linear interpolation from the available test results of Sect. 3. Thus, one value of η_c was defined for each protocol type: $\eta_c = 398$ for a LC-like protocol, and $\eta_c = 1{,}321$ for a HC-like protocol.

Figure 8b shows the time history of δ_B normalized by δ_{By} for one 1st-story BRB under the MCE-IV input motion. It can be concluded from Fig. 8b that demands on BRBs correspond better to a LC-like loading protocol, rather than to a fatigue loading type, and therefore $\eta_c = 398$ was used for the DI calculation shown in the last column of Table 7. Moreover, the force-deformation response obtained in the RGII tests (details can be found in [23]) is also depicted in Fig. 8a for comparison. As seen

Table 7 Maximum earthquake demand and DI calculation

BRB ID	δ_{Bmax} (mm)	δ_{By} (mm)	δ_c (mm)	ε_{pmax} (%)	η_{max}	α	F_1	F_2	DI
1–X	12.99	3.74	54.60	0.54	148.7	0.419	0.238	0.374	0.31
2–X	14.94	3.74	54.60	0.62	149.3	0.407	0.274	0.375	0.33
3–X	13.31	3.74	54.60	0.55	156.3	0.417	0.244	0.393	0.32
4–X	15.24	3.74	54.60	0.64	168.3	0.405	0.279	0.423	0.36
1–Y	12.72	3.74	54.60	0.53	116.9	0.421	0.233	0.294	0.27
2–Y	11.86	3.74	54.60	0.49	107.2	0.426	0.217	0.269	0.25
3–Y	13.02	3.74	54.60	0.54	121.0	0.419	0.238	0.304	0.28
4–Y	11.73	3.74	54.60	0.49	112.8	0.427	0.215	0.283	0.25

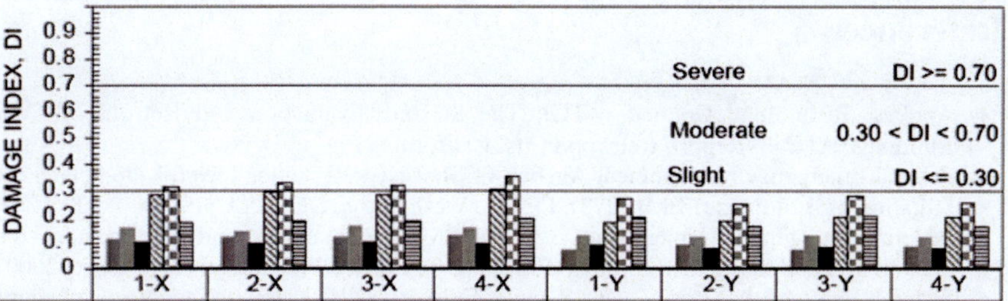

	1-X	2-X	3-X	4-X	1-Y	2-Y	3-Y	4-Y
LP-DE	0.12	0.12	0.13	0.13	0.07	0.09	0.07	0.09
IV-DE	0.15	0.15	0.16	0.16	0.13	0.12	0.13	0.12
A-DE	0.10	0.10	0.11	0.10	0.09	0.08	0.09	0.07
LP-MCE	0.28	0.29	0.29	0.31	0.18	0.19	0.20	0.20
IV-MCE	0.31	0.33	0.32	0.36	0.27	0.25	0.27	0.25
A-MCE	0.17	0.19	0.18	0.19	0.20	0.16	0.20	0.16

Fig. 9 Calculated 1st-story BRB DIs

in Fig. 8a, the normalized force-deformation demand on BRBs is quite below the obtained BRB capacity; so, low values for DI shown in Table 7 agree reasonably well. Based on the calculated DI values and the criteria show in Table 4, the damage level for all eight BRBs is slight, and they can be left on site for future events. Figure 9 shows the damage index for the eight BRBs of the first floor under both seismic levels; the first row indicates the four BRBs located in each direction (X, Y).

5 Concluding Remarks

A methodology for determining the seismic damage index (DI) for BRBs has been introduced and validated through the nonlinear dynamic analysis of a sample building. The proposed damage index was established based on a set of experimental results obtained from a series of tests carried out on a BRB prototype developed in Colombia. From the results of this study, the proposed methodology for the damage index evaluation has proven to be a potential tool for the post-earthquake evaluation of a BRB element, and for determining whether the BRB should be replaced or left on site for future events. Further study for different types of BRB elements, other loading protocols and more building archetypes is needed to gain more insight into this subject, and to validate the applicability of the proposed damage index. Finally, the results of this study are expected to contribute to ongoing efforts on improving PBSD methodologies.

References

1. Applied Technology Council (ATC): The Seismic Evaluation and Retrofit of Concrete Buildings, ATC 40 Report, Redwood City, California (1996)
2. Federal Emergency Management Agency-FEMA: NEHRP Guidelines for the Seismic Rehabilitation of Buildings (FEMA 273), FEMA, Washington, DC, USA (1997)
3. Federal Emergency Management Agency-FEMA: Prestandard and Commentary for the Seismic Rehabilitation of Buildings (FEMA 356), FEMA, Washington, DC, USA (2000)
4. Federal Emergency Management Agency-FEMA: NEHRP Recommended Provisions for Seismic Regulations for New Buildings and other Structures (FEMA 450), FEMA 2003 Edition, Washington, DC, USA (2003)
5. American Society of Civil Engineers: Seismic Evaluation and Retrofit of Existing Buildings, ASCE/SEI 41-17, Reston, Virginia, USA (2017)
6. Park, Y.J., Ang, A.: Mechanistic seismic damage model for reinforced concrete. J. Struct. Eng. **111**(4), 740–775 (1985)
7. Mehanny, S.F., Deierlein, G.G.: Modeling of assessment of seismic performance of composite frames with reinforced concrete columns and steel beams, Report No. 135, the John A. Blume Earthquake Engineering Center, Stanford University (2000)
8. Cao, V.V., Ronagh, H.R., Ashraf, M., Baji, H.: A new damage index for reinforced concrete structures. Earthq. Struct. **6**(6) (2014).
9. Edén, B.M., Gilmore, T., Bojórquez Mora, A., Gómez, J.R., Sonia, E.: Consideración explícita del daño acumulado en el diseño sísmico de estructuras a través de factores de reducción de resistencia por ductilidad. Ingeniería Sísmica **80**, 31–62 (2009)
10. Montiel-Ortega, M.A., Terán-Gilmore, A.: Evaluación y comparación de la confiabilidad de edificios de 24 niveles estructurados con contravientos tradicionales y con contravientos restringidos contra pandeo, XVI Congreso Nacional de Ingeniería Estructural (CD), Veracruz, Veracruz, México (2008)
11. Chang, R.W.K., Albermani, F.: Experimental study of steel slit damper for passive energy dissipation. Eng. Struct. **30**(4), 1058–1066 (2008)
12. Iwata, M., Murai, M.: Buckling-restrained brace using steel mortar planks: performance evaluation as a hysteretic damper. Earthq. Eng. Struct. Dyn. **35**(14), 1807–1826 (2006)
13. Wada, A., Nakashima, M.: From infancy to maturity of buckling restrained braces research. In: The 13th World Conference on Earthquake Engineering, Vancouver, Canada (2004)
14. Whittaker, A.S., Bertero, V.V., Alonso, J., Thompson, C.: Earthquake simulator testing of steel plate added damping and stiffness elements, Report No. UCB/EERC-89/02, Earthquake Engineering Research Center, University of California, Berkeley, California (1989)
15. Oviedo, J.A., Midorikawa, M., Asari, T.: Optimum strength ratio of buckling-restrained braces as hysteretic dissipation devices installed in R/C frames. In: The 14th World Conference on Earthquake Engineering, Vancouver, Canada (2008)
16. Oviedo-A, J.A., Midorikawa, M., Asari, T.: Earthquake response of ten-story story drift-controlled reinforced concrete frames with hysteretic dampers. Eng. Struct. **32**(6), 1735–1746 (2010)
17. Oviedo, J.A.: Influence of the story stiffness of reinforced concrete frame with proportional hysteretic dampers on the seismic response. Revista EIA **17**, 121–137 (2012)
18. Oviedo, J., Buitrago, J., Patiño, J., Hoyos, D.: Evaluación experimental del desempeño de un disipador de energía por deformación tipo riostra. In: VII Congreso Nacional de Ingeniería Sísmica, Bogotá, Colombia (2015)
19. Oviedo, J., Ortiz, N., Blandón, C.: Evaluación experimental del comportamiento de riostras restringidas contra pandeo fabricadas en Colombia. In: VIII Congreso Nacional de Ingeniería Sísmica, Barranquilla, Colombia (2017)
20. Oviedo-Amézquita, J.A., Jaramillo-Santana, N., Blandón-Uribe, C.A.: Evaluación experimental bajo condiciones de fatiga de riostras restringidas contra pandeo fabricadas en Colombia/Fatigue experimental evaluation of buckling-restrained braces fabricated in Colombia. In: IX Congreso Nacional de Ingeniería Sísmica, Cali, Colombia (2019)

21. Speicher, M.S., Harris, J.L.: Collapse prevention seismic performance assessment of new buckling restrained braced frames using ASCE 41. Eng. Struct. **164**, 274–289 (2018)
22. Oviedo-Amézquita, J.A.: Desempeño de un Elemento Estructural de Disipación de Energía por Deformación para F'C Control y Diseño de Estructuras SAS - FASE I, Reporte por F'C SAS, Medellín, Colombia (2015)
23. Oviedo-Amézquita, J.A.: Desempeño de un Elemento Estructural de Disipación de Energía por Deformación para F'C Control y Diseño de Estructuras SAS - FASE II, Reporte por F'C SAS, Medellín, Colombia (2017)
24. Oviedo-Amézquita, J.A.: Desempeño de un Elemento Estructural de Disipación de Energía por Deformación para F'C Control y Diseño de Estructuras SAS - FASE III, Reporte por F'C SAS, Medellín, Colombia (2018)
25. Asociación Colombiana de Ingeniería Sísmica -AIS-: Reglamento Colombiano de Construcciones Sismo Resistente NSR-10. Bogotá D.C., Colombia (2017)
26. American Institute of Steel Construction, Inc.: Seismic Provisions for Structural Steel Buildings, ANSI/AISC 341-05, Chicago, Illinois, USA (2005)
27. Tsai, K., Lai, J., Hwang, Y., Lin, S., Weng, C.: Research and application of double-core buckling restrained braced in Taiwan. In: The 13th World Conference on Earthquake Engineering, Vancouver, Canada (2004)
28. Usami, T., Wang, C., Funayama, J.: Low-cycle fatigue tests of a type of buckling restrained. In: The 12th East Asia-Pacific Conference on Structural Engineering and Construction, Hong Kong, China (2011)
29. American Institute of Steel Construction, Inc.: Seismic Provisions for Structural Steel Buildings, ANSI/AISC 341-16, Chicago, Illinois, USA (2016).
30. American Institute of Steel Construction, Inc.: Seismic Provisions for Structural Steel Buildings, ANSI/AISC 341-10, Chicago, Illinois, USA (2010)
31. Takeuchi, T., Wada, A.: Buckling-restrained Braces and Applications. The Japan Society of Seismic Isolation, JSSI, Tokyo, Japan (2017)
32. Wei, C., Tsai, K.: Local buckling of buckling-restrained braces. In: Proceedings of the 14th World Conference on Earthquake Engineering, Beijing, China (2008)
33. Fahnestock, L.A., Sause, R., Ricles, J.M.: Analytical and large-scale experimental studies of earthquake-resistant buckling-restrained braced frame systems. ATLSS Reports 06-01. https://preserve.lehigh.edu/engr-civil-environmental-atlss-reports/71 (2006)
34. American society of Civil Engineers: Minimum Design Loads and Associated Criteria for Buildings and Other Structures (ASCE/SEI 7-16), Reston, VA, USA (2016)
35. Arias, A.: A Measure of Earthquake Intensity, Seismic Design for Nuclear Power Plants (Hansen, R.J., ed.), pp. 438–483. The MIT Press, Cambridge, MA (1970)
36. Computers & Structures, Inc., ETABS v17.0.1 (2017)

An Energy-Based Prediction of Deformation Demand on Low- to Mid-Rise R/C Buildings with Hysteretic Dampers

J. A. Oviedo-Amezquita, S. Henao-Munoz, and A. M. Bernal-Zuluaga

Abstract This study focuses on frame buildings equipped with hysteretic dampers, and presents an energy-based methodology to estimate the earthquake deformation demand after damper installation. The proposed methodology considers not only structural and dynamic characteristics of the main frame and dampers, but also input ground motion characteristics. For this purpose, five R/C building structures with 2, 4, 6, 8 and 10 stories were designed according to the Colombian seismic code, and further converted into equivalent single-degree-of-freedom (SDOF) system models. Hysteretic dampers with varying mechanical properties were then installed into the SDOF models, and the models were then subjected to a series of 30 input ground motions. The input motions were modified to represent the seismic intensity given in the Colombian code and to grant certain control over the input energy. Furthermore, a discussion over the results of a three-dimension sample R/C frame building is presented. The predicted deformation demand obtained from the proposed methodology was compared with that obtained from the analysis. The methodology was shown to be useful for the preliminary assessment of the earthquake response of frame buildings with dampers.

Keywords Hysteretic damper · Drift demand · Input energy · Reinforced concrete frame · Earthquake response prediction

1 Introduction

Seismic protection techniques such as base isolation and energy dissipation systems focus on the direct control over the seismic deformation demand to building structures, reducing therefore the structural and non-structural seismic damage. Although these systems have been successfully applied to building structures worldwide, the application in Colombia is limited to a few cases. Two of the main reasons of this situation are: (i) an apparent complexity in the design process (including elaborate

J. A. Oviedo-Amezquita (✉) · S. Henao-Munoz · A. M. Bernal-Zuluaga
F'C Control & Design of Structures SAS, Medellin, Colombia
e-mail: joviedo@efeprimace.co

A. Benavent-Climent and F. Mollaioli (eds.), *Energy-Based Seismic Engineering*, Lecture Notes in Civil Engineering 155,
https://doi.org/10.1007/978-3-030-73932-4_13

analytical models and techniques) together with levels of uncertainty on the performance of the devices, and (ii) the lack of an appropriate amount of time allocated in structural design offices for the analysis and design of buildings that incorporate control techniques [1, 2]. In fact, the later reason has become a decision-making aspect when it comes to proposing, to a project owner, the use of a particular seismic control technique; a project rarely has enough time for appropriate design process. Therefore, it is still relevant for design practitioners in many countries to count on simplified methodologies, particularly at the preliminary stage of the design of a building structure [3].

To date, simplified methodologies such as the use of an equivalent single-degree-of-freedom (SDOF) system, play an important role in the engineering practice as a tool for the seismic evaluation of building structures. Several approaches have been reported in the literature for the use of SDOF models [4–7]; here one of the authors proposed an equivalent SDOF system to represent the behavior of reinforced concrete (R/C) frame buildings equipped with hysteretic dampers [3, 8, 9]. Unlike commonly used SDOF system models, the model proposed by Oviedo et al. [8] takes into account the difference of hysteretic behavior between the R/C main frame and the damper system.

Hysteretic dampers are one of the most prevalent energy dissipating devices used for the seismic protection of building structures. These additional structural elements are incorporated into a main frame structure to dissipate part of the vibration energy imposed by ground motions, reducing therefore the seismic demand on the elements of the main frame. In past years, it has been reported a series of efforts aimed at developing hysteretic dampers in Colombia, as well as at introducing and promoting their use in Colombian buildings [10, 11]. These works have certainly gained the attention of the engineering and construction community in Colombia.

Recently, two of the authors studied the behavior of the seismic deformation demand on buildings that incorporate hysteretic dampers. In their work, an analytical procedure for estimating the deformation demand after installing dampers was introduced. The procedure establishes that the input energy imparted to a frame-damper system is, somehow, proportional to the input energy imparted to the main frame prior to damper installation. In other words, the seismic input energy after damper installation equals the input energy before damper installation multiplied by the factor φ. The factor φ represents the variation of the seismic input energy due to the extent of inelastic response, ground motion characteristics and structural characteristics of the building. Results showed a reasonably good correlation between the deformation demand obtained from nonlinear dynamic analysis and the estimated value. However, two main limitations were reported in the study. The first one has to do with the lack of a theoretical expression completely independent of the results of dynamic analysis. The second one has to do with a significant variation in the energy input among the ground motions used for that work; which was mainly because all source records were modified to match the acceleration design spectrum. Therefore, the present study extends this previous work by including (i) a better control over the seismic input energy and deformation demand, (ii) a larger series of ground acceleration records for nonlinear analysis, (iii) an evaluation of the definition of the factor φ, and (iv) a practical expression for estimating the seismic deformation after installing dampers to an R/C main frame.

For this purpose, five R/C frame buildings with 2, 4, 6, 8 and 10 stories were studied as representative of low- to mid-rise building structures. Buildings were designed according to the Colombian seismic code (NSR-10) [12], and further converted into equivalent SDOF systems according to the methodology proposed by Kuramoto et al. [7]. Hysteretic dampers with different mechanical properties were installed into the equivalent SDOF models, and a series of nonlinear dynamic analyses were performed both on the SDOF systems of the building without dampers as well as on the buildings after damper installation. A total of 30 input ground motions were selected and modified to represent the seismic intensity given the in Colombian code, and to grant control over the seismic input energy and deformation demand on the systems; this is a very important aspect for getting adequate and comparable response quantities. After performing the analyses, the behavior of deformation demand and the definition of the factor φ was investigated, and a methodology for predicting the seismic deformation demand was then proposed.

2 Frame Building Models

2.1 R/C Frame System

The five three-dimension moment-resistant, strong-column and weak-beam R/C main frames considered in this study are shown in Fig. 1. The vertical load (dead + live) per unit area is assumed to be the same for all stories with a typical floor load of 9.61 kN/m^2. The vertical load for the roof, however, was set to 7.75 kN/m^2. Prior to damper installation, the structural design of all five buildings was established based on the current Colombian seismic code NSR-10. Details of the structural design and design parameters of the studied buildings can be found in [10]. Table 1 summarizes the structural and dynamic characteristics of the R/C main frames.

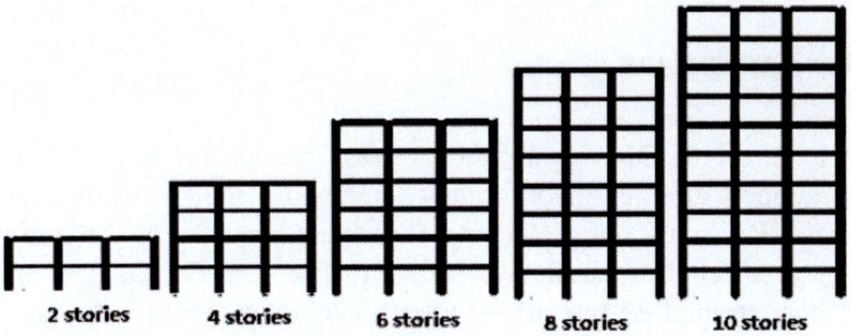

Fig. 1 Elevation of the analyzed R/C main frames

Table 1 Structural properties of analyzed R/C main frames

# Stories	Total height (m)	Period (seg)	Weight (kN)	Beams		Column	
				Cross section (cm)	Concrete strength (MPa)	Cross section (cm)	Concrete strength (MPa)
2	6.60	0.44	4085	30 × 35	21	40 × 40	21
4	13.20	0.73	9720	40 × 40	21	45 × 45	28
6	19.80	0.87	16,458	40 × 45	21	60 × 60	28
8	26.40	1.03	23,977	40 × 50	21	70 × 70	28
10	33.00	1.19	36,464	40 × 50	21	100 × 100	28

Table 2 Structural properties of the SDOF systems

# Stories	W (kN)	Period T (s)	Q_{Fc} (kN)	Δ_{Fc} (cm)	Q_{Fy} (kN)	Δ_{Fy} (cm)
2	3,532	0.41	263	0.31	967	2.26
4	7,984	0.70	529	0.81	1,645	5.37
6	13,048	0.78	665	0.77	2,608	7.18
8	18,680	0.84	891	0.83	3,633	8.11
10	26,788	0.97	991	0.86	4,371	8.47

2.2 3-D R/C Frame Conversion into an Equivalent SDOF System

The five buildings described in the previous section are converted into equivalent SDOF systems models according the methodology proposed by Kuramoto et al. [7]. Table 2 summarizes the structural and dynamic characteristics of the five SDOF models. These five models are later used for nonlinear dynamic analyses to serve as a reference point for comparison with the response of the buildings with dampers.

2.3 Damper System

Having defined the SDOF system models that represent the R/C main frames, the hysteretic dampers are then incorporated. The restoring force characteristics of the entire system (R/C main frame + dampers) is then idealized as the combination of two springs connected in parallel, as shown in Fig. 2. Here, Q_S, Q_{Fy} and Q_{Dy} are the yield shear strength of the entire system, R/C main frame and damper system, respectively. Δ_{Fc}, Δ_{Fy}, Δ_{Dy}, Δ_{max}, μ_F, μ_D are the cracking story drift, the yield story drift of the R/C main frame, the yield story drift of the damper system, the maximum story drift, the ductility of the R/C main frame and the ductility of the damper system,

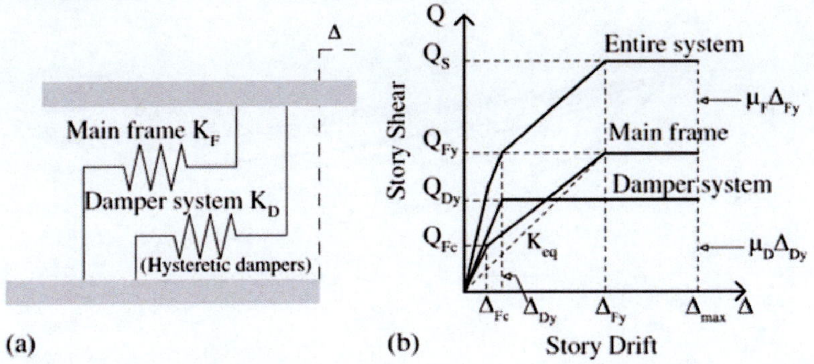

Fig. 2 SDOF system model for an R/C main frame with a damper system: **a** schematic configuration and **b** restoring force characteristics (taken from [3])

respectively. α and ρ define the shear at the cracking point Q_{Fc} and the equivalent stiffness K_{eq} for the R/C main frame, respectively.

The structural characteristics of the damper system, i.e., yield strength and stiffness, are assumed to be proportional to those of the R/C main frame. To determine the yield strength Q_{Dy} and the stiffness K_D of the damper system, the damper strength ratio β (hereafter the strength ratio) and the damper yield drift ratio ν (hereafter the drift ratio) are used. Thus, referring to Fig. 2, Q_{Dy}, Q_{FY}, Δ_{FY} and Δ_{DY} are related by:

$$Q_S = Q_{Fy} + Q_{Dy} \tag{1}$$

$$Q_{Dy} = \beta Q_{Fy} \tag{2}$$

$$\Delta_{Dy} = \nu \Delta_{Fy} \tag{3}$$

To define the restoring force characteristics of the SDOF model for the entire system, the value of ν was varied from 0.1 to 1.0 with an interval of 0.1, and the value of β varied from 0.1 to 0.9 with an interval of 0.1. Finally, the elastic stiffness of the dampers system K_D is determined by:

$$K_D = Q_{Dy}/\Delta_{Dy} \tag{4}$$

With damper installation, the total stiffness and strength of the entire system are increased, and the dynamic properties are then modified. Equation (5) represents the value of the natural period of the entire system T as a function of β, ρ, ν, and the natural period of the R/C main frame T_0. Figure 3 shows the variation of the natural period T of analyzed models after installing hysteretic dampers, and Fig. 4 shows the variation of the stiffness of the entire system K_S. In general, it can be seen that the natural period shortens with increasing values of β and decreasing values of ν.

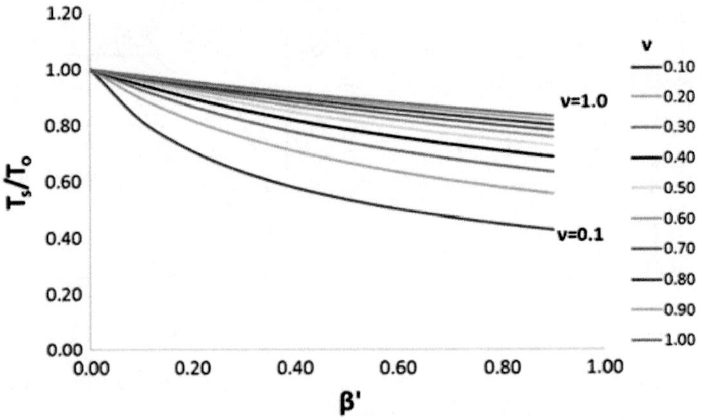

Fig. 3 Variation of natural period of analyzed SDOF models with respect to β and υ

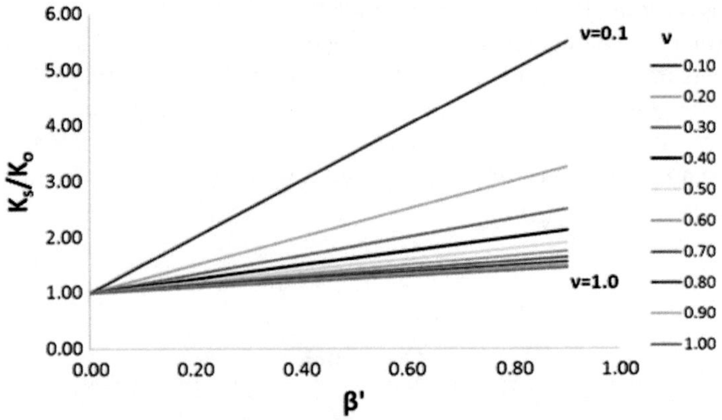

Fig. 4 Variation of total stiffness of analyzed SDOF models with respect to β and υ

$$T = T_0\sqrt{\frac{1}{1+\rho\frac{\beta}{\upsilon}}} \tag{5}$$

3 Earthquake Response Analysis

3.1 Input Earthquake Motions

The input ground motions used for the nonlinear time-history analyses can be found in [13]. All source records represent near-fault type earthquakes, categorized within a type C soil (dense soils or soft rock, 360 m/s $\leq Vs_{30} \leq$ 760 m/s, where Vs_{30} is the average shear wave velocity down to a depth of 30 m) according to NSR-10. Thus, all records have an epicentral distance of no more than 20 km, and present a single velocity pulses of significant amplitude. All acceleration records were obtained from the Pacific Earthquake Engineering Research Center (PEER) website. As mentioned earlier, unlike the previous work done by two of the authors, this study considers a better control over the input energy imparted to the buildings. This is achieved by modifying the source records so that the response spectrum of a source record matches the design velocity spectrum given in the Colombian code. The use of a design velocity spectrum instead of the design acceleration spectrum, leads to a relative uniform energy input among all modified records. Figure 5 shows the velocity spectra and the energy spectra as an equivalent velocity V_{eq}. It can be seen a relatively small variation of the input energy. These 30 modified records are then grouped into Case I records.

Another group of modified records was also studied. Case II ground motions include the 30 records, but modified to match a design energy spectrum. Since the Colombian code does not consider this design spectrum, a design energy spectrum was then obtained through the velocity design spectrum of the Colombian code. More details on the assumed design energy spectrum can be found in [13]. Figure 6 shows the velocity and energy spectra of the Case II records (dashed line indicates the target design spectrum). It is clear after comparing Case I and Case II response spectra that

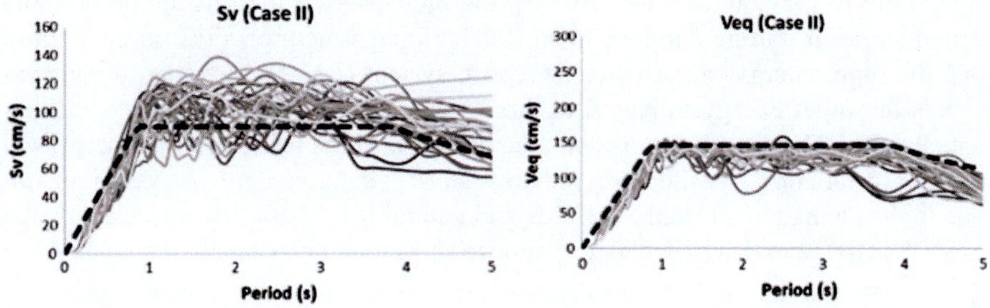

Fig. 5 Response spectra of Case I records. Left: velocity spectra. Right: energy spectra

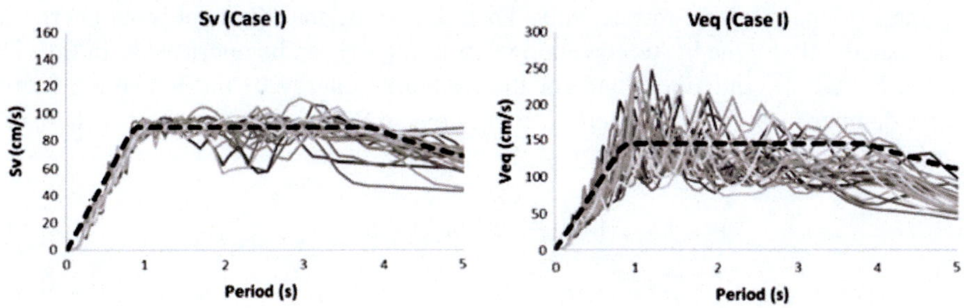

Fig. 6 Response spectra of Case II records. Left: velocity spectra. Right: energy spectra

Case II records grant less variation of the input energy, and therefore, a more uniform deformation demand can be expected among all 30 records. Consequently, the Case II records were used for the nonlinear dynamic analyses.

3.2 Nonlinear Dynamic Analysis

In the numerical analyses of the SDOF systems, a 2-spring SDOF model with the degrading trilinear Takeda model to represent the hysteretic behavior of the R/C main frame and with a bilinear model to represent dampers was used, according to [3, 8]. Damping ratio of 3% of the critical, integration time step of 0.005 s, and a post-elastic stiffness ratio of 0.01 was assumed in all analyses. The series of analyses correspond to the following cases: (1) five numbers of stories (n = 2, 4, 6, 8, 10), (2) nine strength ratios ($\beta = 0.1$–0.9), (3) ten drift ratios ($\nu = 0.1$–1.0), and (4) the 30 modified input ground motions listed [13]. In total more than 27,300 analyses were performed.

4 Prediction of Deformation Demand

According to previous studies [10, 13], the prediction of the seismic deformation demand after installing dampers to an R/C building structure works on the premise that the input energy imparted to the entire system (R/C main frame + dampers) equals the input energy imparted to the R/C main frame multiplied by a factor φ (see Eq. (6)) The factor φ represents the variation in the seismic input energy with damper installation. As shown in previous works [10], the input energy might vary due to the change of strength, stiffness and natural period after damper installation, as well as to the extent of inelastic response and ground motion characteristics.

$$E_{i0} = \varphi E_i \tag{6}$$

where, E_i and E_{i0} stand for the input energy of the entire system and the input energy of the R/C main frame, respectively. Then, assuming that the total input energy is dissipated only by the hysteretic energy work, Eq. (6) can be rewritten as in Eq. (7). Here, E_{HF}, E_{HF0} and E_{HD} stand for the hysteretic energy of the R/C main frame after damper installation, hysteretic energy of the R/C main frame before dampers installation, and hysteretic energy of the damper system.

$$E_{HF0} = \varphi(E_{HF} + E_{HD}) \tag{7}$$

Moreover, this hysteretic energy can be obtained from the work of the entire system under monotonically increasing loads, as shown in Fig. 7. In other words, the hysteretic energy is obtained through the area under the force-deformation curve.

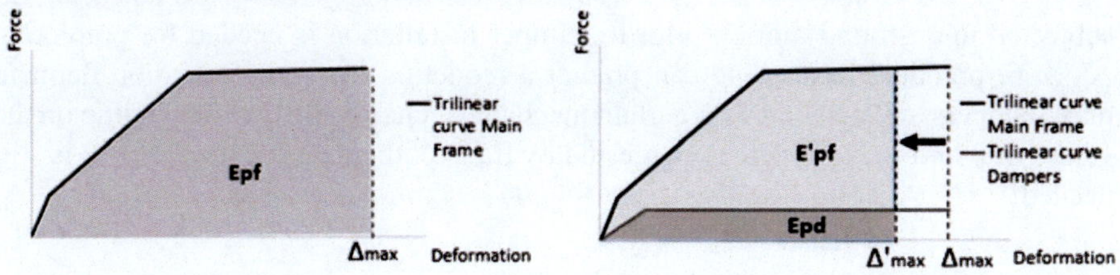

Fig. 7 Force-deformation monotonic curves. Left: w/o dampers, and right: with dampers

Here, it is important to note that the relationship depicted in Fig. 7 is only a reasonable approximation of the comparative behavior of the energy absorbed (Ep) by the interaction of the two systems (R/C main frame and dampers). Thus, since no time-depending response factors can be considered, the use of the factor φ is needed. In Fig. 7, Δ_{max} and Δ'_{max} stands for the maximum deformation of the R/C main frame and the maximum deformation of the entire system after damper installation, respectively. From Fig. 7 two cases apply: Case (1) applies when both the R/C main frame and the damper system behave inelastically ($\Delta'_{max} > \Delta_{Fy}$ and $\Delta'_{max} > \Delta_{Dy}$), and Case (2) applies when only the damper system behaves inelastically $\Delta'_{max} < \Delta_{Fy}$ and $\Delta'_{max} > \Delta_{Dy}$). Therefore, Eq. (7) can be rewritten, depending on the previous cases, as:

For Case (1)

$$\frac{\Delta'_{max}}{\Delta_{max}} = \frac{1}{\varphi(1+\beta)} + \frac{\nu\beta}{2\mu_{F0}(1+\beta)} + \frac{(1-\varphi)(\alpha-\alpha\rho-1)}{2\mu_{F0}\varphi(1+\beta)} \tag{8.1}$$

For Case (2)

$$A\left(\frac{\Delta'_{max}}{\Delta_{max}}\right)^2 + B\left(\frac{\Delta'_{max}}{\Delta_{max}}\right) + C = 0 \tag{8.2}$$

where,

$$A = \varphi\frac{1-\alpha}{1-\alpha\rho}\mu_{F0} \tag{8.2.1}$$

$$B = \varphi\left(1 - \frac{1-\alpha}{1-\alpha\rho} + \alpha - \alpha\rho\frac{1-\alpha}{1-\alpha\rho} + 2\beta\right) \tag{8.2.2}$$

$$C = \frac{1}{\mu_{F0}}\left(\alpha\rho\left(\varphi\frac{1-\alpha}{1-\alpha\rho} - \varphi + 1\right) - 2\mu_{F0} - \alpha - \varphi\nu\beta + 1\right) \tag{8.2.3}$$

The series of equations above presented are thus used for the prediction of the maximum deformation demand Δ'_{max} after installing dampers to an R/C main frame. It is important to note that a reasonable estimate of the response of the R/C main frame subjected to a ground motion prior to damper installation is needed for predicting Δ'_{max}. In practical terms, one can predict a reduction in the deformation demand once dampers are installed with certain mechanical characteristics, depending on the values of β and ν. However, as suggested by Eq. (8), the value of the factor φ is still needed.

4.1 Evaluation of Factor φ

As previously mentioned, the value of factor φ is needed for the prediction of Δ'_{max}, and depends on structural, dynamic and ground motions parameters. To solve this issue, the behavior of the factor φ was investigated through both the variation of the input energy and the variation of the hysteretic energy. The top row of Fig. 8 depicts the relationship between the exact value of φ (hereafter φ_{real}) obtained from the results of the nonlinear analyses through Eq. (6) and the value of φ required to satisfy the series of Eq. (8) (hereafter φ_{req}). On the other hand, the bottom row of Fig. 8 depicts the relationship between φ_{real} obtained through Eq. (7) and the value of φ_{req}. The value of φ_{req} can be readily calculated now that both Δ_{max} and Δ'_{max} are known from the analyses. Figure 8 clearly indicates that it is rather complicated to establish an adequate correlation between both factors (φ_{real}, φ_{req}), no matter the definition used for defining the value of φ.

Another aspect to consider is the evaluation of φ_{real}. At the preliminary stage of a structural design of a building structure equipped with dampers, it is not possible to determine the value of φ_{real} unless several nonlinear analyses are performed. Therefore, a theoretical value for φ_{real} (hereafter φ_{theo}) and a correlation between the two factors is needed (φ_{real}, φ_{theo}). Gómez [13] studied different expressions for input energy evaluation proposed in the literature, and showed that the equation proposed by Housner [14] led to a better correlation, compared to other definitions. Further details can be found in [13]. On the other hand, a theoretical value for φ_{real} seems not available through the hysteretic energy variation, as in Eq. (7).

4.2 Simplified Prediction

From the results of the previous section, it was concluded that a proper expression for the factor φ is somehow complicated to achieve. It was also concluded that determining the value of φ from the input energy or hysteretic energy has a major issue: there is a significant difference between the behavior of both input and hysteretic energy and the behavior depicted in Fig. 7 due to the nature of loading. Moreover, both cases lack a proper theoretical value which one can use in the preliminary stage of a structural design. Thus, after identifying other directions for solving the issue of the factor φ, a simplified prediction is then presented. Equation (7) can be rewritten as in Eq. (9) by using the areas under the force-deformation curve in Fig. 7. Here,

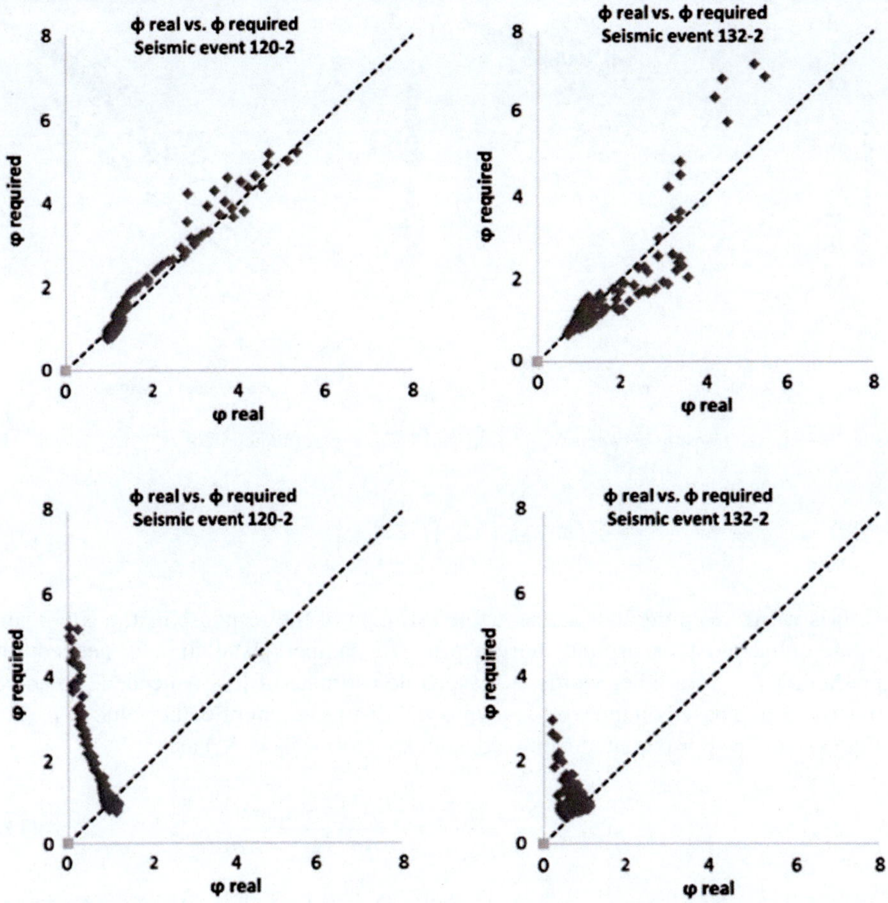

Fig. 8 Relationship between φ_{real} and φ_{req}. Top row: value of φ_{real} obtained through Eq. (6). Bottom row: value of φ_{real} obtained through Eq. (7)

A_F, A_{F0} and A_D stand for the absorbed energy by the R/C main frame after damper installation, absorbed energy by the R/C main frame before damper installation, and the absorbed energy by the damper system.

$$A_{F0} = A_F + A_D \tag{9}$$

Figure 9 shows the relationship between $\Delta'_{max}/\Delta_{max}$ and T/T_0 for the series of ground motions used. The data shown in Fig. 9 correspond to average values among all ground motions. It can be seen that the ratio $\Delta'_{max}/\Delta_{max}$ tends to decrease as the ratio T/T_0 decreases. Thus, the result obtained through either Eq. (8.1) or (8.2), with the value of φ set to unity, is then affected according to Eq. (10) to serve as a correction factor.

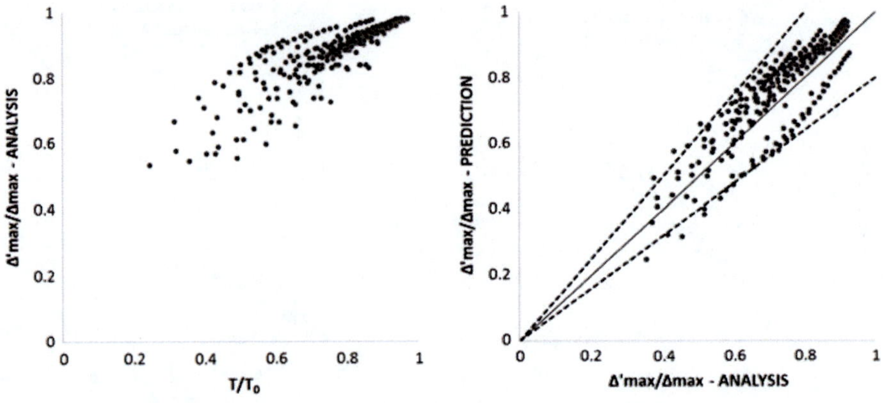

Fig. 9 Left: relationship between $\Delta'_{max}/\Delta_{max}$ and T/T_0. Right: prediction of $\Delta'_{max}/\Delta_{max}$

$$\frac{\Delta'_{max}}{\Delta_{max}} = \left(\frac{T}{T_0}\right)\left(\frac{\Delta'_{max}}{\Delta_{max}}\right)_{Eq.(8)} \tag{10}$$

It is worth recalling that a reasonable estimate of the response of the R/C main frame subjected to a ground motion prior to damper installation is needed for predicting Δ'_{max}. In other words, a reasonable estimate of μ_{F0} is needed. To solve this issue, and based on the well-known equal displacement rule, the value of μ_{F0} in Eq. (8) can be approximated using a design spectrum (S_d or S_a) as:

$$\mu_{F0} \approx \frac{Sd_{(T=T_0)}}{\Delta_{Fy}} = \frac{(T_0)^2 Sa_{(T=T_0)}}{4\pi^2 \Delta_{Fy}} \tag{11}$$

Figure 9 (right) shows the correlation between the exact value of $\Delta'_{max}/\Delta_{max}$ obtained from the nonlinear analyses and the predicted value obtained from Eqs. (8) and (10) (dashed line indicates ±20%); here, the value of μ_{F0} was determined using the Eq. (11). Values shown in Fig. 9 correspond to analysis cases with $\beta \leq 0.5$ and $\nu \leq 1.0$ according to previous studies which show that dampers are more effective within this range [3, 10, 15]. It can be clearly observed that the simplified methodology leads to predictions that can be used in the design practice for the case of SDOF systems.

4.2.1 Application on a MDOF System

It is worth noting that the aforementioned methodology has been proven useful for the case of an equivalent SDOF system that represents the behavior of a whole R/C building equipped with dampers. On the other hand, with regards to estimating the story drift demand on MDOF systems, the authors have been working on this issue through the series of Eq. (8) and the methodologies proposed in [3, 8, 15]. Herein, one six-story building sample is introduced, in which, the mechanical properties of dampers were defined according to the methodology of the constant yield story-drift ratio scheme proposed in [15]. Consequently, Fig. 10 shows the story drift-story shear relationship of the R/C main frame, obtained from a pushover analysis and using cracked concrete cross-sections, and Table 3 lists the idealized story properties of the R/C main frame from Fig. 10. The fundamental elastic periods of the R/C main frame with gross and cracked sections are 1.03 s and 1.59 s, respectively. Thus, having defined the idealized R/C main frame's story restoring-force characteristics and the global story deformation demand (story drift) of the R/C main frame under the studied ground motions, the designer can thus decide the reduction in the story drift response needed to get the structure to comply with a code-base drift limit or with any other drift limit of interest. Figure 11 shows the capacity curve and the envelope of the story drift demand on the R/C main frame (prior to damper installation). Here, it can be seen that if the maximum story drift demand was reduced by about 50%, the story drift would comply with the limit of 1.5% given by the Colombian code when cracked sections are used. A reduction of 50% means that the ratio $\Delta'_{max}/\Delta_{max}$ of the left side of Eq. (10) is set to 0.50.

In the sample building, the behavior of both systems (R/C main frame and damper system) is assumed to be inelastic; thus, the formulation for the Case (1) is used for predicting the deformation demand. Consequently, the following parameters are

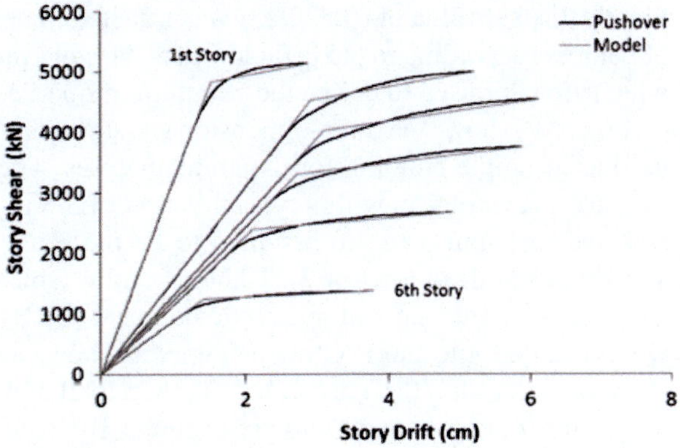

Fig. 10 Story drift-story shear curves of R/C main frame

Table 3 Idealized story properties of R/C main frame

Story	Q_{Fy} (kN)	Q_{Fu} (kN)	Q_{Fc} (kN)	Δ_{Fy} (cm)	K_{eq} (kN/cm)
1	4,830	4,878	48	1.52	3,174
2	4,520	4,565	45	2.92	1,548
3	4,020	4,060	40	3.10	1,295
4	3,310	3,343	33	2.73	1,213
5	2,380	2,404	24	2.10	1,135
6	1,240	1,252	12	1.42	875

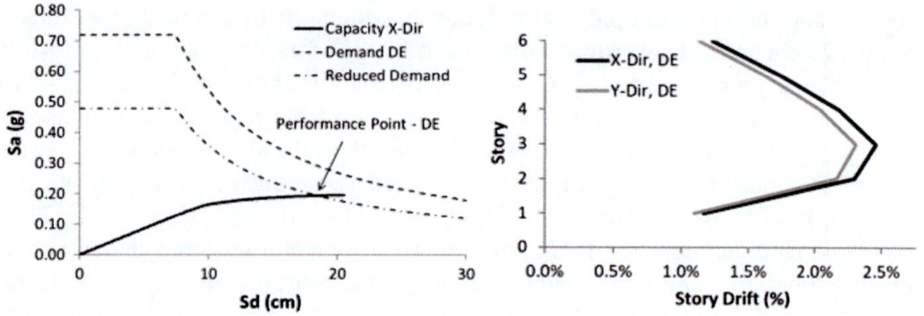

Fig. 11 Capacity curve and envelope of maximum drift demand on the R/C main frame

employed in Eq. (8.1): (i) the factor φ is set to 1.0 for ease of calculation; (ii) the factor μ_{F0} is determined from the capacity curve of the R/C main frame under the methodology of the capacity spectrum method (CSM), in which, T_0 is set to 1.59 s, Δ_{Fy} corresponds to S_{dy} in de acceleration-displacement response spectrum (ADRS) format, and S_a is calculated from the acceleration design spectrum; (iii) the factor α is set to 0.01 since the R/C cracking strength has no relevance in the sample building; and (iv) the factor ρ is taken as the square of the ratio of the elastic period to the elastic period using cracked sections, this is $(1.03/1.59)^2$. Thus, Eq. (10) leads to several couples (β, ν) suitable to obtain a reduction in the story drift demand of about 50%. Finally, the set value of (0.5, 0.25) was selected to define the mechanical properties of dampers according to [15]. Table 4 lists the story properties of the damper system, which are later used to define the structural properties of the BRB elements.

Figure 12 shows the three-dimension six-story reinforced concrete (R/C) structure used as a sample building for dynamic analyses. The symmetric plan consists of 3 by 4 bays each of 7 m with a typical height of 3.5 m. The gravitational loads (dead and live) per unit area are assumed to be the same for all stories, with a typical superimposed dead load of 3.23 kN/m^2 and a typical live load of 4.0 kN/m^2 (for hospital use). Two pairs of hysteretic dampers (BRBs in this case) in each direction were installed into the R/C main frame, as shown in Fig. 12; BRBs of each floor share the same structural properties. Thus, a total of 48 BRBs were installed into the R/C main frame, grouped into six different BRB types. The structural design was

Table 4 Idealized story properties of the damper system

Story	Q_{Dy} (kN)	Δ_{Dy} (kN)	K_D (kN/cm)	K_D/K_{eq}
1	2,415	0.38	6,348	2.0
2	2,260	0.73	3,096	2.0
3	2,010	0.78	2,590	2.0
4	1,655	0.68	2,426	2.0
5	1,190	0.52	2,270	2.0
6	620	0.35	1,750	2.0

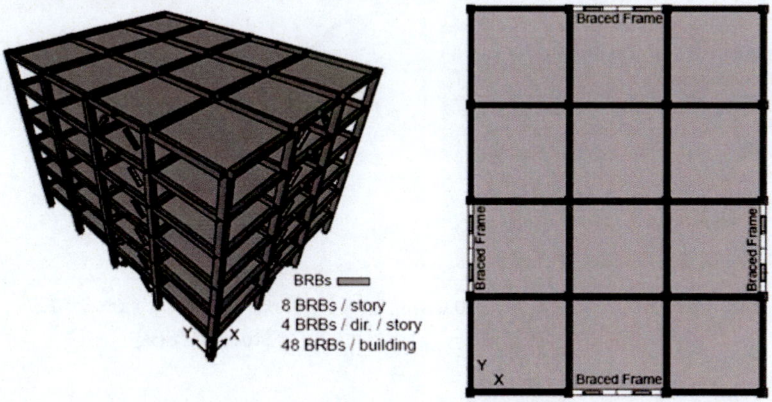

Fig. 12 Six-story sample building with hysteretic dampers (BRBs)

established based on the Colombian seismic code (NSR-10) for R/C braced moment-resisting frames, and BRBs were sized and designed according to the technical advice of the BRB manufacturer in Colombia.

Table 5 summarizes the structural properties of the sample building. In Table 5, P_{ysc}, K_{Bwp} and K_{Bp} correspond to the yield strength (for a nominal yield stress of 250 MPa), workpoint-to-workpoint axial stiffness and post-yield axial stiffness of BRB. L_{Bwp} is the workpoint-to-workpoint length of the brace. The total seismic weight of the three-dimension sample building is 29,742 kN, and the fundamental period of the entire system is 0.84 s (with cracked concrete cross-sections). A series of earthquake nonlinear dynamic analyses (NLDA) was carried out on the sample building in order to investigate the behavior of the proposed set of Eq. (8). Thus,

Table 5 Structural properties of the sample building

Element	Section geom. (m)	Concrete strength (MPa)	ϕP_{ysc} (kN)	K_{Bwp}; K_{Bp} (kN/mm)
1st to 6th story cols.	0.60 × 0.60	28	–	–
1st story braced frame cols.	0.70 × 0.70	28	–	–
1st to 6th floor beams	0.40 × 0.55	21	–	–
1st story BRBs	$L_{Bwp} = 4.95$ $L_B = 4.0$ $L_P = 2.2$	–	932	295; 8.2
2nd story BRBs		–	871	276; 7.6
3rd story BRBs		–	770	244; 6.7
4th story BRBs		–	641	203; 5.6
5th story BRBs		–	459	145; 4.1
6th story BRBs		–	243	79; 2.2

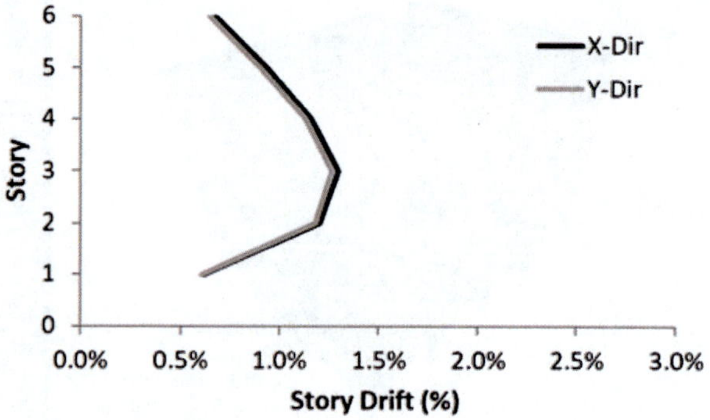

Fig. 13 Envelope of maximum drift demand on the entire system

Fig. 13 shows the envelope of the story drift response on the entire system. Here, it can be seen that the maximum story drift response was effectively reduced in a little more than 50%; this behavior suggests a good correlation between prediction and actual response.

5 Concluding Remarks

The behavior of deformation demand on R/C frame buildings with hysteretic dampers was investigated and a methodology for predicting the seismic deformation demand was introduced. Based on the results of nonlinear dynamic analyses on a set of equivalent SDOF systems and on a 3-D sample frame building, it is concluded that the proposed methodology for predicting the seismic deformation demand after installing dampers into an R/C main frame leads to adequate estimates that can be used by design practitioners at the preliminary design stage. With this estimate, the designer can then set the mechanical properties of dampers required to achieve a desired story drift limit. Finally, the methodology herein proposed is expected to contribute to ongoing efforts for the seismic response control of building structures with hysteretic dampers, and to encourage the use of hysteretic dampers in Colombia.

References

1. Oviedo-Amezquita, J.A., Uribe, P.: Seismic response control systems in building structures. Rev. EIA **6**, 105–120 (2006)
2. Oviedo-Amezquita, J.A., Uribe, P.: Metallic hysteretic dampers as a seismic response control technique in Colombian buildings. Rev. EIA **11**, 51–63 (2009)
3. Oviedo-A, J.A., Midorikawa, M., Asari, T.: An equivalent SDOF system model for estimating the response of R/C building structures with proportional hysteretic dampers subjected to ground motions. Earthq. Eng. Struct. Dyn. **40**, 571–589 (2011)

4. Freeman, S.A., Nicoletti, J.P., Tyrell, J.V.: Evaluation of existing buildings for seismic risk—a case study of Puget Sound Naval Shipyard, Bremerton, Washington. In: Proceedings of the 1st U.S National Conference on Earthquake Engineering, Washington, USA (1975)
5. Applied Technology Council (ATC): The Seismic Evaluation and Retrofit of Concrete Buildings, ATC 40 Report, Redwood City, CA (1996)
6. Federal Emergency Management Agency (FEMA): NEHRP Guidelines for the Seismic Rehabilitation of Buildings (FEMA 273). FEMA, Washington, DC, USA (1997)
7. Kuramoto, H., Teshigawara, M., Okuzono, T., Koshika, N., Takayama, M., Hori, T.: Predicting the earthquake response of buildings using equivalent single degree of freedom system. In: Proceedings of the 12th World Conference on Earthquake Engineering, Auckland (2000)
8. Oviedo-A, J.A., Midorikawa, M., Asari, T.: An equivalent SDOF system model for estimating the earthquake response of R/C buildings with hysteretic dampers. In: Proceedings of the 14th European Conference on Earthquake Engineering, Ohrid, Macedon (2010)
9. Oviedo, J.A., Londoño, C.C.: Modelo inelástico simplificado para la predicción de la respuesta sísmica de pórticos en concreto con disipadores de energía. In: VI Congreso Nacional de Ingeniería Sísmica, Medellín, Colombia (2013)
10. Henao, S., Garza, L., Oviedo, J.A.: Demanda de deformación en edificaciones con y sin disipadores de energía. In: VI Congreso Nacional de Ingeniería Sísmica, Medellín, Colombia (2013)
11. Oviedo, J.A., Buitrago, J.J., Patiño, J.C., Hoyos, D.: Evaluación experimental del desempeño de un disipador de energía por deformación tipo riostra. In: VII Congreso Nacional de Ingeniería Sísmica, Bogotá, Colombia (2015)
12. Asociación Colombiana de Ingeniería Sísmica (AIS): Reglamento Colombiano de Construcciones Sismo Resistente NSR-10. Bogotá D.C., Colombia (2017)
13. Gómez, V.: Seismic Deformation Demand in RC Frame Buildings with and without Energy Dampers. Civil Engineering Graduation Report, EIA University, Colombia (2015)
14. Housner, G.W.: Behavior of structures during earthquakes. J. Eng. Mech. Div. **85**(4), 109–129 (1959)
15. Oviedo-A, J.A., Midorikawa, M., Asari, T.: Earthquake response of ten-story story drift-controlled reinforced concrete frames with hysteretic dampers. Eng. Struct. **32**(6), 1735–1746 (2010)

Energy-Based Topology Optimization Under Stochastic Seismic Ground Motion: Preliminary Framework

Giulia Angelucci, Giuseppe Quaranta, and Fabrizio Mollaioli

Abstract The growing availability of suitable computational resources to support the design of complex and large buildings makes the topology optimization more and more attractive to achieve high structural performances while reducing the use of building materials and thus cutting the total costs. In case of buildings under dynamic loads, displacement- and acceleration-based criteria are most commonly employed in topology optimization for preventing damage in structural components and protecting high-frequency sensitive non-structural components, respectively. The present work introduces the energy-based topology optimization of large structures as a more effective design approach to mitigate damage due to earthquake. The inherent randomness of the seismic excitation is taken into account by means of the random vibration theory, in such a way to avoid the direct integration of the motion equations for a large number of records. Topology optimization is performed via Solid Isotropic Material with Penalization (SIMP) method and resorting to an analytical evaluation of the gradient. A stationary-type stochastic seismic ground motion is considered in the preliminary framework presented in this study, whereas the final case study here discussed is concerned the search of the optimal layout for a lateral resisting system in a multi-story building subjected to earthquake.

Keywords Energy-based design · Random vibrations · Solid isotropic material with penalization method · Stochastic seismic ground motion · Topology optimization

G. Angelucci · F. Mollaioli
Department of Structural and Geotechnical Engineering, Sapienza University of Rome, via Gramsci 53, 00197 Rome, Italy
e-mail: giulia.angelucci@uniroma1.it

F. Mollaioli
e-mail: fabrizio.mollaioli@uniroma1.it

G. Quaranta (✉)
Department of Structural and Geotechnical Engineering, Sapienza University of Rome, via Eudossiana 18, 00184 Rome, Italy
e-mail: giuseppe.quaranta@uniroma1.it

A. Benavent-Climent and F. Mollaioli (eds.), *Energy-Based Seismic Engineering*, Lecture Notes in Civil Engineering 155,
https://doi.org/10.1007/978-3-030-73932-4_14

1 Introduction

The design of lateral bracing systems for multi-story buildings requires the selection of a suitable pattern for the diagonals arrangement, which is commonly carried out through trial-and-error procedures requiring many iteration cycles driven by experts opinion. Despite its widespread use in practice, it is evident that this approach is neither efficient nor able to ensure that the final design solution is optimal in a given sense. In this regard, topology optimization has gained increasing interest in recent years as a powerful preliminary design tool to look for solutions that ensure the best mechanical performance while limiting the structural weight (and thus the material cost) up to a maximum threshold. Within this framework, extensive researches have been conducted in the last decades to develop well-posed topology optimization procedures, and they have been successfully employed to solve the minimum compliance problem of planar [1–3] and three-dimensional domains [4, 5] under static loads.

So far, topology optimization for multi-story buildings has been traditionally framed in a static and deterministic setting considering wind loading scenarios. However, it is well known that lateral resisting systems designed for wind action can exhibit poor performance in case of seismic excitation. In fact, the seismic response of multi-story structural systems is strongly influenced by a complex dynamic behavior in which higher modes can produce significant accelerations with consequent demands for higher performance. Since traditional approaches may fail to address important issues related to specific dynamic loading conditions and their inherent randomness, research efforts have been also spent towards topology optimization of the structures both in frequency and time domain.

Starting from the work by Díaaz and Kikuchi [6], who first dealt with the maximization of natural eigenfrequencies, several authors have developed topological design procedures for frequency domain optimization of structures in order to enhance their dynamic performance [7–9]. It is noted that the frequency domain optimization can be naturally performed via density-based formulations as the design variables operate on both stiffness and self-mass of the elements, thereby allowing the mitigation of the overall dynamic response by tuning the relevant eigenfrequencies straightforwardly. Alternative approaches to formulations based on eigenvalue analysis aim at improving the dynamic performance of structures through minimization of the dynamic compliance [10–12]. Unfortunately, this turns out to be computationally expensive for large-scale domains. To cope with this challenge, model reduction techniques based on modal decomposition methods have been introduced to manage efficiently a huge number of degrees of freedom [13, 14]. An adaptive multi-modal procedure was also developed by Angelucci et al. [15] to incorporate the dynamic response within a standard static and deterministic framework by exploiting the equivalent lateral force procedure through linear combination of higher modes. Furthermore, dynamic topology optimization problems have been successfully addressed using the random vibration theory, in which the seismic ground motion is modeled as stochastic process. In most of the existing works that employ such approach, seismic loading conditions are most commonly simulated using a stationary Gaussian-type stochastic excitation [16–18]. While significant results have been achieved using compliance formulations for dynamic optimization problems, only a few works have explored energy-based design approaches, with emphasis on the strain energy minimization [19]. However, to the best of the authors' knowledge, no previous research has been directed towards the investigation of topology optimization problems by minimizing the total energy dissipated in structural systems under seismic ground motion.

Therefore, this work aims at introducing an energy-based topological design procedure. Specifically, the main goal of the present study is the one of minimizing the energy dissipated by linear elastic systems representative of multi-story buildings subjected to earthquake through the optimization of its topology. To this end, the implemented computational approach merges the Solid Isotropic Material with Penalization (SIMP) method for topological design and the random vibration theory for the dynamic assessment under seismic excitation. It is highlighted that the random vibration theory allows to account for the inherent randomness of the seismic ground motion without the need of performing extensive Monte Carlo simulations based on direct integration of the motion equations, which is especially prohibitive for large structures. This, in turns, allows to perform the optimum design at affordable computational cost and memory savings. The proposed framework employs a gradient-based optimizer, and the relevant derivatives have been estimated analytically to speed up the optimization while ensuring superior convergence properties. The application of the energy-based formulation within topology optimization procedures is demonstrated by discussing a paradigmatic numerical example. Further ongoing developments are briefly outlined in the conclusions.

2 Topology Optimization Under Deterministic Loading Conditions

2.1 Optimum Design Under Deterministic Static Loads

In order to better understand the proposed approach for the topology optimization of structures under dynamic loads due to an earthquake, it is first considered a linear elastic structure discretized into n_d degrees of freedom and n_f finite elements subjected to static loads, whose global stiffness matrix and displacements vector are **K** and **u**, respectively. Moreover, let $\boldsymbol{\rho} = \left\{d_1 \; \ldots \; \rho_e \; \ldots \; \rho_{n_f}\right\}$ be the set of n_f design variables that define the topology of such structural system at the finite element level. Within this framework, the most common formulation of the topology optimization problem aims at minimizing the compliance of the structural system under area or volume constraint. Therefore, under the constraint that the total volume must be equal to a maximum value V_{max}, the compliance minimization problem takes the following form:

$$\min_{\boldsymbol{\rho}} \left\{\mathbf{u}^\top \mathbf{K}\mathbf{u}\right\}$$
$$\text{s.t.}$$
$$\sum_{e=1}^{n_f} V_e = V_{max} \tag{1}$$
$$\boldsymbol{\rho}_\ell \leq \boldsymbol{\rho} \leq \boldsymbol{\rho}_u,$$

where e labels the generic finite element of the mesh whose volume is V_e. The dependence from the design variables $\boldsymbol{\rho}$ is left tacit for the sake of conciseness. In Eq. (1), $\boldsymbol{\rho}_\ell$ and $\boldsymbol{\rho}_u$ are the lower and upper bound of the design variables, respectively. Note that compliance is minimized, or equivalently, stiffness is maximized, in such a way to reduce the displacements undergone by the structure in response to static loading.

2.2 Optimum Design Under Deterministic Seismic Ground Motion

It is now considered a linear viscous elastic structure whose constant global mass matrix and stiffness matrix are **M** and **K**, respectively. As regards the viscous damping matrix, the Rayleigh damping model is adopted, and thus:

$$\mathbf{C} = a_0\mathbf{M} + a_1\mathbf{K}, \tag{2}$$

where a_0 and a_1 are two positive constants calculated by imposing that the damping ratio ξ_s is the same for the first two modes of the structure. Therefore, it is obtained:

$$a_0 = \xi_s \frac{2\omega_1\omega_2}{\omega_1 + \omega_2}, \tag{3a}$$

$$a_1 = \xi_s \frac{2}{\omega_1 + \omega_2}, \tag{3b}$$

where ω_1 and ω_2 are the first two circular natural frequencies of the structure. Under a seismic ground acceleration $\ddot{u}_g$ within the time window $[0, t_f]$, the motion equation reads:

$$\ddot{\mathbf{u}} + \mathbf{M}^{-1}\mathbf{C}\dot{\mathbf{u}} + \mathbf{M}^{-1}\mathbf{K}\mathbf{u} = -\mathbf{r}\ddot{u}_g, \tag{4}$$

where the dependence from the time variable t is—and will be—omitted for the sake of conciseness. In Eq. (4), $\mathbf{u}$ are again the displacements of the structure, $\mathbf{r}$ is the incidence vector and the dots indicate the derivative with respect to time.

The topology optimization problem here proposed in case of seismic ground motion aims at finding the set of design variables $\boldsymbol{\rho}$ that solves the following optimization problem:

$$\begin{aligned} &\min_{\rho} \left\{ \int_0^{t_f} \dot{\mathbf{u}}^\top \mathbf{C} \dot{\mathbf{u}} \mathrm{d}t \right\} \\ &\text{s.t.} \\ &\sum_{e=1}^{n_f} V_e = V_{max} \\ &\boldsymbol{\rho}_\ell \leq \boldsymbol{\rho} \leq \boldsymbol{\rho}_u, \end{aligned} \tag{5}$$

where the dependence from the design variables $\boldsymbol{\rho}$ is again—and will be—left tacit for the sake of conciseness.

The optimization problem in Eq. (5) is the mirror dynamic version of the well-known compliance minimization problem under static loads given by Eq. (1). The engineering interpretation of this analogy is easily inferred by noting that

$$\mathcal{E} = \int_0^{t_f} \dot{\mathbf{u}}^\top \mathbf{C} \dot{\mathbf{u}} \mathrm{d}t \tag{6}$$

denotes the energy dissipated by the structure. Actually, this optimization problem is thus intended to minimize the velocities undergo by the structure in response to dynamic (seismic) loading in order to avoid excessive damage, under the assumption that the overall dissipative mechanism of the structure is idealized through a linear viscous damping. The hypothesis of a pure viscous damping-based dissipative behavior is a reasonable simplification for structures subjected to low-moderate seismic demands. This is acceptable for high-rise and tall buildings whose natural periods corresponding to the dominant vibration modes are far enough from the main period of the seismic ground motion, and when damage limitation is likely the governing design limit state. On the other hand, this simplification is useful to derive a semi-analytical computational approach, and can be further extended to take into account inelastic behaviors.

3 Topology Optimization Under Stochastic Seismic Ground Motion

3.1 Stochastic Seismic Ground Motion Modeling

Time-history analysis based on the deterministic representation of the seismic ground motion is not convenient for the topology optimization of large structures, since many simulations would be required to take into account the variability of the structural response due to the inherent randomness of the seismic events. Conversely, the random vibration theory is here preferred as it provides an efficient, yet rigorous, way to address this issue. In this preliminary study, the seismic ground motion is thus modeled as a stationary filtered white Gaussian noise. In the context of earthquake engineering, two commonly adopted linear filtering techniques are based on the Kanai-Tajimi filter and the Clough-Penzien filter. The Kanai-Tajimi model employs a single linear second-order filter. A refinement of this stochastic modeling is based on the Clough-Penzien filter, which results more appropriate for modeling an earthquake because the pseudo-static component is here canceled.

Assuming a stationary white Gaussian noise filtered through the Clough-Penzien filter, the random seismic ground motion is defined as follows:

$$\ddot{u}_g = \mathbf{a}_p^\top \mathbf{z}_p, \tag{7a}$$

$$\dot{\mathbf{z}}_p = \mathbf{D}_p \mathbf{z}_p + \mathbf{v}_p W, \tag{7b}$$

where

$$\mathbf{a}_p = \begin{Bmatrix} -\omega_p^2 \\ -2\xi_p\omega_p \\ \omega_k^2 \\ 2\xi_k\omega_k \end{Bmatrix}, \tag{8a}$$

$$\mathbf{z}_p = \begin{Bmatrix} u_p \\ \dot{u}_p \\ u_k \\ \dot{u}_k \end{Bmatrix}, \tag{8b}$$

$$\mathbf{D}_p = \begin{bmatrix} 0 & 1 & 0 & 0 \\ -\omega_p^2 & -2\xi_p\omega_p & \omega_k^2 & 2\xi_k\omega_k \\ 0 & 0 & 0 & 1 \\ 0 & 0 & -\omega_k^2 & -2\xi_k\omega_k \end{bmatrix}, \tag{8c}$$

$$\mathbf{v}_p = \begin{Bmatrix} 0 \\ 0 \\ 0 \\ -1 \end{Bmatrix}. \tag{8d}$$

where ω_p and ξ_p, ω_k and ξ_k are the filter parameters. Moreover, W is a zero-mean white Gaussian noise having constant power spectral density S_0. For practical applications, it is possible to evaluate S_0 as function of the peak ground acceleration $\ddot{u}_g^{max}$. In the present study, the following relationship will be employed:

$$S_0 = \frac{\left(\ddot{u}_g^{max}\right)^2}{\gamma^2 \left[\pi \omega_k \left(2\xi_k + \frac{1}{2\xi_k}\right)\right]}, \tag{9}$$

where $\gamma = 2.8$ denotes the peak factor, see for instance Liu et al. [20].

3.2 Covariance Analysis

The state-space representation of the overall dynamics is obtained by assembling the equation of motion in Eq. (4) and the filter equation in Eq. (7) as follows:

$$\underbrace{\begin{Bmatrix} \dot{\mathbf{z}}_s \\ \dot{\mathbf{z}}_p \end{Bmatrix}}_{\dot{\mathbf{z}}} = \underbrace{\begin{bmatrix} \mathbf{A}_s & \mathbf{H}_p \\ \mathbf{0}_{4\times 2n_d} & \mathbf{D}_p \end{bmatrix}}_{\mathbf{A}} \underbrace{\begin{Bmatrix} \mathbf{z}_s \\ \mathbf{z}_p \end{Bmatrix}}_{\mathbf{z}} + \underbrace{\begin{Bmatrix} \mathbf{0}_{2n_d\times 1} \\ \mathbf{v}_p W \end{Bmatrix}}_{\mathbf{f}}, \tag{10}$$

where

$$\mathbf{z}_s = \begin{Bmatrix} \mathbf{u} \\ \dot{\mathbf{u}} \end{Bmatrix}, \tag{11}$$

$$\mathbf{A}_s = \begin{bmatrix} \mathbf{0}_{n_d} & \mathbf{I}_{n_d} \\ -\mathbf{M}^{-1}\mathbf{K} & -\mathbf{M}^{-1}\mathbf{C} \end{bmatrix}, \tag{12}$$

$$\mathbf{H}_p = \begin{bmatrix} \mathbf{0}_{n_d\times 4} \\ -\mathbf{r}\mathbf{a}_p^\top \end{bmatrix}, \tag{13}$$

while $\mathbf{0}$ is a (square or rectangular) null matrix or vector and $\mathbf{I}$ is the (square) identity matrix, respectively (their size is explicated in the corresponding subscripts, where a single number is used for a square matrix). The covariance matrix of the state-space system is:

$$\mathbf{R} = \mathrm{E}\left[\mathbf{z}\mathbf{z}^\top\right] = \mathrm{E}\left[\begin{Bmatrix} \mathbf{z}_s \\ \mathbf{z}_p \end{Bmatrix} \left\{\mathbf{z}_s \ \mathbf{z}_p\right\}\right] = \begin{bmatrix} \mathbf{R}_{\mathbf{z}_s\mathbf{z}_s} & \mathbf{R}_{\mathbf{z}_s\mathbf{z}_p} \\ \mathbf{R}_{\mathbf{z}_p\mathbf{z}_s} & \mathbf{R}_{\mathbf{z}_p\mathbf{z}_p} \end{bmatrix}, \tag{14}$$

in which

$$\mathbf{R}_{\mathbf{z}_s\mathbf{z}_s} = \begin{bmatrix} \mathbf{R}_{\mathbf{u}\mathbf{u}} & \mathbf{R}_{\mathbf{u}\dot{\mathbf{u}}} \\ \mathbf{R}_{\dot{\mathbf{u}}\mathbf{u}} & \mathbf{R}_{\dot{\mathbf{u}}\dot{\mathbf{u}}} \end{bmatrix}. \tag{15}$$

The matrix $\mathbf{R}$, in turn, is the solution of the Lyapunov equation in stationary conditions, which reads:

$$\mathbf{A}\mathbf{R} + \mathbf{R}\mathbf{A}^\top + \mathbf{B} = \mathbf{0}_{2n_d+4\times 2n_d+4}, \tag{16}$$

where $\mathbf{B}$ is a matrix whose elements are equal to zero except the element whose index is $(2n_d+4, 2n_d+4)$, which is equal to $2\pi S_0$.

3.3 *Optimum Design Under Stochastic Seismic Loads*

3.4 *Optimization Problem*

By applying the mean value operator E[·] to Eq. (6), the mean total energy dissipated by the building $\mathrm{E}[\mathcal{E}]$ in case of stochastic seismic ground motion is calculated as follows:

$$\mathrm{E}[\mathcal{E}] = \mathrm{E}\left[\int_0^T \dot{\mathbf{u}}^\top \mathbf{C}\dot{\mathbf{u}} \mathrm{d}t\right] = \int_0^T \mathbf{r}^\top (\mathbf{C} \otimes \mathbf{R}_{\dot{\mathbf{u}}\dot{\mathbf{u}}}) \mathbf{r} \mathrm{d}t, \tag{17}$$

where $\otimes$ is the term-by-term product and $\mathbf{R}_{\dot{\mathbf{u}}\dot{\mathbf{u}}}$ is the covariance matrix of the structural velocity (i.e., the square block of size $n_d \times n_d$ of the matrix $\mathbf{R}$ in Eq. (15) whose elements index lies between $n_d + 1$ and $2n_d$). Hence, by recalling that a stationary ground motion is here assumed, the topology optimization problem in Eq. (5) can be simplified as follows:

$$\begin{aligned} &\min_{\boldsymbol{\rho}} \left\{\mathbf{r}^\top (\mathbf{C} \otimes \mathbf{R}_{\dot{\mathbf{u}}\dot{\mathbf{u}}}) \mathbf{r}\right\} \\ &\text{s.t.} \\ &\sum_{e=1}^{n_f} V_e = V_{max} \\ &\boldsymbol{\rho}_\ell \leq \boldsymbol{\rho} \leq \boldsymbol{\rho}_u \end{aligned} \tag{18}$$

4 Computational Aspects

4.1 *Topology Design via Solid Isotropic Material with Penalization*

A linear elastic and isotropic material is considered in this work, and its mechanical properties are linearly interpolated using the Solid Isotropic Material with Penalization (SIMP) model [21] for the purpose of the topology optimization. Throughout the finite element analysis, therefore, the elastic modulus E_e for each finite element e is manipulated using a convex power function in the following form:

$$E_e = \rho_e^p E_0, \tag{19}$$

where ρ_e is the design variable of the eth finite element and E_0 is the constant elastic modulus of the base material. A penalty factor p greater than zero, typically 3 or 5, is introduced to penalize the presence of intermediate densities in the relaxed setting and to steer the solution to binary 0–1 values. In order to avoid any singularity of the stiffness matrix, a lower bound is set on ρ_e such that $0 < \rho_e(d_j) \leq 1$, where d_j is the initial value of the design variable for all finite elements j within the neighborhood $\mathcal{N}_e$ of the finite element e under consideration. In fact, since topology optimization problems are prone to numerical instabilities, a regularization scheme [22] is adopted to ensure the existence of a solution and avoid computational anomalies, e.g., checker-board patterns and mesh-dependent designs. Based on the application of a convolution product between a linear kernel and the vector of initial design variables, a density filter is applied to obtain ρ_e according to the following relationship:

$$\rho_e = \frac{\sum_{j \in \mathcal{N}_e} H_{ej} d_j}{\sum_{j \in \mathcal{N}_e} H_{ej}}, \tag{20}$$

where H_{ej} are weighting functions defined through a linearly decaying function of fixed radius r_{min} measured from the centroid of the finite element e under consideration and the finite element $j \in \mathcal{N}_e$.

Topology optimization is performed based on the assumption that the stiffness matrix of each finite element is proportional to its fictitious elastic modulus E_e defined according to Eq. (19). Therefore, following the SIMP method, even the stiffness matrix $\mathbf{K}_e$ of the eth element is interpolated as follows:

$$\mathbf{K}_e = \rho_e^p \mathbf{K}_{e,0}, \tag{21}$$

where $\mathbf{K}_{e,0}$ is the element stiffness matrix of the solid material.

A similar approach applies to the mass matrix of the eth element $\mathbf{M}_e$, which is evaluated as follows:

$$\mathbf{M}_e = \rho_e^q \mathbf{M}_{e,0}, \tag{22}$$

where $\mathbf{M}_{e,0}$ is the mass matrix of the solid material and q is another penalty factor.

4.2 Analytical Evaluation of the Gradient

The derivative of the objective function f in Eq. (18) with respect to the design variable ρ_e is:

$$\frac{\partial f}{\partial \rho_e} = \mathbf{r}^\top \left(\frac{\partial \mathbf{C}}{\partial \rho_e} \otimes \mathbf{R}_{\dot{\mathbf{u}}\dot{\mathbf{u}}} \right) \mathbf{r} + \mathbf{r}^\top \left(\mathbf{C} \otimes \frac{\partial \mathbf{R}_{\dot{\mathbf{u}}\dot{\mathbf{u}}}}{\partial \rho_e} \right) \mathbf{r}. \tag{23}$$

The derivative of $\mathbf{C}$ with respect to ρ_e is calculated as follows:

$$\frac{\partial \mathbf{C}}{\partial \rho_e} = \frac{\partial a_0}{\partial \rho_e} \mathbf{M} + a_0 \frac{\partial \mathbf{M}}{\partial \rho_e} + \frac{\partial a_1}{\partial \rho_e} \mathbf{K} + a_1 \frac{\partial \mathbf{K}}{\partial \rho_e}, \tag{24}$$

where

$$\frac{\partial a_0}{\partial \rho_e} = \frac{\partial a_0}{\partial \omega_1}\frac{\partial \omega_1}{\partial \rho_e} + \frac{\partial a_0}{\partial \omega_2}\frac{\partial \omega_2}{\partial \rho_e}, \tag{25a}$$

$$\frac{\partial a_1}{\partial \rho_e} = \frac{\partial a_1}{\partial \omega_1}\frac{\partial \omega_1}{\partial \rho_e} + \frac{\partial a_1}{\partial \omega_2}\frac{\partial \omega_2}{\partial \rho_e}. \tag{25b}$$

The derivatives of the constants a_0 and a_1 with respect to ρ_e are calculated as follows:

$$\frac{\partial a_0}{\partial \omega_1} = \frac{2\xi_s \omega_2^2}{(\omega_1 + \omega_2)^2}, \quad \frac{\partial a_0}{\partial \omega_2} = \frac{2\xi_s \omega_1^2}{(\omega_1 + \omega_2)^2}, \tag{26}$$

$$\frac{\partial a_1}{\partial \omega_1} = \frac{-2\xi_s}{(\omega_1 + \omega_2)^2}, \quad \frac{\partial a_1}{\partial \omega_2} = \frac{-2\xi_s}{(\omega_1 + \omega_2)^2}, \tag{27}$$

whereas

$$\frac{\partial \omega_1}{\partial \rho_e} = \frac{1}{2\omega_1}\frac{\partial \lambda_1}{\partial \rho_e}, \quad \frac{\partial \omega_2}{\partial \rho_e} = \frac{1}{2\omega_2}\frac{\partial \lambda_2}{\partial \rho_e} \tag{28}$$

are the derivatives of the first two circular natural frequencies of the structure. The terms λ_1 and λ_2 are the first two eigenvalues of the following eigenproblem:

$$\boldsymbol{\phi}_k^\top \mathbf{K} \phi_k = \lambda_k \boldsymbol{\phi}_k^\top \mathbf{M} \boldsymbol{\phi}_k, \tag{29}$$

in which $\boldsymbol{\phi}_k$ is the kth eigenvector. The derivatives of λ_1 and λ_2 with respect to ρ_e are evaluated as follows:

$$\frac{\partial \lambda_1}{\partial \rho_e} = \boldsymbol{\phi}_1^\top \left[\frac{\partial \mathbf{K}}{\partial \rho_e} - \lambda_1 \frac{\partial \mathbf{M}}{\partial \rho_e}\right] \boldsymbol{\phi}_1, \quad \frac{\partial \lambda_2}{\partial \rho_e} = \boldsymbol{\phi}_2^\top \left[\frac{\partial \mathbf{K}}{\partial \rho_e} - \lambda_2 \frac{\partial \mathbf{M}}{\partial \rho_e}\right] \boldsymbol{\phi}_2, \tag{30}$$

under the assumption of distinct real eigenvalues and the mass orthonormalization condition.

The calculation of the derivatives $\partial \mathbf{R}_{\dot{\mathbf{u}}\dot{\mathbf{u}}}/\partial \rho_e$, in turn, requires the solution of the following associate stationary Lyapunov equation:

$$\mathbf{A}\frac{\partial \mathbf{R}}{\partial \rho_e} + \frac{\partial \mathbf{R}}{\partial \rho_e}\mathbf{A}^\top + \bar{\mathbf{B}} = \mathbf{0}_{2n_d+4 \times 2n_d+4} \text{ for } e = 1, \ldots, n_f, \tag{31}$$

with

$$\bar{\mathbf{B}} = \frac{\partial \mathbf{A}}{\partial \rho_e}\mathbf{R} + \mathbf{R}\frac{\partial \mathbf{A}^\top}{\partial \rho_e}. \tag{32}$$

It is also noted that:

$$\frac{\partial \mathbf{A}}{\partial \rho_e} = \begin{bmatrix} \dfrac{\partial \mathbf{A}_s}{\partial \rho_e} & \mathbf{0}_{2n_d \times 4} \\ \mathbf{0}_{4 \times 2n_d} & \mathbf{0}_4 \end{bmatrix}, \tag{33}$$

in which

$$\frac{\partial \mathbf{A}_s}{\partial \rho_e} = \begin{bmatrix} \mathbf{0}_{n_d} & \mathbf{0}_{n_d} \\ \mathbf{M}^{-1}\frac{\partial \mathbf{M}}{\partial \rho_e}\mathbf{M}^{-1}\mathbf{K} - \mathbf{M}^{-1}\frac{\partial \mathbf{K}}{\partial \rho_e} & \mathbf{M}^{-1}\frac{\partial \mathbf{M}}{\partial \rho_e}\mathbf{M}^{-1} - \mathbf{M}^{-1}\frac{\partial \mathbf{C}}{\partial \rho_e} \end{bmatrix}. \tag{34}$$

Finally, by virtue of the SIMP method, it is obtained that:

$$\frac{\partial \mathbf{K}}{\partial \rho_e} = \bigwedge_{e=1}^{n_f} \frac{\partial \mathbf{K}_e}{\partial \rho_e} = \bigwedge_{e=1}^{n_f} p\rho_e^{p-1}\mathbf{K}_{e,0}, \tag{35a}$$

$$\frac{\partial \mathbf{M}}{\partial \rho_e} = \bigwedge_{e=1}^{n_f} \frac{\partial \mathbf{M}_e}{\partial \rho_e} = \bigwedge_{e=1}^{n_f} q\rho_e^{q-1}\mathbf{M}_{e,0}, \tag{35b}$$

where $\bigwedge$ stands for standard finite element assembly operator.

5 Case Study

The topology optimization problem formulated in Eq. (18) is solved to find the optimal layout of the lateral resisting system for a multi-story building subjected to earthquake. Specifically, given a stationary seismic ground motion excitation described by means of a filtered white Gaussian noise, the optimal material distribution is attained by minimizing the total energy dissipated in the continuum design domain under a volume constraint. Without loss of generality, the design problem is solved considering a simple case study in order to demonstrate the correctness of the proposed framework. Furthermore, the geometric regularity of the building allows to impose a symmetry constraint with respect to the centerline of the model. This is highly desirable in order to achieve a pattern repetition of the structural components along the elevation, and to minimize manufacturing costs. Additionally, the introduction of a controlled regularization significantly reduces the amount of CPU-time and improves memory efficiency.

The reference model is a regular multi-story building whose height and width are equal to 40 m and 20 m, respectively. The external skin of the building is split into four panels, such that each façade can be analyzed as a continuous optimizable design domain Ω of the material distribution problem, which is discretized using 4-node quadrilateral Lagrangian (Q4) elements with unit mesh size and a uniform thickness equal to 0.15 m. A two-dimensional case study is considered to be accurate enough to capture the main peculiarities of the general problem, but the extension to a three-dimensional case study is straightforward.

The optimization is performed assuming constant penalization factors equal to $p = 3$ and $q = 1$, a projection radius $r_{min} = 1.5$ m and initial volume fraction over the domain equal to 30%. The Rayleigh damping model is assumed with a 5% damping ratio. A seismic ground motion with $\ddot{u}_g^{max} = 0.3\,g$ is assumed, with $\xi_p = \xi_k = 0.6$, $\omega_p = 1.5$ rad/s and $\omega_k = 15$ rad/s. Fixed supports are considered at the base. Concrete decks are assumed as floor slabs, which are placed every 4 m. Their mass is calculated considering dead and live loads equal to 7.0 kN/m^2 and 2.0 kN/m^2, respectively. The resulting seismic mass is lumped at floor level nodes, located at the intersections of an auxiliary frame (also referred to as secondary system). However, the final topology is independent from the specific definition of sub-frame and mass nodes, provided that the auxiliary perimeter framework is not included in the optimizable domain.

The continuum domain is modeled assuming steel as building material, with elastic modulus and mass density equal to 210 GPa and 7864 kg/m^3, respectively. The total mass of the system $\mathbf{M}$ is decomposed as $\mathbf{M} = \mathbf{M}_f + \mathbf{M}_\Omega$, where $\mathbf{M}_f$ is the seismic mass of the concrete slabs estimated by considering the floor tributary areas and $\mathbf{M}_\Omega(\boldsymbol{\rho})$ represents the distributed mass of the optimizable domain, i.e., the mass of the designable lateral resisting system. In the material distribution problem, the floor mass is kept as constant during the optimization routines since it is assumed to be independent of the design variables, which implies that $\partial \mathbf{M}_f / \partial \rho_e$ is a null matrix. Since most of the global mass is due to non-optimizable members (i.e., the floor slabs), it is approximately invariant throughout the optimization procedure. This allows to state that the final topology basically depends on the stiffness properties only of the design domain.

Figure 1 illustrates the reference model and its discretization, together with the results of the topology optimization.

It can be noted in Fig. 1 that the material distribution within the design domain assumes a specific arrangement, which consists of two lateral columns and two pairs of full-width diagonal braces, the number of which is closely related to the aspect ratio of the numerical model (i.e., 2). Additionally, it can be observed that the lateral columns on the top of the optimized domain are interrupted before reaching the last module. This is due to a very low value of intermediate densities and, of course, produces an unfeasible result. Because corner columns cannot be removed from the final braced system, they are introduced as very thin members in the final discrete

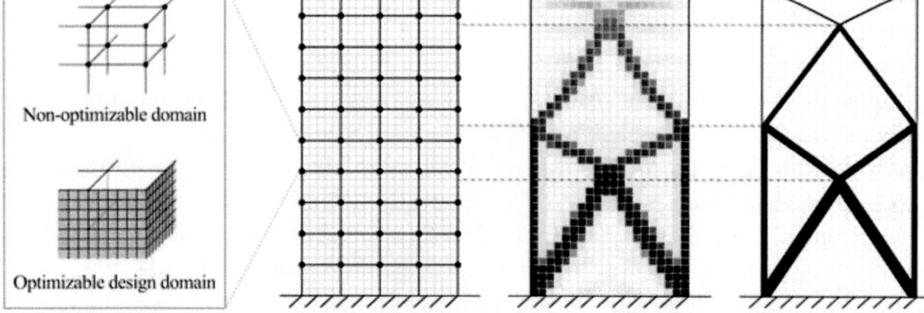

Fig. 1 Reference model and its discretization (**a**), result of the topology optimization (**b**) and post-processed refined optimum layout (**c**)

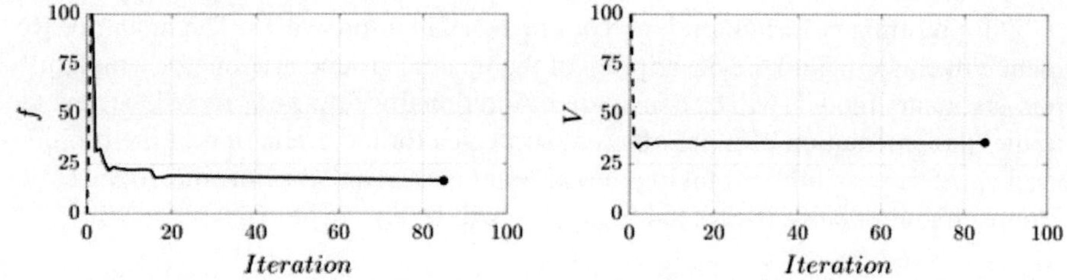

Fig. 2 Convergence of objective function (left) and constraint function (right)

layout. This latter representation is the result of a post-processing operation on the optimized continuum domain. In order to appreciate the results of the topology optimization and correctly identify the localization of the working points of brace-to-brace and brace-to-column nodes, all the pixels are subjected to an image repair process for obtaining the refined final layout. Furthermore, it is worth noticing that the optimal topology does not include the presence of a secondary framed system, which is constituted of intermediate beams and columns, and thus it can be omitted from the final layout.

The evolution of the optimal solution during the optimization routines can be inferred from Fig. 2, where the convergence of objective function (total dissipated power) and constraint function (material volume) is plotted with respect to the number of iterations (the iterative procedure stops when a tolerance equal to or lower than 1% is met). The convergence to the optimal solution is remarked by a filled circle. The efficiency of the topology optimization framework proposed in the present study is confirmed by the steady convergence as well as the limited number of optimization cycles needed to achieve the final optimal solution.

6 Conclusions

The present work has introduced the energy-based approach in the topology optimization of large structures under earthquake. Due to the huge computational effort required for this task (even for the case of linear elastic systems), the classical time-history dynamic analysis was deemed unsuitable, as it would require the direct integration of the motion equations for a large number of records in order to account for the inherent randomness of the seismic excitation. Therefore, the use of the random vibration theory together with the Solid Isotropic Material with Penalization (SIMP) method has been explored in this study to deal with the energy-based topology optimization of large buildings under seismic loads. A gradient-based technique has been employed to solve the resulting optimization problem, and the relevant derivatives have been evaluated analytically. Finally, a paradigmatic case study has been discussed to demonstrate the effectiveness of the proposed framework.

This preliminary formulation will be enhanced in future works. The first improvement will concern with the description of the seismic ground motion, for which fully non-stationary models will be considered. Additionally, future efforts will be devoted to the implementation of more efficient strategies for the evaluation of the problem gradient. The possibility of taking into account nonlinear behaviors due to the structure itself and/or seismic protection devices will be also explored.

References

1. Sigmund, O.: A 99 line topology optimization code written in Matlab. Struct. Multidiscip. Optim. **21**(2), 120–127 (2001)
2. Stromberg, L.L., Beghini, A., Baker, W.F., Paulino, G.H.: Topology optimization for braced frames: combining continuum and beam/column elements. Eng. Struct. **37**, 106–124 (2012)
3. Talischi, C., Paulino, G.H., Pereira, A., Menezes, I.F.M.: PolyTop: a Matlab implementation of a general topology optimization framework using unstructured polygonal finite element meshes. Struct. Multidiscip. Optim. **45**(3), 329–357 (2012)
4. Liu, K., Tovar, A.: An efficient 3D topology optimization code written in Matlab. Struct. Multidiscip. Optim. **50**(6), 1175–1196 (2014)
5. Angelucci, G., Spence, S.M.J., Mollaioli, F.: An integrated topology optimization framework for three-dimensional domains using shell elements. Struct. Des. Tall Special Build. **30**(1), 1–17 (2021)
6. Díaaz, A.R., Kikuchi, N.: Solutions to shape and topology eigenvalue optimization problems using a homogenization method. Int. J. Numer. Methods Eng. **35**(7), 1487–1502 (1992)
7. Seyranian, A.P., Lund, E., Olhoff, N.: Multiple eigenvalues in structural optimization problems. Struct. Optim. **8**(4), 207–227 (1994)
8. Pedersen, N.L.: Maximization of eigenvalues using topology optimization. Struct. Multidiscip. Optim. **20**(1), 2–11 (2000)
9. Yoon, G.H.: Structural topology optimization for frequency response problem using model reduction schemes. Comput. Methods Appl. Mech. Eng. **199**(25–28), 1744–1763 (2010)
10. Min, S., Kikuchi, N., Park, Y.C., Kim, S., Chang, S.: Optimal topology design of structures under dynamic loads. Struct. Optim. **17**(2–3), 208–218 (1999)
11. Maeda, Y., Nishiwaki, S., Izui, K., Yoshimura, M., Matsui, K., Terada, K.: Structural topology optimization of vibrating structures with specified eigenfrequencies and eigenmode shapes. Int. J. Numer. Methods Eng. **67**(5), 597–628 (2006)
12. Filipov, E.T., Chun, J., Paulino, G.H., Song, J.: Polygonal multiresolution topology optimization (PolyMTOP) for structural dynamics. Struct. Multidiscip. Optim. **53**(4), 673–694 (2016)
13. Greene, W.H., Haftka, R.T.: Computational aspects of sensitivity calculations in linear transient structural analysis. Struct. Optim. **3**(3), 176–201 (1991)
14. Zhao, J., Wang, C.: Dynamic response topology optimization in the time domain using model reduction method. Struct. Multidiscip. Optim. **53**(1), 101–114 (2016)
15. Angelucci, G., Mollaioli, F., AlShawa, O.: Evaluation of optimal lateral resisting systems for tall buildings subject to horizontal loads. Procedia Manuf. **44**, 457–464 (2020)
16. Yang, Y., Zhu, M., Shields, M.D., Guest, J.K.: Topology optimization of continuum structures subjected to filtered white noise stochastic excitations. Comput. Methods Appl. Mech. Eng. **324**, 438–456 (2017)
17. Zhu, M., Yang, Y., Guest, J.K., Shields, M.D.: Topology optimization for linear stationary stochastic dynamics: applications to frame structures. Struct. Saf. **67**, 116–131 (2017)
18. Chun, J., Song, J., Paulino, G.H.: System-reliability-based design and topology optimization of structures under constraints on first-passage probability. Struct. Saf. **76**, 81–94 (2019)

19. Kang, B.S., Park, G.J., Arora, J.S.: A review of optimization of structures subjected to transient loads. Struct. Multidiscip. Optim. **31**(2), 81–95 (2006)
20. Liu, Z., Liu, W., Peng, Y.: Random function based spectral representation of stationary and non-stationary stochastic processes. Probab. Eng. Mech. **45**, 115–126 (2016)
21. Bendsøe, M.P., Sigmund, O.: Material interpolation schemes in topology optimization. Arch. Appl. Mech. **69**(9), 635–654 (1999)
22. Bourdin, B.: Filters in topology optimization. Int. J. Numer. Methods Eng. **50**(9), 2143–2158 (2001)

Energy-Based Seismic Design Method for Coupled CLT Shear Walls

Selamawit Dires, Thomas Tannert, and Solomon Tesfamariam

Abstract Ductile cross-laminated timber (CLT) shear walls can be achieved by vertically joining a series of CLT panels with ductile connectors. When such multi-panel systems have a well-defined center of rotation, the resulting kinematic behavior is termed as coupled-panel (CP). In this paper, an iterative energy-based design (EBD) method is proposed for CLT shear walls based on energy balance established on their CP kinematic. Holz-Stahl-Komposit (HSK) connectors were utilized for both hold-downs and vertical joints. The seismic energy demands were estimated from constant ductility hysteretic energy spectra established for elastic-perfectly-plastic single-degree-of-freedom oscillators. The lateral force-deformation characteristics were derived considering the CP behavior in elastic and plastic ranges. Subsequently, the ductility demand was evaluated from these force-deformation relations. The story-wise hysteric seismic energy demands were balanced by the cyclic energy supply. While the lateral yield resistances were attributed to the hold-downs and vertical joints, the lateral plastic deformations were attributed to the vertical joints. The proposed EBD method accounts for the preferred failure mode together with performance criteria derived from either target deformation limit-states or local deformation capacities of the energy dissipative components.

Keywords Cross-laminated timber · Rocking structures · Coupled wall kinematic · HSK connectors · Seismic energy balance

S. Dires (✉) · S. Tesfamariam
The University of British Columbia, Kelowna, Canada
e-mail: selamawit.dires@alumni.ubc.ca

S. Tesfamariam
e-mail: Solomon.Tesfamariam@ubc.ca

T. Tannert
University of Northern British Columbia, Prince George, Canada
e-mail: Thomas.Tannert@unbc.ca

A. Benavent-Climent and F. Mollaioli (eds.), *Energy-Based Seismic Engineering*, Lecture Notes in Civil Engineering 155,
https://doi.org/10.1007/978-3-030-73932-4_15

1 Introduction

1.1 Background

Rapid population growth and the subsequent urban densification have fueled the need for infrastructure developments. Constructing mid- to high-rises with steel, concrete, or the combination thereof have been resulting in enormous carbon footprints [1, 2]. To mitigate the fast-growing urban density demands sustainably, interest in building with wood is increasing, as wood is a natural carbon sink [1, 2]. The development of engineered wood products, in particular cross-laminated-timber (CLT), and related construction technologies are enabling mid- to high-rise wood structures [3].

CLT is a plate-like engineered mass timber product consisting of dimensional lumber layers that are glued together orthogonally so to have strength and stiffness in both principal axes [4]. This process of forming CLT allows for dimensional stability and increased fire resistance compared to dimensional lumber. Because of the in- and out-of-plane resistances, CLT has a versatile application as structural floor and wall panels. CLT can be prefabricated a length of up to 18.0 m, enabling design flexibility, esthetics, and faster erection of buildings. Similar to traditional light-frame wood systems, CLT can be used to make all the components of buildings in platform-type constructions. The high strength-to-weight ratio coupled with CLT's versatile applications attracted builders to use CLT in seismic-prone areas across the world [5, 6].

Several seismic-resistant CLT structural systems have been proposed for mid- to high-rises [7–11]. Integrated experimental programs from component to full-scale tests have been conducted to enhance knowledge regarding the seismic performances of platform-type CLT structures [12–14] and post-tensioned CLT structures [10, 15]. Accordingly, while CLT lateral load resisting systems can provide sufficient lateral stiffness, well-designed ductile connections can provide ductility required to dissipate seismic input energy. After high intensity shaking tests, CLT structures were observed to remain intact except for local damages concentrated around the hold-downs, vertical joints, and shear connections. In consequence, building codes such as the National Building Code of Canada (NBC 2020) [16] incorporated CLT structural systems for mid-rise constructions.

Accounting for proper damage measures, among other steps, has been deemed as an indispensable step for reliable performance-based seismic design [17]. The common force-based design procedure only checks maximum deformation limit-state exceedance by the end of seismic designs, whereas in displacement-based design procedures, limit-state deformations are taken explicitly at the beginning of seismic designs. However, both methods neglect damages associated with cumulative cyclic deformations. Energy-based seismic design (EBD) lends itself to integrate maximum and accumulated damages during the strong ground motions. For this reason, EBD procedures are regarded as advanced methods for performance-based earthquake engineering [18–21]. Beyond as a seismic design method, energy-based approaches are useful to characterize ground motions [22], develop optimal intensity measures

[23], and develop collapse assessment procedures based on explicit consideration of dynamic instability [24].

Despite the potential of using EBD approaches, only few methods have been developed for timber applications; Lopez-Almansa et al. discussed an EBD approach for fussed timber platform frame buildings [25], and Goertz et al. developed EBD for steel-timber hybrid structures [26]. In this paper, an energy-based seismic design procedure for rocking coupled CLT shear walls is proposed.

1.2 Rocking CLT Wall Structures

Platform-type CLT buildings, consisting of CLT wall and floor panels to transfer all the vertical and lateral loads, gained popularity for their assembly ease. The CLT panels have very high in-plane rigidity that the walls go through nearly rigid wall kinematics [12, 14]. Hence, the hold-downs, shear connections, and vertical joints between wall panels introduce flexibility and ductility to platform constructions. To enhance the ductility in such structures, the walls are usually segmented into multi-panels and connected using fasteners (such as nails or self-tapping screws) [12, 14, 27].

The kinematics of CLT walls, when subjected to earthquakes, play a crucial role in seismic energy dissipation. Depending on its design details, CLT walls could rock, slide, or exhibit a combination of both while anticipating minimal flexural and shear deformations [12]. The prevalent kinematic behavior is governed by the relative stiffness and strength between the hold-downs and vertical wall joints, wall panels aspect ratios, gravity loads carried by the walls, the shear connectors tension to shear resistance interaction, the wall to ceiling connections besides the magnitude of lateral actions [28]. Experimental results showed that multi-panel rocking CLT walls possess higher ductility levels than equivalent single panel walls [12, 14]. When such multi-panel CLT walls have a well-defined rotation about the toe of the walls, the resulting kinematic mode is known as coupled-panel (CP) behavior [28] (see Fig. 1).

Some shear connectors between the CLT wall panels and the floors (or ground level for the first story) are known to supply both in-plane shear and uplift resistance [14]. Such coupling of resistances presents intricate circumstances for the tension-shear capacity interactions are not radially available. Therefore, in the case of shear connections with high uplift resistance, it would be tenuous to dimension the hold-down as well as vertical joint connectors without knowing the approximate uplift resistance provided by the shear connectors. In platform-type construction, the floor diaphragms also constrain the walls' rocking movement, and thus proper connection detailing among the walls and CLT floors would be essential.

Gravity loads resisted by the CLT walls reduce the seismic uplift force, decreasing the demands for hold-down designs. However, the presence of a large magnitude of gravity load may hinder the seismic performance by intervening the multi-panel walls from attaining a rocking kinematic. Thus, the whole system becomes very stiff, resulting in a brittle mode of failure. For an earthquake-resistant design, gravity

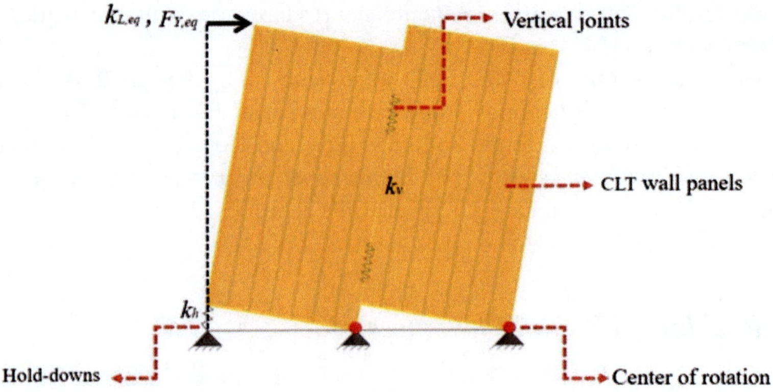

Fig. 1 Coupled-panel kinematic

loads sustained by lateral force resisting CLT walls have to be moderated to allow the desirable rocking kinematic.

For the earthquake-resistant design of multi-panel coupled CLT shear walls, yielding of the vertical joints is the preferred mode of failure while capacity-protecting CLT panels, hold-downs, shear and other non-dissipative connections [29]. The vertical joints can be engaged in energy dissipation by designing multi-panel Wall systems to realize CP behavior when subjected to seismic events. The CP behavior across elastic and plastic ranges can be maintained by designing hold-downs to remain elastic until the vertical joints fail in flexure [30], resulting in stringent design requirements to meet.

1.3 Modified HSK Connectors

The modified Holz-Stahl-Komposit (HSK) connectors consist of perforated steel plates (Fig. 2a) glued inside CLT slots (Fig. 2b) [31]. The connection was developed for high-capacity hold-downs (Fig. 2c) but can also be used to couple wall panels (Fig. 2b). The steel segments between the holes, termed as steel links (SL), provide the strength, stiffness, and ductility. While plates' embedment in CLT prevented local buckling of the plates, enabling the connectors to reach their ultimate strength under reversed shear loading, the failure was assured to occur by yielding of SLs covered by duct tape (highlighted with dark gray in Fig. 2b and c).

The reversed cyclic shear tests exhibited full force-deformation hysteresis depicting large energy dissipation [31, 32] (see Fig. 3). Shear and Hold-down experimental tests using the modified HSK connectors showed consistency in the recorded capacities per SLs [32].

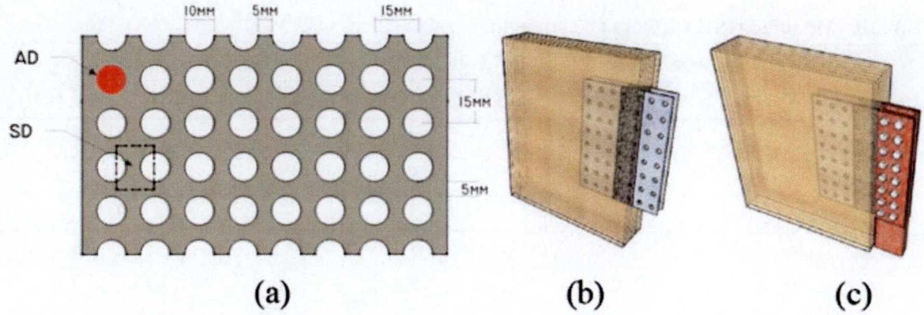

Fig. 2 Modified HSK connector [31]: **a** perforated steel plate; **b** vertical joint connector; **c** hold-down

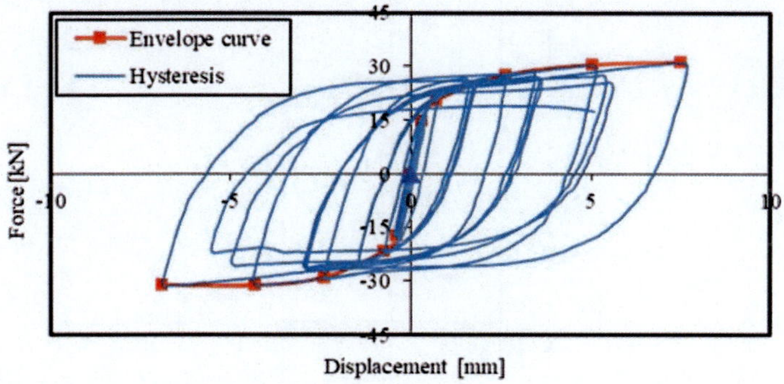

Fig. 3 Mid-scale shear test result of modified HSK system [31]

1.4 Proposed Coupled CLT Shear Walls

The proposed EBD method was implemented on coupled CLT shear wall systems that utilized modified HSK connectors for hold-downs and vertical joints. The number of SLs was adapted as a design parameter to meet the proposed system's stiffness and strength requirements. Adapting the same detail for the HSK plates as in the available experimental tests [32], the mean stiffness, strength and deformation capacities were taken from the mid-scale and full-scale tests, respectively, for the vertical joints and hold-downs. The 5th and 95th percentile strengths were evaluated using ASTM D5055 [33], considering a 75% confidence for one-sided tolerance limits of normally distributed resistances with a 20% coefficient of variation. Table 1 summarizes the stiffness and strength values relevant to the design of the hold-downs and vertical joints.

The proposed EBD is demonstrated for a 3-story two equal width coupled CLT wall structures (see Fig. 4). The building height H, the wall height h, and the panel width b were taken to be 9.6 m, 3.0 m, and 1.5 m, respectively. A seismic weight of 3.0 kN/m per wall length in each story was assumed. An idealized CP kinematic

Table 1 Modified HSK stiffness and strength

	For vertical joints design			For hold-downs design	
	$k_{e,v,SL}$ (kN/mm)	$f_{y,v,SL}$ (kN)	$f_{u,v,SL}$ (kN)	$k_{e,h,SL}$ (kN/mm)	$f_{y,h,SL}$ (kN)
Mean	4.6	2.0	3.0	3.2	2.0
5th percentile	–	1.2	1.8	–	0.8
95th percentile	–	2.8	4.2	–	3.2

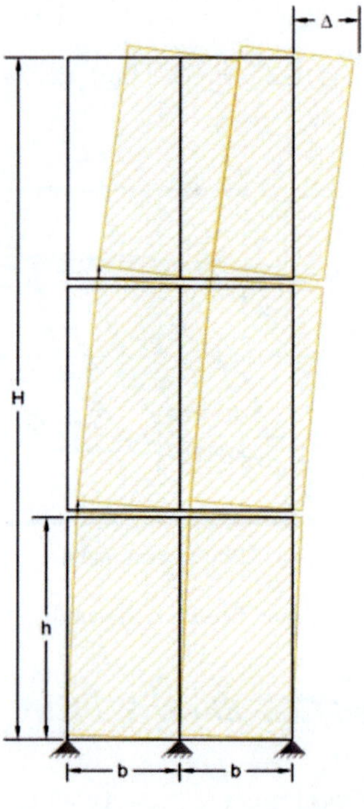

Fig. 4 An idealized schematic of CP behavior of a 3-story CLT wall

is shown in Fig. 4, whereby the walls were constrained from sliding, and the in-plane bending and shear deformations of the CLT panel were neglected. The panels' top was considered to go through the same lateral deformation, assuming a rigid floor diaphragm would distribute the deformations evenly. The system's lateral load-deformation was approximated by elastic-perfectly plastic (EPP) hysteretic behavior, consistent with the load-deformation response of the mid-scale tests of modified HSK connectors under reversed shear (see Fig. 3).

2 Energy-Based Design of Coupled CLT Walls

2.1 Proposed EBD Method

An iterative EBD method was proposed for coupled CLT walls by formulating the energy balance between the story-wise hysteretic energy demand and energy supply on the CP kinematic deformed shape (see Fig. 5). The CP behavior was considered to be maintained until the ductile vertical joints fail. Hence, while vertical joints were designed as energy dissipative components, hold-downs, CLT walls, shear and other non-dissipative connections were capacity-protected.

Herein, shear connectors were considered to resist only shear and the hold-downs to resist seismic uplift forces. The decoupling of the shear and tensile strengths of shear connections can be feasible by using special type connectors [34]. For simplicity, the contribution of gravity loadings to impede uplifting was excluded from the design example.

The energy input to a multiple-degree-of-freedom (MDOF) system was estimated from constant ductility hysteretic energy spectra developed for EPP SDOF oscillators. The target ductility was defined as a ratio of a lateral target to yield deformation. The lateral yield and target displacements were derived from the lateral force-deformation behavior. The proposed systems' lateral force-deformation response was characterized considering the CP kinematic, preferred hierarchy of failure modes, and imposed maximum deformation criteria.

The lateral yield deformations were evaluated from the ratio of the lateral strength to elastic stiffness, which were computed analytically considering the wall panel dimensions, initial hold-downs' and vertical joints' capacities. The maximum plastic deformation was taken as the minimum of the lateral drift limit-states prescribed in building design codes (or prescribed to suffice chosen performance objectives) and maximum lateral deformation capacities evaluated from jointly using experimental results and analytical expressions derived from the CP behavior.

The total hysteretic energy demand was distributed across the stories adapting linear distribution schemes. The story-wise energy demands were equilibrated by the energy supply provided by coupled panel CLT shear walls. The elastic yield strengths were attained jointly from the hold-downs and vertical joints; the lateral plastic deformations were achieved from inelastic yield excursions of the vertical joints. If the initial hold-downs and vertical joints size are not adequate to meet the hysteretic energy demands, larger-sized connectors should be assigned, and the design must be repeated until the seismic energy demands are fulfilled.

Finally, capacity-protected components' elastic resistances were checked to be greater than demands derived based on the vertical joints' probable capacities. This way, failures can be ensured to occur at the covered rows of SLs across the vertical joints while the non-dissipative components remain elastic. Hence, the vertical joints' 95th percentile ultimate resistances were used to compute their probable capacities.

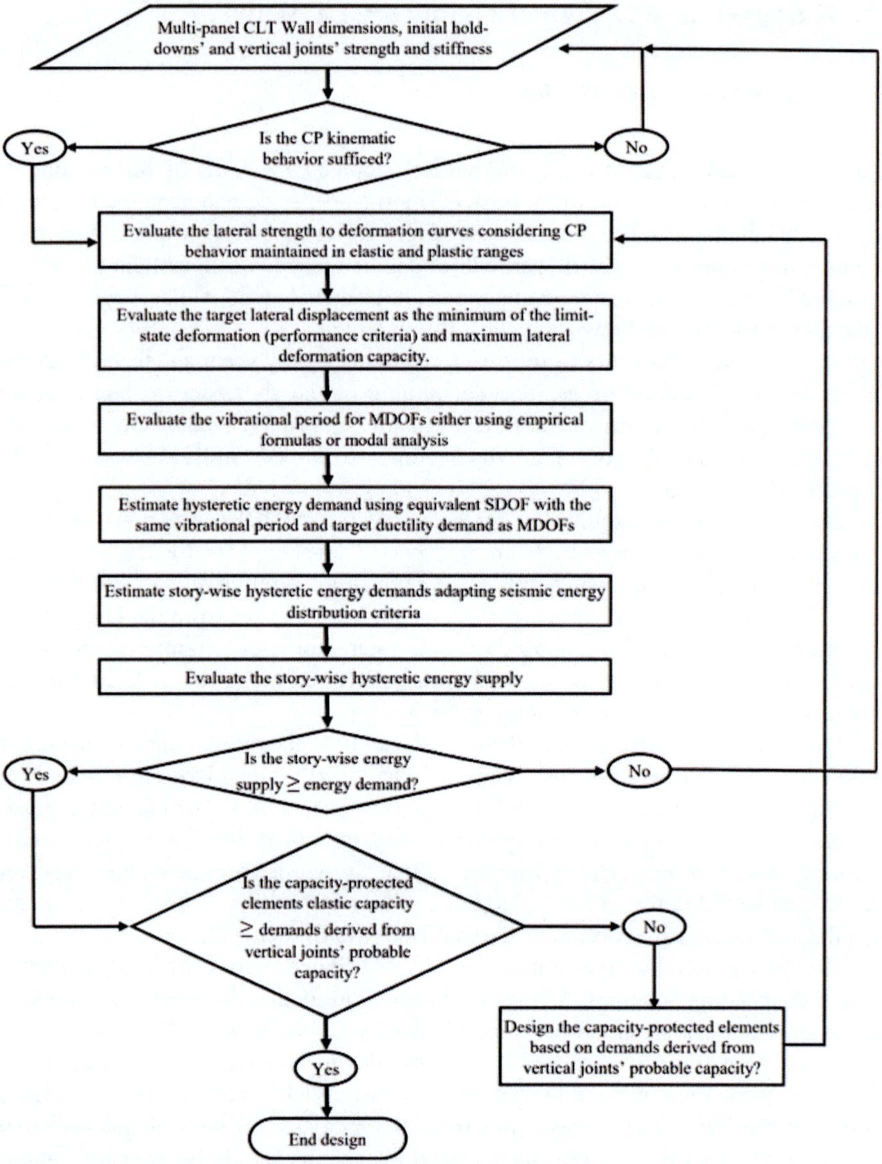

Fig. 5 Proposed energy-based design method flowchart

2.2 *Estimation of Lateral Force-Deformation Relations*

The hold-downs and vertical joints could be viewed as springs in parallel undergoing the same lateral drifts upon the realization of CP kinematic mode (see Figs. 1 and 4).

The story-wise equivalent lateral yield strength and elastic stiffness were determined analytically from the CP kinematic. Accordingly, the equivalent hold-down tensile stiffness $k_{h,eq}$ was computed by using Eqs. (1)–(3) [28].

$$k_{h,eq} = k_h + k_v \tag{1}$$

$$k_h = n_h k_{e,h,SL} \tag{2}$$

$$k_v = n_v k_{e,v,SL} \tag{3}$$

where k_h and k_v = elastic stiffness of hold-downs and vertical joints, respectively, n_h and n_v = number of SLs in hold-downs and vertical joints, respectively, $k_{e,h,SL}$ and $k_{e,v,SL}$ = elastic stiffness per SL in hold-downs and vertical joints, respectively, see Table 1.

The story-wise equivalent lateral stiffness $k_{L,eq}$ (see Fig. 1) was evaluated considering the equivalent hold-down stiffness and wall panels geometry (Eq. 4) [28].

$$k_{L,eq} = \frac{k_{h,eq} b^2}{h^2} \tag{4}$$

The elastic lateral yield strength of the CP system ($F_{Y,eq}$) (see Fig. 1) was governed by the vertical joints' strength, assuming the hold-downs are designed stronger than the vertical joints and was estimated using Eq. (5) [28].

$$F_{Y,eq} = \frac{f_{y,v,SL} k_{h,eq} b}{k_{e,v,SL} h} \tag{5}$$

where $f_{y,v,SL}$ and $k_{y,v,SL}$ = elastic strength and stiffness of vertical joints per SL (see Table 1). Therefore, the lateral yield displacement can be computed from the known lateral yield strength and stiffness, Eq. (11).

$$\Delta_y = \frac{F_{Y,eq}}{K_{L,eq}} \tag{6}$$

The lateral target displacement was evaluated considering drift limit states and the ultimate deformation capacities of the vertical join connectors. The ultimate deformation limit state can be defined using either building code inter-story drift ratios (ISDR) or desired performance criteria. Herein, a 2.5% ISDR limit was adapted; hence, the target maximum inter-story plastic drift was evaluated to 80 mm, resulting in a total roof displacement Δ of 240 mm (Fig. 4). The average ultimate capacity of the modified HSK connectors when tested under shear was approximately 9.4 mm. The corresponding lateral displacement was evaluated analytically, considering the plastic range CP behavior, using Eq. 7 [30], which yielded 18.8 mm.

Table 2 Example design summary

	Story		
	1	2	3
n_v	50	40	40
n_h	425	350	330
$E_{p,d}$ (kN m)	8.2	5.5	2.7
$E_{p,c}$ (kN m)	12.7	10.4	10.0
$F_{u,v,95th}$ (kN)	333	270	263
$F_{y,h,5th}$ (kN)	340	280	264

$$d_L = d_{u,v}\frac{h}{b} \tag{7}$$

where d_L= ultimate lateral displacement, $d_{u,v}$ = ultimate deformation capacity of the vertical joints.

Hence, the target displacement was governed by the vertical joints' ultimate deformation. The corresponding target displacement was calculated accordingly using Eq. (8). For the design example, which was summarized in Table 2, the target ductility at the first level was evaluated to 43.

$$\mu = \frac{\Delta_t}{\Delta_y} \tag{8}$$

2.3 Hysteretic Energy Demand

Considering the seismicity of south-western (SW) Canada, constant-ductility energy spectra were established for generalized SDOF oscillators with EPP hysteretic curves and 5% critical damping ratios. The details regarding the hysteretic energy spectra can be found in [35]. The SW Canada seismicity can result from three source mechanisms: crustal, in-slab, and interface [36]. Herein, ensembles of five ground motions represented each of the seismic events [37]. Figure 6 shows the mean constant-ductility hysteretic energy spectra for the selected seismic records. Seismic energy imparted to a MDOF system, dominated by the fundamental vibration mode, could be reasonably estimated from an equivalent SDOF system with the same fundamental vibration period and ductility demand as that of the corresponding MDOF system [38].

The initial vibration period of MDOF systems can be evaluated either using empirical formulas for the construction types or modal analysis. For the sample design, the period was predicted from the NBCC 2015 empirical formula for shear wall systems using Eq. 9, which yielded approximately 0.3 s. Once the first round of

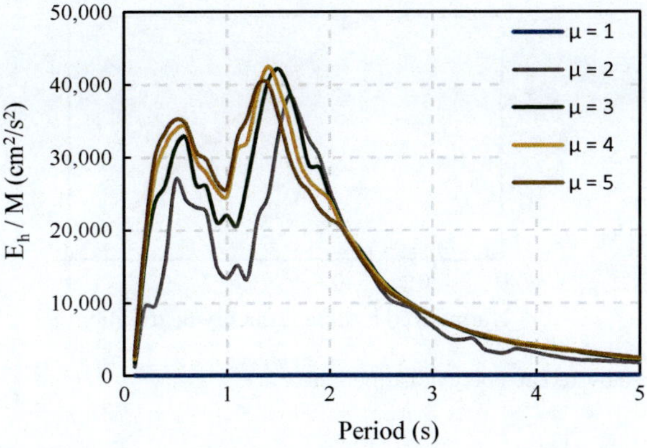

Fig. 6 Constant ductility hysteretic energy spectra

design is completed, a more accurate period can be estimated from a modal analysis procedure, and the design could be updated.

$$T = 0.05(h_n)^{0.75} \tag{9}$$

where h_n = building height in m.

The total hysteretic energy demand of a MDOF system per unit mass was estimated from the energy spectra (Fig. 6) with known period and ductility demand. The target ductility was estimated based on the first story kinematic response. Even though the computed target ductility was 43, the hysteretic energy demand per unit mass was estimated reasonably using the spectra for ductility level 5 as the ductility effect on energy demand diminishes for highly inelastic systems [38].

2.4 *Formulation of Story-Wise Energy Balance*

Hysteretic energy distributions depend on the distribution of strength and stiffness across the stories [38]. Hence, a thorough understanding of the energy distribution schemes requires substantial parametric nonlinear analyses. In the proposed design, linear hysteretic energy distribution (Fig. 7) was considered to yield a reasonable approximation based on a comparable previous study for low-rise buildings [19].

In the proposed design, a structure would be considered unfit whenever the energy demand is more than the energy absorption capacity of a structure. The total number of yield excursions and yield amplitudes undergone by structures are highly dependent on the ground motion characteristics [39]. The number and amplitude of yield cycles, in turn, affect the total hysteretic energy supply by a system [40]. Substantial testing subordinated by numerical analyses would be required to examine structural capacities in terms of energy [41]. For the proposed system, which was characterized

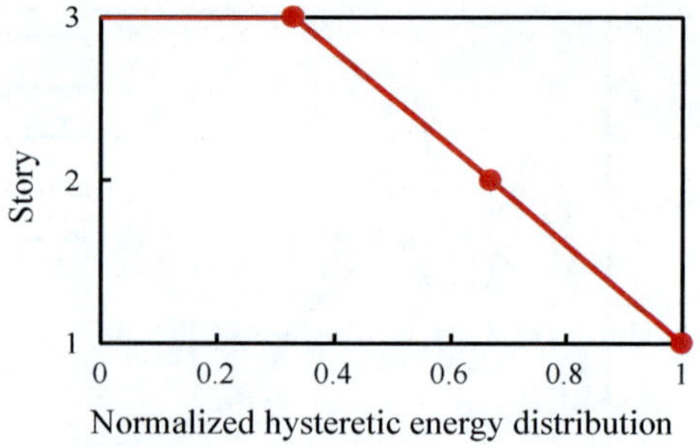

Fig. 7 Normalized hysteretic energy distribution

by a fuller hysteretic curve (Fig. 3), the plastic energy supply was estimated with Eq. 10. Similar approximations were adopted for structures that own fuller hysteretic behavior [21, 42].

$$E_{p,c} = 4F_{Y,eq}(\Delta_t - \Delta_y) \tag{10}$$

2.5 Design of Capacity-Protected Components

The vertical joints and hold-downs were dimensioned to achieve the hysteretic energy demands while maintaining the CP behavior and prescribed performance hierarchies. The vertical joints' probable capacities based on their 95th percentile ultimate strengths were evaluated to check the adequacy of elastic hold-down capacities. Likewise, all the other capacity-protected components must be designed for sufficient strength and deformation capacities based on the vertical joints' probable capacities to avoid premature failures. The hold-downs elastic demand based on the probable capacities can be estimated using Eqs. (11) and (12) [30].

$$F_{p,eq} = F_{Y,eq} + (F_{v,p} - F_{v,e})\frac{b}{h} \tag{11}$$

$$F_h = \left(\frac{F_{Y,eq}h^2}{b^2}\right)\frac{k_{,h}b}{k_{h,eq}h} + \left(F_{p,eq} - F_{Y,eq}\right)\frac{h}{b} \tag{12}$$

where F_h = hold-down tensile force, $F_{v,p}$ and $F_{v,e}$ = vertical joints probable and yield capacities, respectively, $F_{p,eq}$ = equivalent lateral plastic strength.

The design example is summarized in Table 2 for each story. The stringent requirements imposed on hold-downs necessitated more SLs than required to maintain the CP kinematic and energy demands.

3 Conclusion

In this paper, an EBD method was proposed for multi-story coupled-panel CLT wall systems. The proposed EBD was demonstrated with coupled CLT walls that utilized the modified HSK hold-down and vertical joint connectors. Considering the shear and hold-down experimental test results [32], the force-deformation response was approximated with EPP curves. However, it is vital to validate this force-deformation approximation experimentally by conducting full-scale tests on coupled-panel CLT walls connected using the modified HSK system.

The CP kinematic was maintained in elastic and plastic ranges to assure failure to occur in the vertical joints while capacity-protected components remain elastic. Such failure criteria introduced stringent design requirements for the hold-down designs. The target failure mode seems to require stronger hold-down solutions with higher elastic deformation capacities. The CP mode up until failure of the vertical joints can be granted if the hold-downs' elastic deformation capacities exceed that of the ultimate deformation capacities of the vertical joints. Likewise, all the other capacity-protected connections' deformation capacities should be assessed to avoid premature brittle failures. The availability of fewer experimental data manifested in evaluating the 5th and 95th percentile SL strengths, which further made the hold-down designs conservative. More experiments would assure a better estimation of strengths.

The shear connections' tensile-shear interaction resistances, gravity loadings, floor diaphragms, and other connections may intervene in the anticipated kinematic behavior of multi-panel systems. Hence, elaborated studies including other intervening mechanisms are important as such systems' seismic energy dissipations depend on the kinematic behavior.

The hysteretic energy demands of MDOF systems were estimated using equivalent SDOF systems with the same period and ductility as MDOF systems. While previous studies [38] showed the adequacy of such approximations for shear-type buildings dominated by the fundamental modes of vibration, nonlinear time history analyses would be essential to validate for rocking systems. Moreover, energy distributions across high-rises necessities rigorous investigations for higher mode contributions.

The story-wise hysteretic energy supply equilibrated the story-wise seismic energy demands. Hence, while the lateral yield capacities were attributed to the hold-down and vertical joint connectors, the lateral plastic deformations were attributed to the yielding of the vertical joints. The story-wise energy supplies were conservatively evaluated considering the fuller hysteresis of the proposed system. Further experimental and numerical studies of multi-panel systems are underway to investigate the multi-story kinematic and energy supply capacities.

References

1. Green, M., Karsh, E.: The Case for Tall Wood Buildings. Wood Enterprise Coalition, Vancouver, BC, Canada (2012)
2. Skidmore, Owings, Merrill, L.L.P.: Timber tower research project. Final report, Greenbuild 2013, WoodWorks Education Lab, Chicago, IL, USA (2013)
3. Karacabeyli, E., Lum, C.: Technical guide for the design and construction of tall wood buildings in Canada. FPInnovation, Special Publication SP-55E (2014). ISBN: 978-0-86488-555-5
4. Brandner, R., Flatscher, G., Ringhofer, A., Schickhofer, G., Thiel, A.: Cross laminated timber (CLT): overview and development. Eur. J. Wood Wood Prod. **74**(3), 331–351 (2016)
5. Izzi, M., Casagrande, D., Bezzi, S., Pasca, D., Follesa, M., Tomasi, R.: Seismic behaviour of cross-laminated timber structures: a state-of-the-art review. Eng. Struct. **170**, 42–52 (2018)
6. Tannert, T., Follesa, M., Fragiacomo, M., Gonzalez, P., Isoda, H., Moroder, D., van de Lindt, J.: Seismic design of cross-laminated timber buildings. Wood Fiber Sci. 3–26 (2018)
7. Dickof, C., Stiemer, S.F., Bezabeh, M., Tesfamariam, S.: CLT–steel hybrid system: ductility and overstrength values based on static pushover analysis. J. Perform. Constr. Facil. **28**(6), A4014012 (2014)
8. Zhang, X., Fairhurst, M., Tannert, T.: Ductility estimation for a novel timber-steel-hybrid system. J. Struct. Eng. **142**(4), E4015001 (2015)
9. Tesfamariam, S., Bezabeh, M., Skandalos, K., Martinez, E., Dires, S., Bitsuamlak, G., Goda, K.: Wind and earthquake design framework for tall wood-concrete hybrid system. Technical report, University of British Columbia, BC, Canada (2019)
10. Pei, S., van de Lindt, J.W., Barbosa, A.R., Berman, J.W., McDonnell, E., Dolan, D.J., Blomgren, H.E., Zimmeran, R.B., Huang, D., Wichman, S.: Experimental seismic response of a resilient 2-story mass-timber building with post-tensioned rocking walls. J. Struct. Eng. **145**(11), 04019120 (2019)
11. Hashemi, A., Quenneville, P.: Seismic performance of timber structures using rocking walls with low damage hold-down connectors. Structures **27**, 274–284 (2020)
12. Popovski, M., Karacabeyli, E.: Seismic behaviour of cross-laminated timber structures. In: World Conference on Timber Engineering, pp. 335–344. Auckland, New Zealand (2012)
13. Ceccotti, A., Sandhaas, C., Okabe, M., Yasumura, M., Minowa, C., Kawai, N.: SOFIE project—3D shaking table test on a seven-story full-scale cross-laminated timber building. Earthq. Eng. Struct. Dyn. **42**(13), 2003–2021 (2013)
14. Gavric, I., Fragiacomo, M., Ceccotti, A.: Cyclic behavior of CLT wall systems: experimental tests and analytical prediction models. J. Struct. Eng. **141**(11), 04015034 (2015)
15. Ganey, R., Berman, J., Akbas, T., Loftus, S., Dolan, D.J., Sause, R., Ricles, J., Pei, S., van de Lindt, J., Blomgren, H.E.: Experimental investigation of self-centering cross-laminated timber walls. J. Struct. Eng. **143**(10), 04017135 (2017)
16. National Research Council (NRC) Canada: National Building Code of Canada. National Research Council of Canada, Ottawa, Ontario (2020)
17. Priestley, M.J.N.: Myths and fallacies in earthquake engineering—conflicts between design and reality. Bull. N. Z. Natl. Soc. Earth Quake Eng. **26**(3), 329–341 (1993)
18. SEAOC: Vision 2000: performance-based seismic engineering for buildings. Report prepared by Structural Engineers Association of California, Sacramento, CA, USA (1995)
19. Choi, H., Kim, J.: Energy-based seismic design of buckling-restrained braced frames using hysteretic energy spectrum. Eng. Struct. **28**(2), 304–311 (2006)
20. Benavent-Climent, A.: An energy-based method for seismic retrofit of existing frames using hysteretic dampers. Soil Dyn. Earthq. Eng. **31**(10), 1385–1396 (2011)
21. Mezgebo, M.G., Lui, E.M.: A new methodology for energy-based seismic design of steel moment frames. Earthq. Eng. Eng. Vib. **16**(1), 131–152 (2017)
22. Mollaioli, F., Donaire-Avila, J., Lucchini, A., Benavent-Climent, A.: On the importance of energy-based parameters. In: Papadrakakis, M., Fragiadakis, M. (eds.) COMPDYN 2019 7th ECCOMAS Thematic Conference on Computational Methods in Structural Dynamics and Earthquake Engineering, Crete, Greece (2019)

23. Donaire-Ávila, J., Mollaioli, F., Lucchini, A., Benavent-Climent, A.: Intensity measures for the seismic response prediction of mid-rise buildings with hysteretic dampers. Eng. Struct. **102**, 278–295 (2015)
24. Deniz, D., Song, J., Hajjar, J.F.: Energy-based sidesway collapse fragilities for ductile structural frames under earthquake loadings. Eng. Struct. **174**, 282–294 (2018)
25. López-Almansa, F., Segués, E., Cantalapiedra, I.R.: Energy-based design of a seismic protection system of timber platform frame buildings using energy dissipators. In: Second European Conference on Earthquake Engineering and Seismology, Istanbul, Turkey (2015)
26. Goertz, C., Mollaioli, F., Tesfamariam, S.: Energy based design of a novel timber-steel building. Earthq. Struct. **15**(4), 351–360 (2018)
27. Hossain, A., Danzig, I., Tannert, T.: Cross-laminated timber shear connections with double-angled self-tapping screw assemblies. J. Struct. Eng. **142**(11), 04016099 (2016)
28. Casagrande, D., Doudak, G., Mauro, L., Polastri, A.: Analytical approach to establishing the elastic behavior of multi-panel CLT shear walls subjected to lateral loads. J. Struct. Eng. **144**(2), 04017193 (2018)
29. CSA Standard 086-19, Engineering Design in Wood. Canadian Standard Association, Ottawa, Ont., Canada (2019)
30. Nolet, V., Casagrande, D., Doudak, G.: Multi-panel CLT shearwalls: an analytical methodology to predict the elastic-plastic behaviour. Eng. Struct. **179**, 640–654 (2019)
31. Zhang, X.: Seismic design of timber steel hybrid high-rise buildings. Ph.D. Dissertation, University of British Columbia, BC, Canada (2017)
32. Zhang, X., Popovski, M., Tannert, T.: High-capacity hold-down for mass-timber buildings. Constr. Build. Mater. **164**, 688–703 (2018)
33. ASTM D5055—19: Standard Specification for Establishing and Monitoring Structural Capacities of Prefabricated Wood I-Joists. ASTM International, West Conshohocken, PA, USA (2019)
34. Hashemi, A., Zarnani, P., Masoudnia, R., Quenneville, P.: Seismic resilient lateral load resisting system for timber structures. Constr. Build. Mater. **149**, 432–443 (2017)
35. Dires, S., Tannert, T.: Input and hysteretic energy demands for the seismicity of south-western Canada. In: 17th World Conference on Earthquake Engineering. Sendai, Japan (2020)
36. Atkinson, G.M., Goda, K.: Effects of seismicity models and new ground-motion prediction equations on seismic hazard assessment for four Canadian cities. Bull. Seismol. Soc. Am. **101**(1), 176–189 (2011)
37. Pieper, C.G.: Seismic analysis and design of hybrid concrete timber structures with 2015 National Building Code of Canada. Master's thesis, The University of British Columbia, BC, Canada (2018)
38. Akiyama, H.: Earthquake-Resistant Limit-State Design for Buildings. University of Tokyo Press (1985)
39. Zahrah, T.F., Hall, W.J.: Earthquake energy absorption in SDOF structures. J. Struct. Eng. **110**(8), 1757–1772 (1984)
40. Manfredi, G.: Evaluation of seismic energy demand. Earthq. Eng. Struct. Dyn. **30**(4), 485–499 (2001)
41. Fardis, M.N.: From force- to displacement-based seismic design of concrete structures and beyond. In: Pitilakis, K. (ed.) Recent Advances in Earthquake Engineering in Europe. ECEE 2018. Geotechnical, Geological and Earthquake Engineering, vol. 46, pp. 101–122. Springer, Cham
42. Choi, H., Kim, J., Chung, L.: Seismic design of buckling-restrained braced frames based on a modified energy-balance concept. Can. J. Civ. Eng. **33**(10), 1251–1260 (2006)

Energy-Based Design Process for Passive Control Structures Considering Torsional Effect

Sanghoon Oh, Seunghoon Shin, and Bahador Bagheri

Abstract The seismic performance of structures associated with dampers, called as passive vibration-controlled structures, depends considerably on the damper's mechanical characteristics, which is expected to dissipate most of seismic energy by their nonlinear hysteretic behavior. Force-based design procedure presented in ASCE7, mostly referred for the design of structures with dampers, has limitations in quantifying the plastic deformation capacity required for dampers. Moreover, there is a lack of knowledge in formulation for torsional effect, which is potential by irregularities according to variations in damper characteristics and irregular arrangement of the dampers. Therefore, it inevitably requires repetitive analysis to verify the seismic performance in the design. In this research, a design method is proposed based on energy-based design approach to consider the torsional effect on passive control structures. The proposed method is the prediction of seismic response of structures, which is obtained throughout a large number of nonlinear analysis. The yield strength distribution of dampers through the height of structure was found be vital parameter for energy dissipation. Therefore, the optimum yield strength distribution of dampers is suggested to evenly distribute the accumulated plastic deformation ratio in torsional systems. Also, the yield deformation ratio between dampers and structural members is suggested in order to induce the most of damage concentrated to dampers at each floor. The analysis shows that by selecting the appropriate range of damper's characteristics such as strength and stiffness, the response of structures can be controlled in accordance with design objectives.

Keywords Energy-based design · Torsional effect · Passive control structure

S. Oh · B. Bagheri
Pusan National University, Busan, South Korea

S. Shin (✉)
LERA Consulting Structural Engineers, LLC, Seoul, Korea
e-mail: seunghoon.shin@lera.com

A. Benavent-Climent and F. Mollaioli (eds.), *Energy-Based Seismic Engineering*, Lecture Notes in Civil Engineering 155,
https://doi.org/10.1007/978-3-030-73932-4_16

1 Introduction

Energy dissipating systems are commonly applied to improve the seismic performance of a structure. Therefore, the seismic performance of the structure is highly dependent on the design of the damper. However, current design standards do not propose systematic damper design methods to maximize the seismic performance of structures with damping system. ASCE7-16, which has been commonly used as design criteria for structures with damping system, specifies seismic load and design requirements based on strength. This force-based design involves the following limitations in the design of structures with damping system. The acceleration response spectrum cannot quantify the plastic deformation capacity required for the damper. In addition, the behavior of the member after yielding cannot be evaluated because the nonlinear behavior of the structure is evaluated by the response modification factor. Therefore, the effects of the difference in the characteristics of dampers with various energy dissipation characteristics cannot be clearly elucidated. To design a structure with efficient energy dissipation capacity, it is necessary to perform repeated nonlinear time history analysis for evaluating the seismic performance.

Meanwhile, in a structure associated with dampers, it is appropriate to arrange dampers such that to minimize the torsional effect of the structure. However, this may cause eccentricity because the center of the mass of the floor becomes inconsistent with the center of stiffness owing to the addition of the dampers depending on the structural plan. In addition, the damper characteristics may change depending on the environment, manufacturing deviations, and the inherent material characteristics which are not clearly addressed in the prototype test. Therefore, the torsional effect is likely to occur in the structure owing to the deviation in the damper characteristics and the irregular arrangement of the dampers. Chapter 15 of ASCE41-17 presents an analysis procedure to consider the accidental eccentricity of a structures with damping system. When a time history analysis is conducted and the response amplification ratio caused by the eccentricity exceeds 10%, the torsional effect should be considered in a design. The load and displacement multiplied with the amplification ratio should be used for member design. As mentioned, ASCE41-17 specifies that the torsional effect of a structures with damping system should be considered in the design, but it does not present a damper design method that considers the torsional effect.

To improve the above limitations of the force-based design and to propose a design procedure based on the performance of a structures with damping system, an energy-based design was applied in this study. The energy-based design of a structure using the relationship between the total input energy of the structure due to an earthquake and the hysteretic energy dissipated by the hysteretic behavior of the structure has been evaluated as a reasonable method for design based on the performance of the structure. Benavent-Climent [1], Oh and Shin [2], and Bagheri et al. [3] applied an energy-based design to structures with damping system with hysteretic dampers and evaluated their seismic performance.

In previous studies related to the energy-based design of structures with damping system, it was assumed that the main structure, excluding the dampers, exhibited elastic behavior. However, considering the possibility of large earthquakes in the future, it is also necessary to consider the significant plastification of the main structure. In addition, the range of design parameters required for the design of structures with damping system was limited, and the parameters for generalizing the design equations were not sufficiently analyzed. Also, structures with dampers installed on a plane frame have been mainly researched heretofore. Thus, further research is required to identify the response characteristics of structures with torsional effect in terms of energy-based design.

When a structure is affected by torsion, damage is highly likely to be concentrated on the members located in areas where the response is amplified by torsional motion. Considering a case in which the dampers and main structure exceed elastic deformation by the amplification due to torsional motion, a design method that minimizes the energy absorbed by the main structure and causes most of the energy to be absorbed by the dampers is required. At positions where the influence of the response amplification is relatively small, sufficient energy dissipation owing to the hysteretic behavior of the dampers cannot be expected. This concentration of damage on the dampers due to the torsional effect is likely to significantly reduce the energy absorption capacity of the entire structure. As dampers are used in a structure to improve the seismic performance of the entire structure, it is necessary to maximize the energy absorption capacity of the entire structure through the hysteretic behavior of all dampers. To this end, a design method that evenly distributes the damage to the dampers on all floors instead of concentrating such damage on a specific floor is required.

2 Literature Review

The energy-based seismic design theory was systematically proposed by Housner [4] and Akiyama [5]. In the seismic design based on the energy balance proposed by Akiyama [5], the load effect due to an earthquake is considered as the energy input to the structure. In addition, for the seismic structure modeled using the plane frame, the seismic design method was proposed through identifying the distribution of the total input energy of the structure to each floor and each member and the design parameters governing the energy distribution in terms of energy. Subsequently, studies on the seismic design of structures have been conducted using the concept of energy. Most of these studies focused on the design of seismic structures, such as plane steel moment-resisting frames [6–8].

Benavent-Climent [1] applied hysteretic dampers to the seismic retrofit and suggested a structural design procedure based on energy-based design. Oh and Shin [2] suggested the design procedures for an energy dissipation device in which dampers were installed in all floors and installed only on the first floor with respect to the plane steel frame with hysteretic dampers. Habibi et al. [7] applied the design method proposed by Chou and Uang [9] to structures with hysteretic dampers and evaluated its seismic performance. They calculated the hysteretic energy requirements for each floor based on the mode function for the plane frame with dampers placed on all floors, and used them for the design of dampers. Benavent-Climent [1] proposed a design energy response spectrum considering the site characteristics of low-to-moderate seismicity regions. Bagheri and Oh [10] proposed a design equation to quantify of the hysteretic energy required for dampers according to the natural vibration period and yield base shear force coefficient of the structure for shear wall structural systems with coupling dampers. Oh et al. [11] analyzed the stability of the acceleration response spectrum and energy response spectrum according to the mass, stiffness, and strength distributions for each floor of a structure as well as the torsional effect, hysteretic characteristics, and plastification of its members, and verified the validity of the energy response spectrum.

The design story shear force distribution suggested in most design standards is based on the response of the elastic system. As seismic design allows with the performance of collapse prevention, it is reasonable that the distribution of the story shear force is based on the response of the elasto-plastic system. This is equivalent to considering that the distribution of story shear force is related to the damage on each floor due to the plastic deformation of the floor. Akiyama [5] evaluated the damage on each floor of the frame using the accumulative plastic deformation ratio, and reported that the ratio is dependent on the mass, yield strength, and stiffness distributions on the floor and it is mainly affected by the yield strength distribution. Through the analysis of variables, the yield strength distribution that evenly distributes the damage across the floors was proposed. This represents the strength distribution that evenly distributes the damage across the floors without concentrating the damage on a particular floor until the structure reaches its limit state. Oh and Jeon [12] pointed out that the damage to the upper story tends to be increased under the influence of high-order modes, and proposed the yield strength distribution as a function of the natural period of the structure.

In the event of an earthquake, structures vibrate due to the simultaneous translational and torsional modes. This is due to the accidental eccentricity present in symmetric structures, in addition to the irregularity present in asymmetric structures. According to Chopra and De la Llera [13], symmetric structures could be influenced by the torsional mode to some extent because the foundation may have rotational motion, and perfectly symmetrical structures are rare. ASCE41-17 suggests that analysis should be conducted by considering the eccentricity corresponding to 5% of the diaphragm in the direction of the principal axis or the response amplification factor should be applied to the design to consider the accidental eccentricity of a structures with damping system. The amplification factor is the ratio of the response when the torsional effect by eccentricity is considered to the response when the eccentricity is not considered. If the amplification factor is 1.1 or higher, a nonlinear dynamic procedure should be performed.

Elms [14] suggested the response amplification factor to reflect the torsional effect by considering three torsional effects for seismic structures: accidental eccentricity, ground motion accompanied by the torsional mode, and the combination of the torsional mode and translational mode. De la Llera and Chopra [15] analyzed the response amplified by the rotational displacement added to the foundation of a single-degree-of-freedom (SDOF) structure, and suggested a analysis procedure considering the torsional effect of the structure. Many studies have been conducted to identify the torsional effect on seismic structures [16, 17], including the aforementioned studies, and they have been used as the basis for developing design standards.

The story damage distribution according to the distribution of the story shear force in the dampers has not been adequately researched. Structures with damping system should be designed in a way that most damage is concentrated on the dampers. Therefore, it is necessary to propose that the story shear force that induces damage evenly distributed among the dampers placed on all floor to maximize the energy absorption capacity of the dampers, without concentrating the damage on the dampers on a certain floor. In addition, considering the cases in which the frame exceeds elastic limit, the effect of the distribution of story shear force in the frame on the story damage distribution of the dampers should be researched. Because it is highly likely that structures with damping system will be subjected to the torsional effect through the deviation caused by fluctuations in the damper characteristics and the irregular arrangement of the dampers, damper design methods should be suggested considering the influence of the response amplification due to the torsional effect.

3 Response Characteristics of Structures with Dampers Without Torsional Effect

In this section, the distribution of story shear force that evenly distributes the damage to the dampers on all floors, and the appropriate design range of dampers to minimize the damage of the frame are proposed for a structure with dampers without considering the torsional effect.

3.1 Damage Distribution Between the Frame and Dampers

The analysis model was constructed as a three-dimensional structure with a 1 × 1 span, as shown in Fig. 1. The widths of the analysis model, L_1 and L_2, were 6 m and 2 m, respectively, and the height of each floor, h_i, was 4 m. The uniform vertical load on the floor was 12 kPa. Nonlinearity was simulated based on the hysteretic behavior of the members. The hysteretic characteristics of the flexural and shear springs of the columns were set to be perfectly elasto-plastic, and the beams were modeled as a rigid beam to feasible structural analysis. The inverted-V-shaped dampers were installed in the frame in the same direction as that of the input seismic wave. The dampers were modeled as axial link elements, and their ends were modeled as pins. The dampers were set to have the hysteretic characteristics of the buckling-restrained brace specified in AISC 341-16. According to the experimental research on the buckling-restrained brace by Merritt et al. [18], the compressive strength adjustment factor β and the tensile strength adjustment factor ω were set to 1.95 and 1.5, respectively.

The yield story shear force coefficient of the frame for the first floor, ${}_f\alpha_{Y1}$, is assumed to be 0.15, and the yield story drift ${}_f\theta_{Yi}$ is 1/150. ${}_f\alpha_{Yi}$ is the value obtained by dividing the yield story shear force of the frame for the i-th floor, ${}_fQ_{Yi}$, by the cumulative weight of the upper story, and is expressed as ${}_f\alpha_{Yi} = {}_fQ_{Yi}/\sum_{j=i}^{N} m_j g$. The distribution of story shear force coefficient in the frame, ${}_f\overline{\alpha}_i$, was calculated in accordance with ASCE7-16, and the distribution of story shear force coefficient in the dampers, ${}_s\overline{\alpha}_i$, was calculated in accordance with Akiyama [5]. ${}_f\overline{\alpha}_i$ and ${}_s\overline{\alpha}_i$ are the values obtained by dividing the yield story shear force coefficients of the frame and dampers of the i-th floor, ${}_f\alpha_{Yi}$ and ${}_s\alpha_{Yi}$, by those of the first floor, ${}_f\alpha_{Y1}$ and ${}_s\alpha_{Y1}$, respectively, and are expressed as $\overline{\alpha}_i = \alpha_{Yi}/\alpha_{Y1}$. r_q is the ratio of the yield strength of the frame, ${}_fQ_{Yi}$, to the yield strength of the dampers for the i-th floor, ${}_sQ_{Yi}$. Nine cases were considered in the range of $0.6 < r_q < 7.5$. The cases in which r_q is smaller than 1.0 are the cases where the yield strength of the dampers, ${}_sQ_{Yi}$, was designed to be higher than the yield strength of the frame, ${}_fQ_{Yi}$. r_δ is the ratio of the yield deformation of the frame, ${}_f\delta_{Yi}$, to the yield deformation of the dampers, ${}_s\delta_{Yi}$, and 12 cases were considered in the range of $0.2 < r_\delta < 15$. In the analysis, $P - \delta$ effect was considered, and the damping ratio was set as zero.

Five input waves, that is, El Centro NS, Hachinohe NS/EW, Kobe NS, and Tohoku NS, were used, and the analysis results were derived as the average response for the five input waves. The intensity of input acceleration was scaled by trial-and-error procedure to obtain the responses at the limit state of the structure.

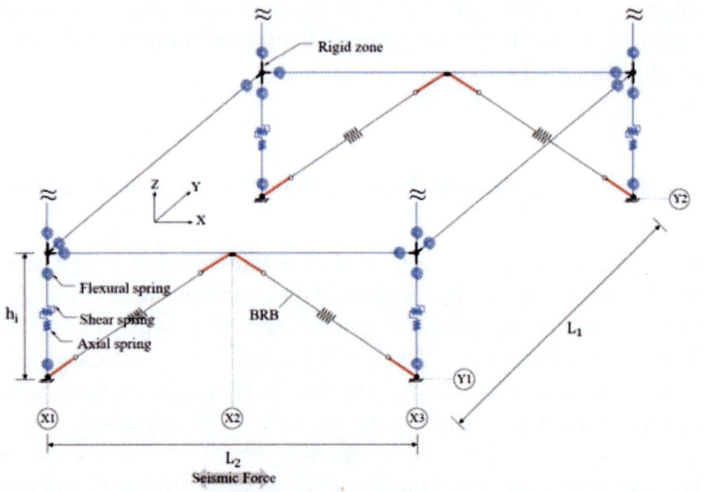

Fig. 1 Analysis model without torsional effect

The limit state is the case in which (1) the accumulative plastic deformation ratio of any frame member, η_f, reaches 10, (2) η_s of any damper reaches 500, or (3) the story drift, θ_i, reaches 5.0% rad. The limit state of the analysis model was determined by any of the conditions (1) to (3) that occurred first. CANNY (v12.0) software was used for the nonlinear time history analysis.

In this study, the damage was evaluated using the accumulative plastic deformation ratio η, which is the cumulative damage by non-dimensionalizing the cumulative plastic strain energy W_p by the yield story shear force Q_y. and the yield deformation δ_y, as $\eta = W_p/(Q_y\delta_y)$. Figure 2 shows the accumulative plastic deformation ratio of the frame according to r_δ. η_f/η in the vertical axis is the ratio of the damage of the frame, η_f, to the total damage of the frame and dampers, $\eta = \eta_f + \eta_s$. η_f/η decreased as the yield deformation ratio r_δ increased, and it was less than 5% for $r_\delta > 4.0$. This is because the energy dissipation caused by the plastification of the dampers increased as the yield deformation of the dampers decreased in comparison with that of the frame.

Akiyama [5] proposed a damage distribution evaluation equation for the shear-type multi-story frame with elements presenting perfectly elasto-plastic hysteretic characteristics having different stiffness on the same floor as follows.

$$\eta_s/\eta_f = r_\delta^2 \tag{1}$$

Equation (1) evaluates the ratio of the damage of stiff and flexible elements, η_s/η_f, as proportional to the square of r_δ. An attempt was made to verify the applicability of Eq. (1) to cases other than perfectly elasto-plastic elements, such as the hysteretic characteristics of the buckling-restrained brace—asymmetric hysteresis characteristics on both the compression and tension sides with secondary stiffness. The analysis results were compared with Eq. (1) and are shown in Fig. 3. The vertical axis represents the ratio of the damage of the dampers to the frame, η_s/η_f. The range of $\eta_s/\eta_f > 19$ corresponds to the range of $\eta_f/\eta < 5\%$ in which the damage to the frame becomes less than 5% of the total damage.

Equation (1) was found to conservatively evaluate the damage of the frame in the range of $\eta_s/\eta_f < 19$, excluding a partial range of the yield strength ratio ($r_q < 1.0$ corresponding to the case in which the yield strength of the dampers is set to be higher than that of the frame). Excluding this range, the damage of the frame was evaluated to be on the safe side in the typical design range of dampers, i.e., $1.5 < r_q < 2.5$ [2].

When the energy absorbed by the frame and dampers were evaluated according to the yield deformation ratio and yield strength ratio, it was found that the yield strength ratio in the range of $1.5 < r_q < 2.5$ had little influence on the amount of energy absorbed by the dampers. Meanwhile, when the yield deformation ratio was 10 or higher, the amount of energy absorbed by the dampers decreased, as the limit state of the entire structure was determined by that of the dampers. Based on the above results, the design ranges of the yield deformation ratio and yield strength ratio for transferring most of the total damage to the dampers and maximizing the

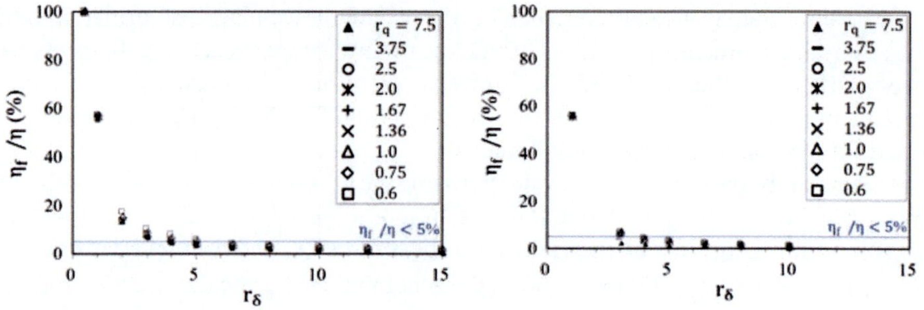

(a) SDOF model (b) MDOF model (5 stories)

Fig. 2 Accumulative plastic deformation ratio of the frame according to the yield deformation ratio

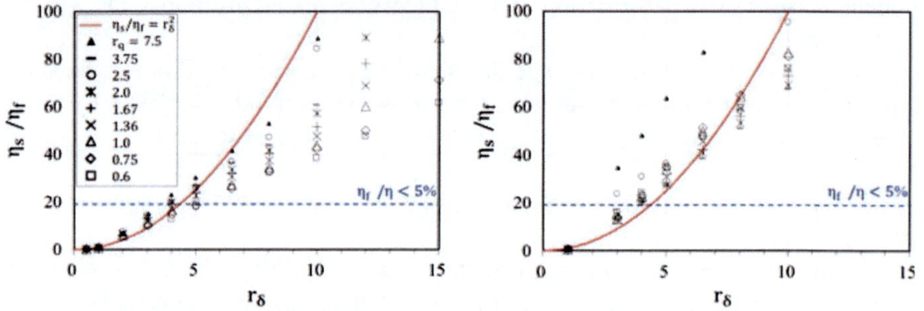

(a) SDOF model (b) MDOF model (5 stories)

Fig. 3 Accumulative plastic deformation ratio between the frame and dampers

energy absorption capacity of the entire structure by using dampers were found to be $4.0 < r_\delta < 10$ and $1.5 < r_q < 2.5$, respectively.

3.2 *Story Damage Distribution of the Dampers*

For the distribution of story shear force coefficient ${}_s\overline{\alpha}_i$ in the dampers, four design standards (ASCE7-16, UBC97, NZS1170, AIJ96) and the distribution by Akiyama [5] were considered. The distribution by Akiyama [5] was proposed to evenly distribute the damage to the frame on each floor in a frame structure. The dampers are the main energy absorption elements in the target design range, that is, the design range in which the damage of the frame is less than 5% of the total damage; however, it is necessary to evenly distribute the generated damage to the dampers on each floor. Therefore, the distribution by Akiyama [5] was applied to the design of dampers to evaluate its applicability. Meanwhile, the aforementioned five distributions were applied to the design of the frame to evaluate the effect of the distribution of story shear force coefficient in the frame, ${}_{f}\overline{\alpha}_{i}$, on the damage distribution of the dampers throughout the height of the structure.

The yield story shear force coefficient of the frame for the first floor, ${}_{f}\alpha_{Y1}$, was 0.15, and the yield story drift ${}_{f}\theta_{Yi}$ was 1/150. The yield strength ratio and yield deformation ratio of the frame and dampers, r_q and r_δ, were set to 2.0 and 8.0, respectively, which are the average values of the design range derived in Sect. 3.1. For the analysis, the limit state, damping ratio, input waves, and hysteretic characteristics of the dampers and frame considered in Sect. 3.1 were applied. To consider the influence of high-order modes, 3, 6, 12, and 18 stories were considered in the analysis model.

Story damage distribution of the dampers according to the distribution of story shear force in the dampers. Figure 4 shows the distribution of story shear force coefficient in the dampers, ${}_{s}\overline{\alpha}_{i}$, calculated in accordance with the five design standards for the 6- and 18-story models. x′ in the vertical axis represents the stories of the model by non-dimensionalizing them in the range between zero and one, and is expressed as $x' = (i - 1)/N$. Here, i denotes the corresponding floor and N the total number of stories. When the dampers were designed in accordance with ASCE7-16, ${}_{s}\overline{\alpha}_{i}$ varied almost linearly according to the height of the structure. In the distributions according to UBC97, NZS1170, and AIJ96, ${}_{s}\overline{\alpha}_{i}$ exhibited a similar tendency as ASCE7-16 up to the middle story, but showed a larger increase relatively at the upper story. For the top floor, it was approximately 1.5 times larger than that under ASCE7-16. On the other hand, the value of ${}_{s}\overline{\alpha}_{i}$ by Akiyama [5] showed relatively a smaller value up to the two third of the total height and increased to a greater extent in the upper story. The frame was designed according to the distribution of story shear force in ASCE7-16.

The story damage distribution of the dampers according to the distribution of story shear force was evaluated based on the distribution of the accumulative plastic deformation ratio, and the results are shown in Fig. 5. η_{si}/η_s in the vertical axis represents the accumulative plastic deformation ratio of the dampers for the i-th floor, η_{si}, to the accumulative plastic deformation ratio for all dampers, η_s. The damage, evaluated based on the standard code design was more concentrated on the upper story, as the story shear force of the upper story was evaluated to be smaller. As a result, the degree of damage concentration on a certain floor of the upper story showed an increase tendency. In addition, as the number of stories increased, the degree of damage concentration increased. For instance, in the 18-story model designed in accordance with ASCE7-16, the damage was concentrated on the roof story, which accounted for 80% of the total damage. However, in the models designed in accordance with UBC97, NZS1170, and AIJ96, the concentration of damage on the upper story was slightly smaller. In the models designed in accordance with Akiyama [5], the damage was evenly distributed among the dampers on each floor regardless of the number of stories.

Figure 6 shows the energy absorption capacity of the entire structure in the limit state according to the distribution of the story shear force in the dampers. The vertical axis represents the ratio of the energy equivalent velocity V_D when the dampers are designed in accordance with the four design standards to the energy equivalent velocity $V_{D(\overline{\alpha}_i)}$ when the dampers are designed according to the distribution of the story shear force coefficient $\overline{\alpha}_i$ of Akiyama [5]. V_D is obtained by $E_D = MV_D^2/2$. E_D is the plastic strain energy of the frame and dampers. M is the effective mass of the structure. For the models designed in accordance with ASCE7-16, the damage to the dampers was insignificant in the lower story, because the damage was concentrated on the roof story or the dampers did not contribute to the energy adsorption, as they remained in the elastic region. Therefore, the energy absorption capacity of the entire structure was significantly reduced. The energy absorption capacity of the 18-story model designed in accordance with ASCE7-16 was approximately 70% lower than that of the model designed in accordance with Akiyama [5]. The energy absorption capacity of the models designed in accordance with UBC97, NZS1170, and AIJ96 was higher than that of the models designed with ASCE7-16, but it remained approximately 40–80% of that of the models designed in accordance with Akiyama [5].

Story damage distribution of the dampers according to the distribution of story shear force in the frame. Figure 7 shows the story damage distribution of the dampers according to the distribution of story shear force in the frame, η_{si}/η_s. The dampers were designed according to the distribution of story shear force by

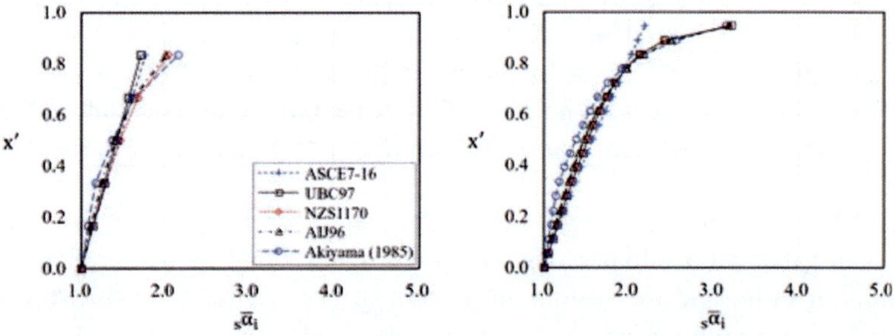

(a) 6 stories (b) 18 stories

Fig. 4 Distribution of story shear force coefficient in the dampers according to the design standards

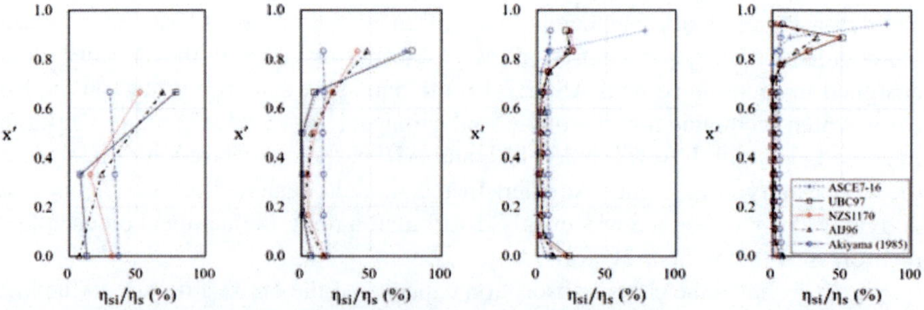

(a) 3 stories (b) 6 stories (c) 12 stories (d) 18 stories

Fig. 5 Story damage distribution of the dampers according to the distribution of story shear force

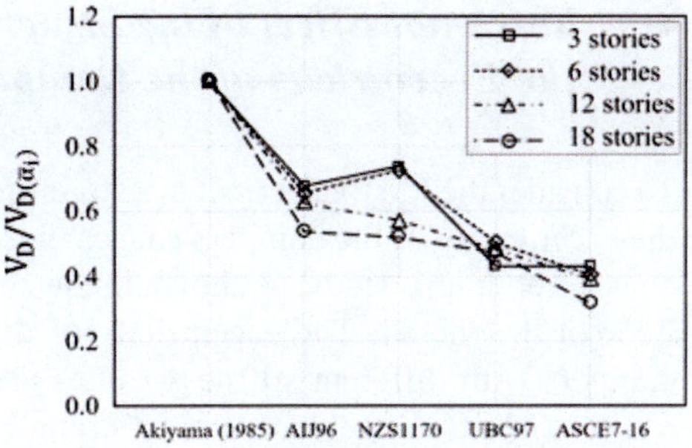

Fig. 6 Energy absorption capacity of the entire structure according to the distribution of story shear force in the dampers

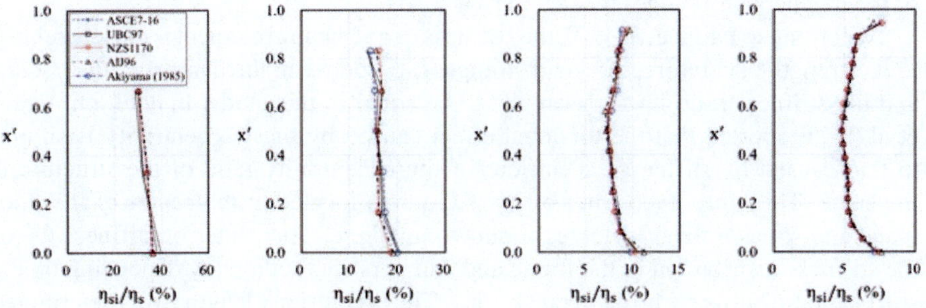

(a) 3 stories (b) 6 stories (c) 12 stories (d) 18 stories

Fig. 7 Story damage distribution of dampers according to the distribution of story shear force in the frame

Akiyama [5]. It was found that the distribution of story shear force in the frame had an insignificant influence on the story damage distribution of the dampers. therefore, the distribution of story shear force in the frame hardly affected the energy absorption capacity of the entire structure in the limit state.

4 Response Characteristics of Structures with Dampers Considering the Torsional Effect

In this section, the energy distribution characteristics of structures with combined translational and torsional modes are analyzed. In Sects. 4.2 and 4.3, damage distribution between the frame and dampers and the characteristics of the story damage distribution are analyzed. Based on the results, the design range of the parameters of the dampers for minimizing the damage to the frame and inducing most of the damage to be concentrated on the dampers, even in the presence of the torsional effect, is proposed. In Sect. 4.4, the distribution of story shear force in the dampers in consideration of the torsional effect is proposed, and its validity is verified.

4.1 Torsional Effect of the Structure Considering the Eccentricity of the Dampers

To consider the torsional effect in a structure with hysteretic dampers, a case in which the eccentricity of the dampers causes the torsional effect of the entire structure was considered in this study, as shown in the analysis model in Fig. 8, which depicts one floor of the model. The eccentricity of the dampers was considered by designing dampers with different stiffness values on the same floor. The dampers located in row Y1 (flexible edge) have a yield strength of ${}_sQ_{Y1}$ and stiffness of k_{s1}. These values were set to be lower than ${}_sQ_{Y2}$ and k_{s2}, which were the yield strength and stiffness of the dampers (stiff edge) located in row Y2.

According to Fajfar et al. [17], the stiffness of a structural element is closely related to its strength. Therefore, the yield strength ratio between the dampers, ${}_sQ_{Y2}/{}_sQ_{Y1}$, and the stiffness ratio k_{s2}/k_{s1} were set to be equal in this study. In addition, Fajfar et al. [17] reported that the torsional effect caused by mass eccentricity is similar to that caused by stiffness eccentricity if the eccentricity ratio of the structure is the same. Therefore, the center of mass C_M was located at the center of the floor plane and set as a fixed variable, as shown in Fig. 8. The center of stiffness C_R of the stiffness distribution in the frame and dampers was set to vary depending on the stiffness ratio between the dampers, k_{s2}/k_{s1}. The eccentricity length e was determined using the difference between the center of mass and the center of stiffness, and the eccentricity ratio λ was calculated using the eccentricity length e and the diaphragm area $\sqrt{L_1^2 + L_2^2}$, as shown in Eq. (2).

$$\lambda = e'/\sqrt{12},\ e' = e\sqrt{12}/\sqrt{L_1^2 + L_2^2},\ e = \frac{(L_1 k_{s2}/k_{s1})}{(1 + k_{s2}/k_{s1})} - L_1/2 \quad (2)$$

The eccentricity ratio λ was assigned a value of up to 40% according to the ratio between the plane widths, L_2/L_1, and the stiffness ratio between the dampers, k_{s2}/k_{s1}. This was intentionally set to derive the general tendency of the torsional response according to the degree of eccentricity of the structure by analyzing the response characteristics for the wide range of eccentricity from realistic to significantly large eccentricity.

4.2 Damage Distribution Between the Frame and Dampers Considering the Torsional Effect

The analysis model was designed as a five-story model that has a torsional effect due to the difference in damper stiffness as shown in Fig. 8. The dampers were designed based on the distribution of story shear force according to Akiyama [5], which evenly distributes the damage in the damper on each floor of a structure without eccentricity,

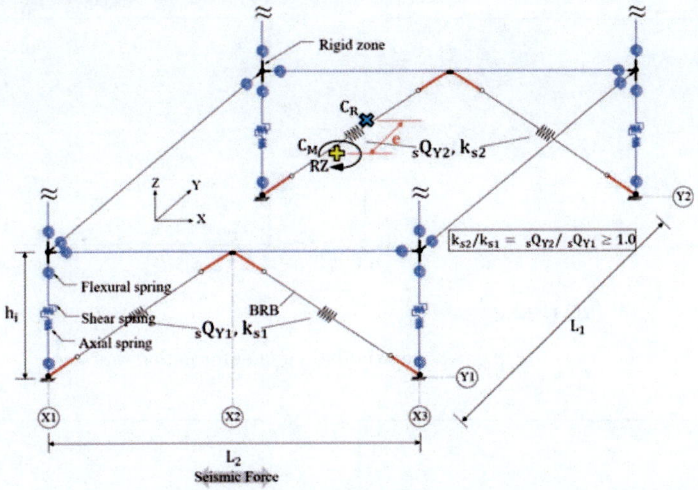

Fig. 8 Analysis model with torsional effect

as mentioned in Sect. 3.2. The frame was designed based on the distribution of the story shear force specified in ASCE7-16. The yield story shear force coefficient of the frame for the first floor, ${}_f\alpha_{Y1}$, was 0.15, and the yield story drift, ${}_f\theta_{Yi}$, was 1/150. For the yield-strength ratio between the dampers and frame, r_q, nine cases were considered with ranging from 0.6 to 7.5. For the yield deformation ratio between the dampers and frame, r_δ, 12 cases were considered with ranging from 0.2 to 15. L_2/L_1 was 0.333, and five cases were considered for k_{s2}/k_{s1} ranging from 1.5 to 11. Accordingly, the eccentricity ratio λ was assigned a value of up to 40%. $P-\delta$ effect was considered in the analysis, and the damping ratio was set as zero. The hysteretic characteristics of the dampers and frame, the input waves, and the definition of the limit state are as described in Sect. 3.

Figure 9 shows the accumulative plastic deformation ratio of the frame according to the yield deformation ratio r_δ and yield strength ratio r_q. ${}_f\eta_\lambda/\eta_\lambda$ in the vertical axis is the ratio of the damage to the frame, ${}_f\eta_\lambda$, to the total damage to the dampers and frame, $\eta_\lambda = {}_f\eta_\lambda + {}_s\eta_\lambda$. As shown in Fig. 9a, the damage of the frame was dominated by the yield deformation ratio, but it also clearly showed a tendency to change depending on the eccentricity ratio. When r_δ was 10 or higher, the influence of the eccentricity ratio was small. However, when the eccentricity ratio was 20% or higher and r_δ was less than 10, the damage of the frame showed a tendency to increase as the eccentricity ratio increased. In addition, Fig. 9b shows that the damage of the frame increased as the yield strength ratio decreased. In the symmetric system, the effect of the yield strength ratio on the damage of the frame was negligible, and the appropriate design range of the yield deformation ratio for limiting the damage to the frame to less than 5% of the total damage was simply $r_\delta > 4.0$. However, in the asymmetric system, it was found that the eccentricity ratio and yield strength ratio as well as the yield deformation ratio affected the damage of the frame, and the damage

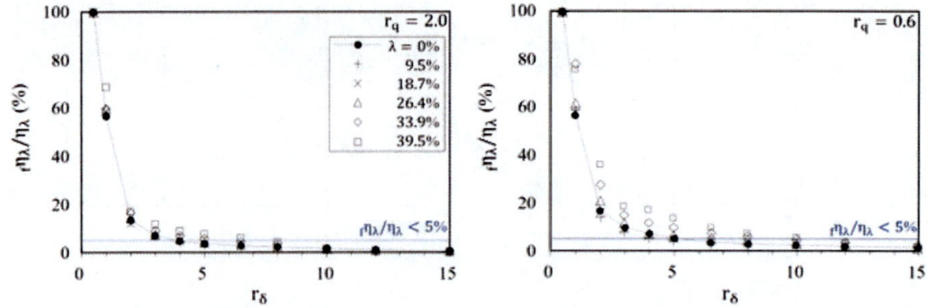

(a1) r_q=2.0 (a2) r_q=0.6

(a) $_f\eta_\lambda/\eta_\lambda$ according to yield deformation ratio r_δ

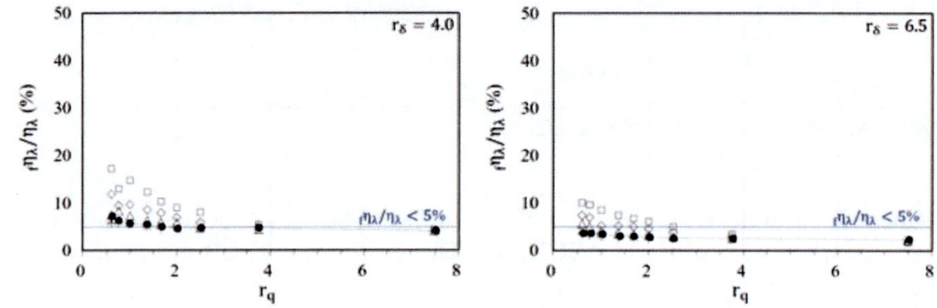

(b1) r_δ=4.0 (b2) r_δ=6.5

(b) $_f\eta_\lambda/\eta_\lambda$ according to yield deformation ratio r_q

Fig. 9 Accumulative plastic deformation ratio of the frame according to the yield deformation ratio, yield strength ratio, and eccentricity ratio

to the frame increased within the specific range of the eccentricity ratio and yield strength ratio.

Figure 10 shows the ratio of the accumulative plastic deformation ratio between the frame and dampers, $_s\eta_\lambda/_f\eta_\lambda$. The range of $_s\eta_\lambda/_f\eta_\lambda > 19$ corresponds to the range in which the damage to the frame is less than 5% of the total damage ($_f\eta_\lambda/\eta_\lambda < 5\%$). When the eccentricity ratio was 20% or higher, the damage of the frame increased as the eccentricity ratio increased because $_s\eta_\lambda/_f\eta_\lambda$ decreased. In addition, as the yield strength ratio decreased, the change in $_s\eta_\lambda/_f\eta_\lambda$ according to the eccentricity ratio increased. For $r_\delta = 7$ and $\lambda = 39.5\%$, the $_s\eta_\lambda/_f\eta_\lambda$ values were 22.7 and 5.8, respectively, at $r_q = 3.75$ and 0.6. The corresponding values for the damage of the frame, $_f\eta_\lambda/\eta_\lambda$, were 4% and 15%. As the yield strength ratio changed, the damage of the frame increased by more than three times.

To evaluate whether Eq. (1) is also effective for an asymmetric system, Eq. (1) was compared with the responses, as shown in Fig. 10. When the eccentricity ratio was less than 20%, Eq. (1) was effective because the eccentricity ratio and yield strength ratio had little influence on the damage distribution between the dampers

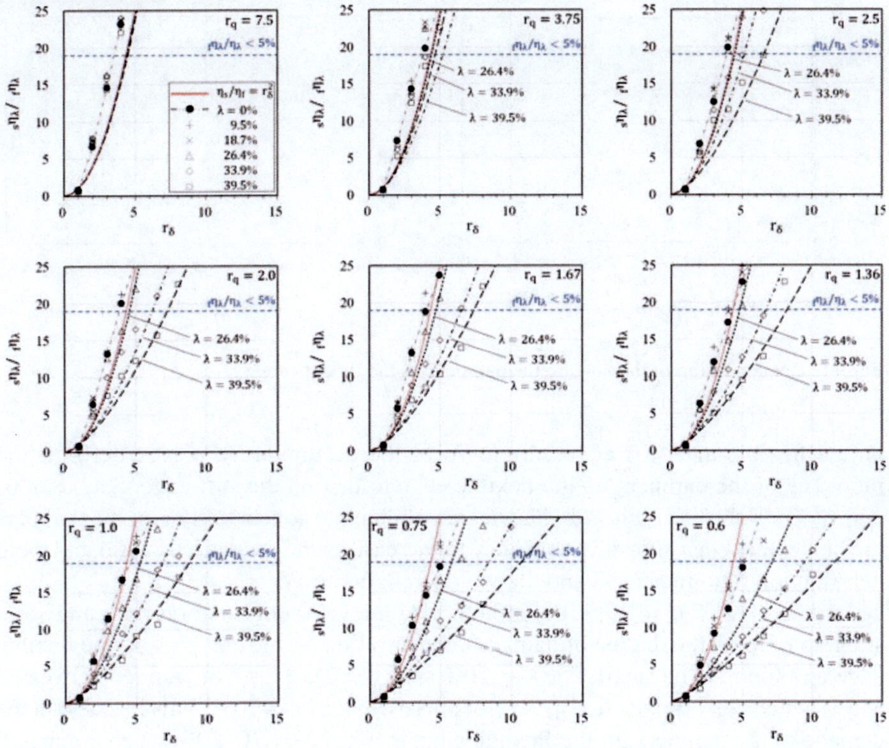

Fig. 10 Evaluation of damage distribution between the frame and dampers considering the torsional effect

and frame. When the eccentricity ratio was 20% or higher, the damage of the frame was underestimated. For $r_\delta = 5$, $r_q = 1.36$, and $\lambda = 39.5\%$, the response was 9.5, but the value obtained from Eq. (1) was 25. When these were calculated as the damage of the frame, ${}_f\eta_\lambda/\eta_\lambda$, the results were 10% and 4%, respectively, thus underestimating the damage to the frame by more than two times. Therefore, for an asymmetric system, Eq. (1) is modified to Eq. (3) so that the influence of the yield strength ratio and eccentricity ratio can be considered.

$$ {}_s\eta_\lambda/{}_f\eta_\lambda = r_\delta^{f(r_q,\lambda)}, \; f(r_q, \lambda) = 2 - (6e - 9)\lambda^a, \quad a = 5.37e^{-0.037 r_q} \tag{3} $$

${}_s\eta_\lambda/{}_f\eta_\lambda$ of the asymmetric system was evaluated using Eq. (3), and the results are shown in Fig. 10. Equation (3) was found to appropriately predict the damage distribution between the frame and dampers according to the eccentricity ratio λ and yield strength ratio r_q when the eccentricity ratio was 20% or higher.

In a structure with the torsional effect, the damage is concentrated on the members around the flexible edge due to the amplification of the displacement of the flexible edge with relatively small stiffness. To compare the damage to the dampers located

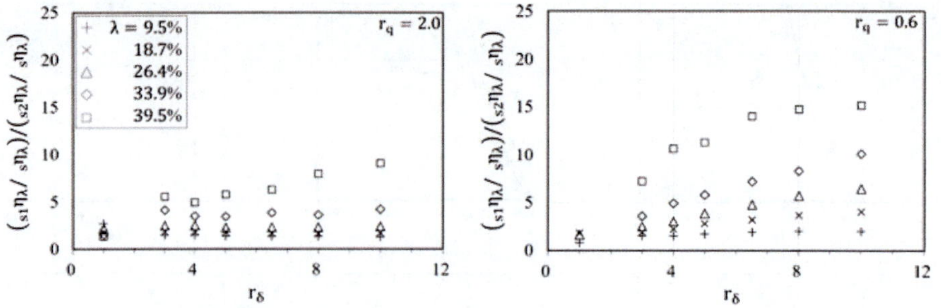

(a) r_q=2.0 (b) r_q=0.6

Fig. 11 Concentration of damage on the dampers on the flexible edge

on the flexible and stiff edges, the ratio of the accumulative plastic deformation ratio ${}_{s1}\eta_\lambda$ of the dampers on the flexible edge to that on the stiff edge, ${}_{s2}\eta_\lambda$, that is, ${}_{s1}\eta_\lambda/{}_{s2}\eta_\lambda$, is shown in Fig. 11. The degree of damage concentration on the dampers on the flexible edge showed a tendency to increase as the eccentricity ratio and yield deformation ratio increased and the yield strength ratio decreased.

In the range of $\lambda < 20\%$, the influence of the yield deformation ratio and yield strength ratio on the degree of damage concentration on the dampers on the flexible edge was found to be small. For $\lambda < 20\%$ and $r_q = 2.0$, ${}_{s1}\eta_\lambda/{}_{s2}\eta_\lambda$ ranged between 2 and 3, which corresponds to ${}_{s1}\eta_\lambda/{}_{s}\eta_\lambda$ of approximately 60–70%. This means that the damage to the dampers on the flexible edge increased by 10–20% when compared with that in the symmetric system. In the range of $\lambda > 20\%$, the concentration of damage clearly showed a tendency to increase as the yield deformation ratio increased and the yield strength ratio decreased. For $\lambda > 20\%$, $r_q = 2.0$, and $6 < r_\delta < 10$, ${}_{s1}\eta_\lambda/{}_{s}\eta_\lambda$ was found to be approximately 85%. This indicates that most of the damage was concentrated on the dampers on the flexible edge.

4.3 *Story Damage Distribution of the Dampers Considering the Torsional Effect*

The analysis model was designed such that the yield strength ratio r_q is 2.0 and the yield deformation ratio r_δ is 8.0. These correspond to the average values in the design range in which the damage of the frame ${}_{f}\eta_\lambda/\eta_\lambda$ is less than 5% at the maximum eccentricity ratio of 40% based on the result derived in Sect. 4.2. Accordingly, the story damage distribution of the dampers was analyzed under the condition that negligible damage occurs in the frame. L_2/L_1 was 0.333, and six cases were considered for k_{s2}/k_{s1} ranging from 1.5 to 11.7. Therefore, the eccentricity ratio λ was assigned a value of up to 40%. $P - \delta$ effect was considered for the analysis, and the damping ratio was set as zero. To consider the characteristics of the story damage distribution according to the natural period, 3, 6, 12, and 18 stories were set for the analysis

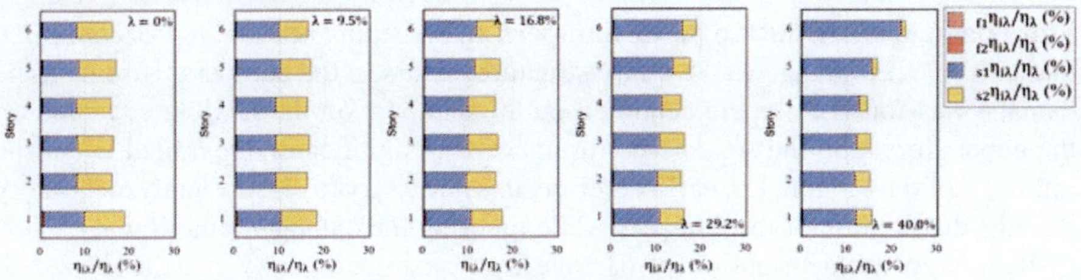

(a) 6-story model

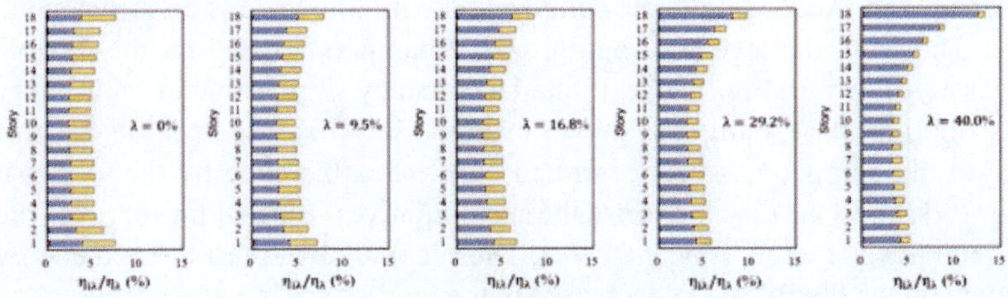

(b) 18-story model

Fig. 12 Story damage distribution of the frame and dampers

model. The dampers and frame were designed based on the distribution of story shear force by Akiyama [5] and ASCE7-16, respectively. Other design parameters and analysis conditions were considered in the same manner as in Sect. 4.2.

Story damage distribution of the frame and dampers. Figure 12 Shows the story damage distribution $\eta_{i\lambda}/\eta_{\lambda}$ for the 6- and 18-story models. The horizontal axis represents the ratio of ${}_{s1}\eta_{i\lambda}$ of the dampers and ${}_{f1}\eta_{i\lambda}$ of the frame on the flexible edge, ${}_{s2}\eta_{i\lambda}$ of the dampers and ${}_{f2}\eta_{i\lambda}$ of the frame on the stiff edge, on the i-th floor to the total accumulative plastic deformation ratio of the entire structure, η_{λ}. Therefore, the damage to the frame on the i-th floor, ${}_{f}\eta_{i\lambda}/\eta_{\lambda}$, is the sum of ${}_{f1}\eta_{i\lambda}/\eta_{\lambda}$ and ${}_{f2}\eta_{i\lambda}/\eta_{\lambda}$. The damage to the dampers on the i-th floor is the sum of ${}_{s1}\eta_{i\lambda}/\eta_{\lambda}$ and ${}_{s2}\eta_{i\lambda}/\eta_{\lambda}$. The damage to the frame ${}_{f}\eta_{\lambda}/\eta_{\lambda}$ was found to range from 0.71 to 2.0% regardless of the number of stories, as intended in the design, thereby meeting the condition of ${}_{f}\eta_{\lambda}/\eta_{\lambda} < 5\%$.

In the symmetric system ($\lambda = 0\%$), the damage to the dampers, ${}_{s}\eta_{i\lambda}/\eta_{\lambda}$, was found to be almost evenly distributed on all floors, as verified in Sect. 3.2. In the asymmetric system, the damage to the dampers tended to be concentrated on the upper story as the eccentricity ratio increased. In addition, the damage tended to be concentrated on the dampers on the flexible edge by the torsional motion as the eccentricity ratio increased, regardless of the number of stories. When the eccentricity ratio was 30% or higher, the damage to the dampers on the flexible edge when compared with the damage to all dampers, ${}_{s1}\eta_{\lambda}/{}_{s}\eta_{\lambda}$, was found to be between 80 and 90%. Figure 11 shows that ${}_{s1}\eta_{\lambda}/{}_{s2}\eta_{\lambda}$ is approximately 8.0 for $\lambda = 39.5\%$, $r_q = 2.0$, and $r_\delta = 8.0$, which corresponds to ${}_{s1}\eta_{\lambda}/{}_{s}\eta_{\lambda}$ of approximately 90%.

Story damage distribution of the dampers. In preceding section, it was confirmed that most of the damage to the entire structure occurs in the dampers. However, the damage was found to be concentrated on the dampers on the flexible edge and on the upper story, and the seismic performance was significantly degraded under the influence of the torsional effect. Therefore, it is necessary to closely analyze the story damage distribution of the dampers while ignoring the damage to the frame.

Regarding the distribution of damage in the dampers, ${}_{s}\eta_{i\lambda}/{}_{s}\eta_{\lambda}$, the degree of damage concentration on the upper story increased as the eccentricity ratio increased, as shown in Fig. 13a. Accordingly, the damage to the middle and lower story tended to decrease. The story damage distribution of the dampers located on the flexible edge, ${}_{s1}\eta_{i\lambda}/{}_{s1}\eta_{\lambda}$, shown in Fig. 13b, exhibited a tendency similar to that of the story damage distribution of all dampers, shown in Fig. 13a. This was because most of the damage to the dampers was concentrated on the flexible edge by the torsional motion. Fig. 13d shows the distribution of the accumulative plastic strain energy in the dampers on the flexible edge, ${}_{s1}W_{pi\lambda}/{}_{s1}W_{p\lambda}$. The accumulative plastic strain energy of the dampers on the flexible edge decreased in the lower story but was concentrated on the upper story as the eccentricity ratio increased. This can be identified as the cause for the concentration of damage on the dampers on the flexible edge of the upper story in Fig. 13b.

Figure 13c shows the story damage distribution of the dampers on the stiff edge, ${}_{s2}\eta_{i\lambda}/{}_{s2}\eta_{\lambda}$. The damage to the dampers on the stiff edge tended to decrease in the upper story and be concentrated on the lower story, contrary to the damage to the dampers on the flexible edge. The torsional effect of the structure induced damage distributions with opposite tendencies on the dampers located on the flexible and stiff edges on the same floor.

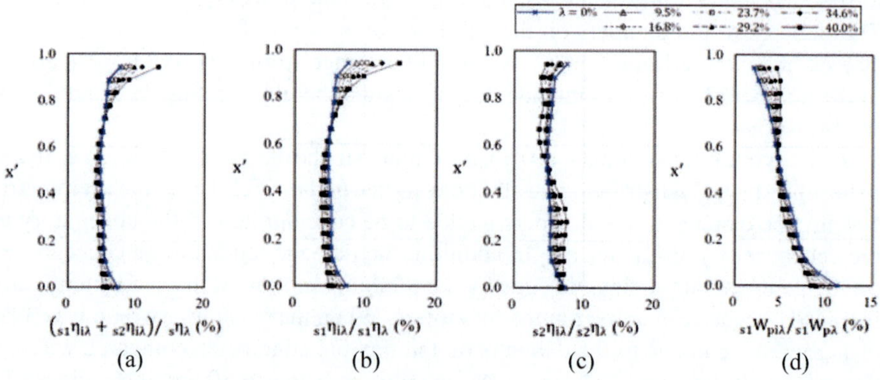

Fig. 13 Story damage distribution of the dampers (3, 6, 12, and 18 stories from the left). **a** Story damage distribution of the dampers; **b** Story damage distribution of the dampers on the flexible edge; **c** Story damage distribution of the dampers on the stiff edge; **d** Distribution of accumulative plastic strain energy in the dampers on the flexible edge

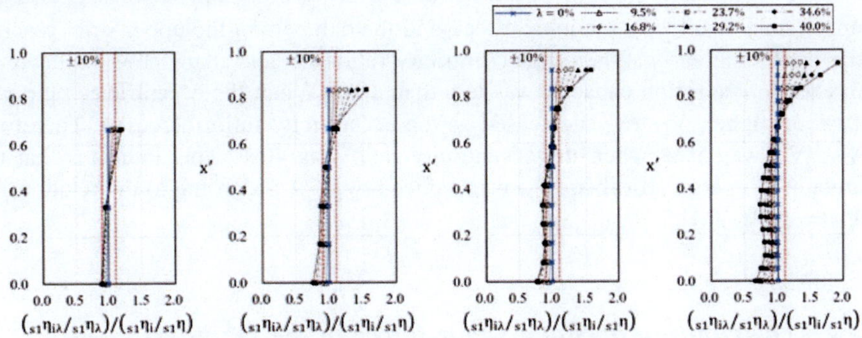

(a) 3 stories (b) 6 stories (c) 12 stories (d) 18 stories

Fig. 14 Difference in the story damage distribution between the asymmetric and symmetric systems (3, 6, 12, and 18 stories from the left)

Evaluation of damage concentration of the dampers. Figure 14 shows the ratio of the damage distribution in the dampers on the flexible edge, ${}_{s1}\eta_{i\lambda}/{}_{s1}\eta_{\lambda}$, to the damage distribution in the models without eccentricity, ${}_{s1}\eta_i/{}_{s1}\eta$. When the eccentricity ratio was less than 10%, the difference from the damage distribution in the symmetric system was found to be less than +/−10%. When the eccentricity ratio was 10% or higher, the damage in the asymmetric system, ${}_{s1}\eta_{i\lambda}/{}_{s1}\eta_{\lambda}$, was gradually concentrated to the upper and roof story, and decreased in the middle and lower story with increase in the eccentricity ratio, when compared with the damage distribution in the symmetric system, ${}_{s1}\eta_i/{}_{s1}\eta$. This tendency was more noticeable as the number of stories increased.

Figure 15 shows the energy absorption capacity according to the eccentricity ratio. V_D and $V_{D\lambda}$ are the energy dissipated at the limit state in the symmetric and asymmetric systems, respectively. $V_{D\lambda}/V_D$ differed depending on the input wave, but its average value was found to decrease as the eccentricity ratio increased regardless of the number of stories. This is because the dampers in the middle and lower story could not dissipate sufficient energy, as the damage was concentrated on the dampers in the

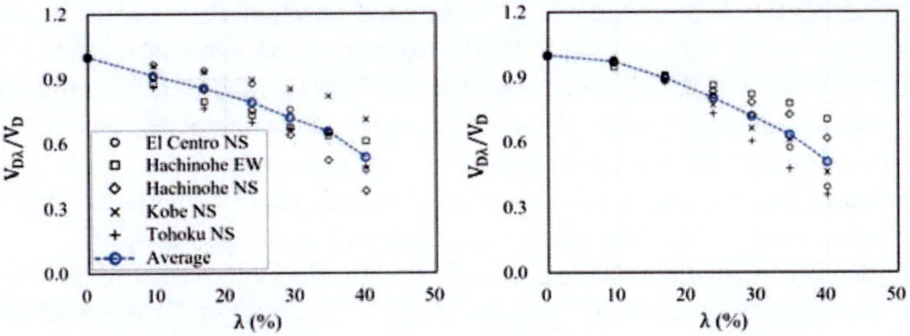

(a) 3 stories (b) 18 stories

Fig. 15 Comparison of energy absorption capacity between the asymmetric and symmetric systems upper story due to the torsional effect, and the dampers in the upper story reached the limit state early. When the eccentricity ratio was less than 10%, the decrease in energy absorption capacity was less than 10%. When the eccentricity ratio was 10% or higher, $V_{D\lambda}/V_D$ decreased as the eccentricity ratio increased. Therefore, $V_{D\lambda}/V_D$ was 50% when the eccentricity ratio was 40%. This indicates that the amount of energy absorbed, E_D, was reduced by 75% according to the relationship $V_D = \sqrt{2E_D/M}$.

4.4 Distribution of Story Shear Force in the Dampers Considering the Torsional Effect

By torsional effect, a considerable proportion of the damage to the dampers was concentrated on the dampers on the flexible edge in the upper story, and thus the dampers in the middle and lower story could not absorb sufficient energy. Accordingly, the energy absorption capacity of the entire structure was significantly degraded. Therefore, it is necessary to deduce the distribution of story shear force that can evenly distribute damage to all floors. Because the response amplification due to the torsional motion is relatively small on the stiff edge, the damage to the dampers on the stiff edge is not directly related to the limit state of the structure. Therefore, an attempt was made to propose the distribution of story shear force in the dampers to control the damage distribution on the flexible edge.

Design story shear force distribution in the dampers considering the torsional effect. An analysis was conducted on structures with an eccentricity ratio of 10% or higher, for which the difference from the damage distribution in the symmetric system was more than 10%, as shown in Fig. 14, and the decrease in energy absorption capacity in comparison with the symmetric system exceeded 10%, as shown in Fig. 15. The analysis to derive the distribution of story shear force in the dampers considering the torsional effect was conducted using the following procedure. (1) The yield strength of the dampers, on a floor where the damage was concentrated, was increased. (2) At the limit state, the story damage distribution was examined to check whether the damage was evenly distributed across all floors. If the distribution was less than 10% compared with the distribution in the symmetric system, it was considered satisfactory. Otherwise, (3) Analysis was conducted repeatedly until the damage was evenly distributed across all floors by increasing or decreasing the yield strength. (4) Steps (1) to (3) were independently performed for 3-, 6-, 12-, and 18-story models, five eccentricity ratios ranging between 10–40%, and five input waves. (5) The determined yield strengths were averaged for each input wave.

${}_s\overline{\alpha}_{i\lambda}$, derived through the above procedure is shown in Eq. (4) and Fig. 16. Figure 16 shows that the yield strength of the upper story, ${}_s\overline{\alpha}_{i\lambda}$, increased ${}_s\overline{\alpha}_i$ as the eccentricity ratio increased. This was because the yield strength of the dampers was increased in the upper story to reduce the concentration of damage due to the torsional effect. ${}_s\overline{\alpha}_i$ is the distribution of the story shear force coefficient that evenly distributes the

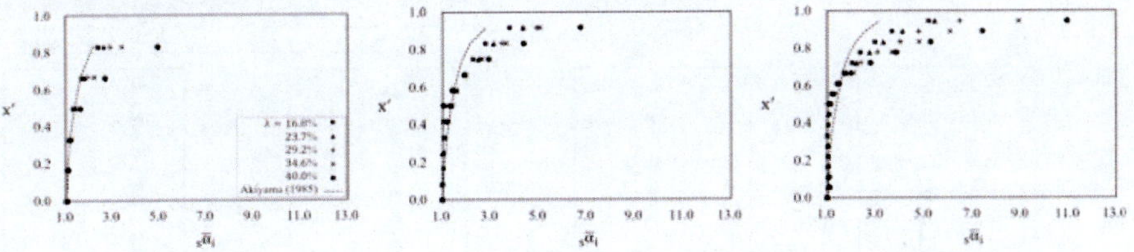

(a) 6 stories (b) 12 stories (c) 18 stories

Fig. 16 Distribution of story shear force in the dampers according to the eccentricity ratio

damage to the damper on each floor in the symmetric system, proposed by Akiyama [5]. On the other hand, ${}_{s}\overline{\alpha}_{i\lambda}$ was evaluated to be smaller than ${}_{s}\overline{\alpha}_{i}$ in the middle and lower story. Figure 17 shows ${}_{s}\overline{\alpha}_{i\lambda}$ according to the number of stories for the same eccentricity ratio. As the number of stories increased, that is, as the natural period increased, the strength of the upper story was evaluated to be larger. This was because the degree of damage concentration on the dampers in the upper story increased as the number of stories increased, as shown in Fig. 14.

$$ {}_{s}Q_{Yi} = {}_{s}\alpha_{Y1} \cdot {}_{s}\overline{\alpha}_{i\lambda} \cdot \sum_{j=i}^{N} m_j g \tag{4} $$

$$ {}_{s}\overline{\alpha}_{i\lambda} = y_0 + a\left(1 - e^{-bx'}\right) + c\left(1 - e^{-dx'}\right), \quad y_0 = 1.0, $$

$$ a = 0.15\lambda^2 + 1.5\lambda - 0.7, \; b = -0.42T^2 + 0.9T + 0.1 $$
$$ c = \lambda^2 - 1.3\ln(T + 3.7x'), \; d = 4.1\lambda + e^{(\lambda^2 + 0.6\lambda - 0.9)} $$

Verification of the proposed distribution of story shear force. To verify the validity of Eq. (4), dampers were designed in accordance with Eq. (4), and the story damage distribution was evaluated. The design parameters and analysis model, excluding the distribution of story shear force in the dampers, were the same as in Sect. 4.3. Figure 18 shows the story damage distribution of the dampers, ${}_{s1}\eta_{i\lambda}/{}_{s1}\eta_{\lambda}$. The concentration of damage on the dampers in the upper story or in the roof story

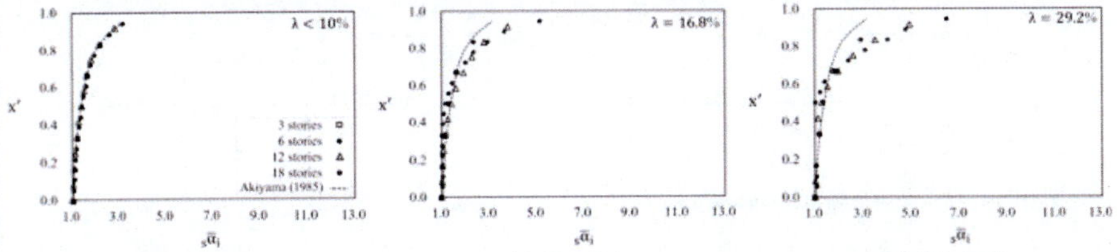

(a) 6 stories (b) 12 stories (c) 18 stories

Fig. 17 Distribution of story shear force in the dampers according to the number of stories

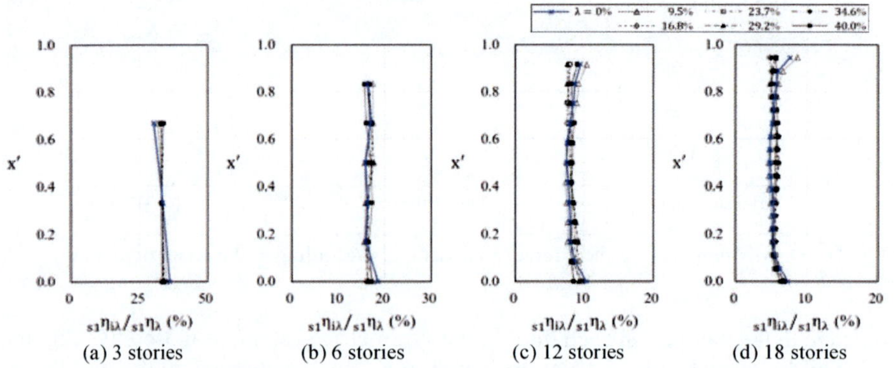

Fig. 18 Story damage distribution of the dampers upon the application of the proposed distribution of story shear force considering the torsional effect

was mitigated, and the damage to the dampers in the middle and lower story was increased, thereby causing the damage to be almost evenly distributed among all floors. Even when the eccentricity ratio was significantly high, differences of less than 10% were observed compared with the damage distribution in the symmetric system. by applying the proposed distribution of story shear force, the dampers exhibited a damage distribution favorable for seismic performance.

Figure 19 shows the energy absorption capacity with $V_{D\lambda({}_s\overline{\alpha}_{i\lambda})}/V_{D\lambda({}_s\overline{\alpha}_i)}$. $V_{D\lambda({}_s\overline{\alpha}_i)}$ and $V_{D\lambda({}_s\overline{\alpha}_{i\lambda})}$ are the energy dissipated by the structure, E_D, converted into V_D when ${}_s\overline{\alpha}_i$ and ${}_s\overline{\alpha}_{i\lambda}$ are applied to the design of the dampers, respectively. Although there were differences depending on the input wave, the maximum average of $V_{D\lambda({}_s\overline{\alpha}_{i\lambda})}/V_{D\lambda({}_s\overline{\alpha}_i)}$ was 1.5. When this was converted into the amount of energy absorbed, an increase of more than two times was observed. As the concentration of damage on the dampers was reduced, almost all dampers could contribute to the energy absorption, thereby significantly improving the energy absorption capacity of the entire structure.

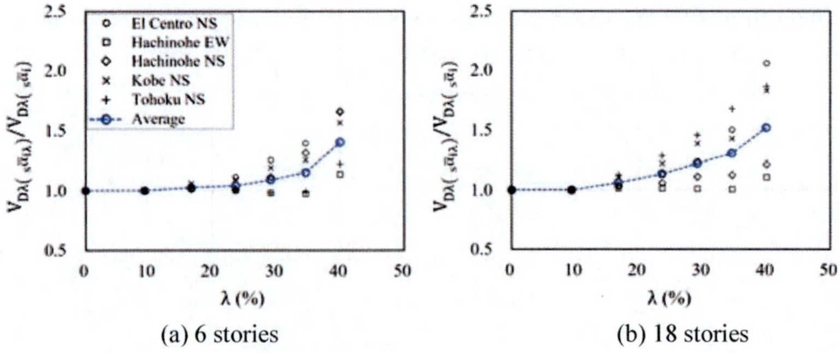

Fig. 19 Evaluation of energy absorption capacity upon the application of the proposed distribution of story shear force considering the torsional effect

5 Suggestion of Energy-Based Design Procedure for Structures with Damping System Considering the Torsional Effect

Based on the results derived in Sects. 3 and 4, a simplified procedure for designing a structure with dampers with consideration of the torsional effect is presented in Table 1. The procedure can be utilized for determining the characteristics of hysteretic dampers, such as the yield strength and the stiffness for each floor, and their distribution throughout the height of the structures without a huge number of analysis.

Step 1. The importance factor of the structure and the target performance level are determined based on ASCE7-16, ASCE41-17, or other standards. The allowable story drift δ_{lim} corresponding to the structural system and the target performance level are determined.

Table 1 Proposed procedure for energy-based design of a structure with dampers considering the torsional effect

Step 1	Target performance objective
	↓
Step 2	Design V_E and S_A response spectrum
	↓
Step 3	Determination of design V_D and base shear
	↓
Step 4	Main structure design by gravity load
	↓
Step 5	Check for deformation range of main structure
	↓
Step 6	Calculation of eccentric ratio
	↓
Step 7	Determination of yield deformation ratio
	↓
Step 8	Yield shear strength coefficient of damper at first story
	↓
Step 9	Calculation of natural fundamental period
	↓
Step 10	Story distribution of yield shear strength
	↓
Step 11	Damper's characteristics considering installation direction
	↓
Step 12	Evaluation of seismic performance by nonlinear dynamic analysis

Step 2. The design acceleration response spectrum and energy response spectrum are calculated. The former is calculated to determine the minimum strength required for the dampers, and the strength of the dampers is determined based on the energy response spectrum.
Step 3. The V_D response spectrum is calculated by applying the damping ratio considered for the design. The base shear force V is calculated. When the frame is considered as a seismic-force-resisting system, the distribution of story shear force is calculated.
Step 4. The frame is designed for the gravity load combination, and the yield strength, stiffness, and yield deformation for each floor of the frame are calculated. When the frame is considered as a seismic-force-resisting system, it is designed based on the story shear force calculated in Step 3.
Step 5. The allowable story drift is compared with the yield deformation for each floor of the frame, ${}_f\delta_{Yi}$. Because the frame should exhibit elastic behavior at the design basis earthquake (DBE) level, the relationship of ${}_f\delta_{Yi} > \delta_{lim,DBE}$ should be satisfied. The plastification of the frame is allowed at the maximum considered earthquake (MCE) level, but $\delta_{lim,MCE}$ should not exceed 2.0 times the value of ${}_f\delta_{Yi}$ to prevent excessive plastic deformation and to maintain some lateral stiffness for safety.
Step 6. The eccentricity distance is calculated from the center of mass and center of stiffness, and it is divided by the diaphragm area to calculate the eccentricity ratio.
Step 7. The yield deformation ratio r_δ is determined using Eq. (3) in the ranges of $4.0 < r_\delta < 10$ and $1.5 < r_q < 2.5$. This is the condition to induce less than 5% of the damage to the entire structure to the frame.
Step 8. The yield story shear force coefficient of the dampers for the first floor, ${}_s\alpha_{Y1}$, should be determined by satisfying the following conditions.

1. Minimum strength according to the base shear force: ${}_s\alpha_{Y1} \geq V/W - {}_f\alpha_{Y1}$
2. Condition to maximize the energy dissipation of the dampers: $r_q = {}_f\alpha_{Y1}/{}_s\alpha_{Y1} < 2.5$
3. Condition for the dampers to absorb the entire input energy: ${}_sW_p > V_D^2M/2$
4. Condition for the damage to the frame being less than 5% of the damage to the entire structure: Eq. (3).

   Here, ${}_f\alpha_{Y1}$ is the yield story shear force coefficient of the frame for the first floor, which is calculated in Step 4. If the eccentricity ratio is less than 20%, it is not necessary to satisfy the condition (4).

Step 9. The natural period with consideration of the torsional mode is calculated.
Step 10. If the eccentricity ratio exceeds 10%, it should be calculated using Eq. (4). The yield strength is obtained from the distribution of the story shear force coefficient of the dampers, and it is divided by the yield deformation ratio to obtain the stiffness.
Step 11. The yield strength, stiffness, and yield deformation are calculated in the installation direction of the dampers, and the corresponding details of the dampers are designed.

Step 12. Nonlinear dynamic analysis is conducted to examine whether the accumulative plastic deformation ratio of the frame is equal to or lower than the limit state. The accumulative plastic strain energy of the dampers is examined to determine whether it is equal to or higher than the energy response spectrum. The distribution of the accumulative plastic deformation ratio of the dampers should be almost even across all floors.

6 Conclusions

In this study, a simplified procedure for the design of structures with hysteretic dampers considering the torsional effect was proposed based on an energy-based design. Considering the torsional motion of the structure, the design methods to ensure the dissipation of most of the seismic energy to the dampers, and the distribution of the story shear force to evenly distribute the damage to the dampers on each floor were proposed and verified. The main conclusions are as follows:

1. As a result of evaluation for the damage distribution between the dampers and frame according to the combination of the design parameters, when there was no torsional effect, it was found that the yield deformation ratio should be 4.0 or higher to induce the accumulative plastic deformation ratio of the frame to be less than 5% of that of the entire structure. When there was a torsional effect, it was found that damage distribution was affected by the yield strength ratio and eccentricity ratio as well as the yield deformation ratio, and the damage to the frame increased as the eccentricity ratio increased. Therefore, an equation for evaluating the damage distribution was proposed using the yield strength ratio, eccentricity ratio, and yield deformation ratio.
2. In the absence of the torsional effect, the analysis of the story damage distribution according to the distribution of the story shear force in the dampers revealed that the damage tended to be concentrated on the dampers located in the upper story when the dampers were designed in accordance with ASCE7-16. The degree of damage concentration increased as the number of stories increased, and maximum 80% of the total damage was concentrated on the dampers in the upper story. When the dampers were designed in accordance with the distribution of yield story shear force proposed by Akiyama [5], the damage concentration on certain dampers was significantly mitigated. Therefore, it was found that the energy absorption capacity of the structure designed in accordance with ASCE7-16 was up to 70% lower than that of the structure designed in accordance with Akiyama [5]. Moreover, the distribution of story shear force in the frame was found to have little influence on the story damage distribution of the dampers.
3. As a result of analyzing the story damage distribution of the dampers with consideration of the torsional effect, it was found that the damage to the dampers tended to be concentrated on the upper story under the influence of torsional motion. Accordingly, the damage to the dampers located in the middle and

lower story was relatively small. In addition, as the response was amplified on the flexible edge with relatively small stiffness, most of the damage was concentrated on the dampers on the flexible edge. Therefore, the damage concentration significantly degraded the energy dissipation capacity of the structure.

4. The distribution of story shear force in the dampers, considering the torsional effect, was proposed to reduce the damage concentration on a certain floor, thereby improving the energy absorption capacity of the structure. When the dampers were designed using the proposed equation and the story damage distribution was evaluated, the damage concentration on the dampers in the upper story and roof story was significantly reduced and the damage to the dampers located in the middle and lower story was increased, thereby causing the damage to be almost evenly distributed on all floors. In addition, the application of the proposed equation increased the amount of energy absorbed by more than two times. Therefore, upon the application of the proposed distribution of story shear force, the dampers exhibited a damage distribution favorable for seismic performance improvement.

References

1. Benavent-Climent, A.: An energy-based method for seismic retrofit of existing frames using hysteretic dampers. Soil Dyn. Earthq. Eng. **31**(10), 1385–1396 (2011)
2. Oh, S.H., Shin, S.H.: A proposal for a seismic design process for a passive control structural system based on the energy equilibrium equation. Int. J. Steel Struct. **17**(4), 1285–1316 (2017)
3. Bagheri, B., Oh, S.H., Shin, S.H.: Distribution of optimum yield-strength and plastic strain energy prediction of hysteretic dampers in coupled shear wall buildings. Int. J. Steel Struct. **18**(4), 1107–1124 (2018)
4. Housner, G.W.: Limit design of structures to resist earthquakes. In: Proceedings of the First World Conference on Earthquake Engineering, vol. 5, pp. 1–13 (1956)
5. Akiyama, H.: Earthquake-Resistant Limit-State Design for Buildings. University of Tokyo (1985)
6. Akbas, B., Shen, J., Hao, H.: Energy approach in performance-based seismic design of steel moment resisting frames for basic safety objective. Struct. Des. Tall Build. **10**(3), 193–217 (2001)
7. Habibi, A., Chan, R.W., Albermani, F.: Energy-based design method for seismic retrofitting with passive energy dissipation systems. Eng. Struct. **46**, 77–86 (2013)
8. Leelataviwat, S., Goel, S.C., Stojadinović, B.: Energy-based seismic design of structures using yield mechanism and target drift. J. Struct. Eng. **128**(8), 1046–1054 (2002)
9. Chou, C.C., Uang, C.M.: A procedure for evaluating seismic energy demand of framed structures. Earthq. Eng. Struct. Dyn. **32**(2), 229–244 (2003)
10. Bagheri, B., Oh, S.H.: Seismic design of coupled shear wall building linked by hysteretic dampers using energy based seismic design. Int. J. Steel Struct. **18**(1), 225–253 (2018)
11. Oh, S.H., Shin, S.H., Bagheri, B.: Stability evaluation of the acceleration and energy response spectra. Soil Dyn. Earthq. Eng. **123**, 124–143 (2019)
12. Oh, S.H., Jeon, J.S.: A study on optimum distribution of story shear force coefficient for seismic design of multi-story structure. Int. J. High-Rise Build. **2**, 121–145 (2014)
13. Chopra, A.K., De la Llera, J.C.: Accidental and natural torsion in earthquake response and design of buildings. In: Eleventh World Conference on Earthquake Engineering (1996)

14. Elms, D.G.: Seismic torsional effects on buildings. Bull. NZ Nat. Soc. Earthq. Eng. **9**(1), 79–83 (1976)
15. De la Llera, J.C., Chopra, A.K.: Accidental torsion in buildings due to base rotational excitation. Earthq. Eng. Struct. Dyn. **23**(9), 1003–1021 (1994)
16. Chopra, A.K., Goel, R.K.: A modal pushover analysis procedure to estimate seismic demands for unsymmetric-plan buildings. Earthq. Eng. Struct. Dyn. **33**(8), 903–927 (2004)
17. Fajfar, P., Marušić, D., Peruš, I.: Torsional effects in the pushover-based seismic analysis of buildings. J. Earthq. Eng. **9**(06), 831–854 (2005)
18. Merritt, S., Uang, C.M., Benzoni, G.: Subassemblage testing of star seismic buckling restrained braces. Report No. TR-2003/04, Final Report to Star Seismic, LLC (2003)

Efficiency of Viscous Damping in Seismic Energy Dissipation and Response Reduction

Fırat Soner Alıcı and Halûk Sucuoğlu

Abstract Input energy accumulates at a specific rate, and viscous damping dissipates the accumulated input energy at a slower pace. The difference between the two energy time histories at a time t is the vibration energy $E_v(t)$, which is the sum of kinetic and potential energies at time t. Maximum displacement occurs shortly after E_v attains its maximum value during the following cycle when potential energy is maximum and kinetic energy is zero. An efficient damping produces lower E_v, accordingly lower maximum displacement. We choose to define the damping efficiency as the ratio of dissipated energy E_D to input energy E_I at the time t_{max} when $E_v(t)$ attains its maximum value for a SDOF system with period T. The influence of earthquake magnitude, fault distance, soil type and fault type on damping efficiency are assessed here under a large set of earthquake ground motions that represent the distribution of such characteristics effectively. A large set of free-field strong motion records are selected from the NGA database. Damping ratio, soil class, distance to epicenter (R_{epi}), moment magnitude (M_w), and fault mechanism are selected as the basic parameters in order to characterize source and site properties of ground motions. Based on the employed GM database, it has been found that damping efficiency is affected most by the earthquake magnitude, soil type, and expectedly by the damping ratio.

Keywords Damping efficiency · Viscous damping · Input energy · Dissipated energy · Vibration energy

F. S. Alıcı (✉)
Başkent University, Ankara 06790, Turkey
e-mail: fsalici@baskent.edu.tr

H. Sucuoğlu
Middle East Technical University, Ankara 06800, Turkey

A. Benavent-Climent and F. Mollaioli (eds.), *Energy-Based Seismic Engineering*, Lecture Notes in Civil Engineering 155,
https://doi.org/10.1007/978-3-030-73932-4_17

1 Introduction

Response spectra of earthquake ground motions associated with damping levels higher than the conventional 5% critical become important for the seismic design, and evaluation of structures equipped with energy-dissipating devices or seismic isolation systems. Therefore, high-damping response spectra and corresponding modification methods generally derived for the conversion of 5% damped response spectra (acceleration or displacement) to high damping conjugate are important ingredients in seismic design and analysis of structures equipped with these type of systems as well as in displacement-based seismic design. However, from the seismic energy point of view, damping effect is not so prominent for input energy spectra when compared to response spectra of ground motions [1, 2]. Evaluation of viscous damping efficiency in terms of energy may be crucial for structures equipped with supplemental damping devices.

Various approaches have been proposed in the literature to approximate spectral response values of systems with high damping. In general, high damping response spectra are derived from a reference 5% damped spectrum by using proper damping scaling (modification) factors, which depend on the equivalent viscous damping value of the structure (>5%). There are several studies proposing different models for damping scaling. Most of them are built on damping ratio, together with spectral period T [3–12]. Some researchers also considered the effects of ground motion duration and site class [13–19].

A number of analogous studies have been conducted in the literature to evaluate the effects of ground motion characteristics and structural properties on damping scaling factors [20–25]. There are common observations as well as disagreements among these studies about the effect of parameters on damping scaling. Earthquake magnitude strongly affects damping scaling; however, the effect of distance to fault is in lesser extent. Based on these observations, they revealed that damping scaling factors decrease with increasing magnitude and slightly decrease with increasing distance, but the effect of site class on damping scaling is not so clear. Furthermore, they also indicated that damping efficiency shows variations based on the response type (acceleration, velocity and displacement).

Damping scaling factors have been most frequently used in displacement based design, and recent research in this field has resulted in new approaches for scaling factors. Few studies have recently appeared in the literature that directly estimate the damping scaling factor for various levels of damping by developing a prediction equation with a more comprehensive worldwide GM dataset [26–31]. In these studies, different parameters are used in the prediction equations. In general, damping ratio, spectral period, earthquake magnitude, distance to fault and site class are the parameters considered in these studies. Recently, damping scaling factors have been estimated for converting the maximum displacements of inelastic displacements to that of elastic systems with equivalent damping [32].

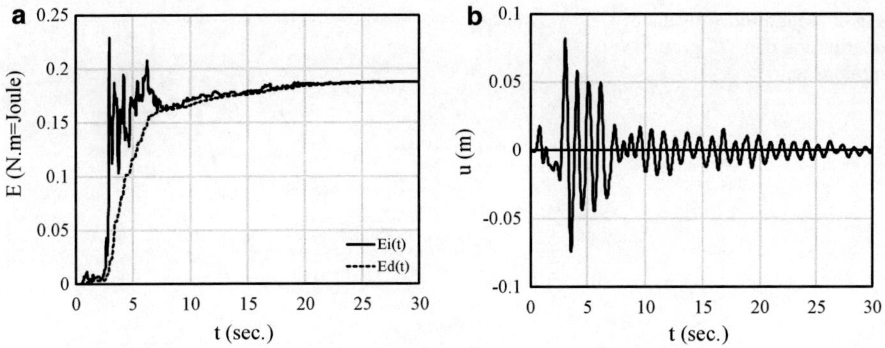

Fig. 1 **a** Energy, **b** displacement time histories. $T = 1$ s, 5% damping

Input energy and dissipated energy time histories of a 5% damped linear elastic SDOF system with a period of 1 s are shown in Fig. 1a. The ground motion excitation was recorded during the 1992 Cape Mendocino M_w 7.01 earthquake, at a fault distance of 10.4 km, on soil type C. Input energy accumulates at a specific rate, and viscous damping dissipates the accumulated input energy at a slower pace. The difference between the two energy time histories at a time t is the vibration energy $E_v(t)$, which is the sum of kinetic and potential energies at time t. Maximum displacement occurs after E_v attains its maximum value during the subsequent cycles, when potential energy is maximum and kinetic energy is zero. This is shown in Fig. 2b. An efficient damping produces lower E_v, accordingly smaller maximum displacement. Hence, higher damping efficiency leads to faster dissipation of accumulated input energy imposed on the system by the ground excitation.

We choose to define the damping efficiency, denoted by e_ξ as a spectral parameter in Eq. (1), which is the ratio of dissipated energy E_D to input energy E_I at the time t_{max} when E_v(t) attains its maximum value for a SDOF system with period T.

$$e_\xi(T) = \frac{E_D(T)}{E_I(T)} \tag{1}$$

t_{max} usually occurs at the earlier stages of ground motion duration, shortly before the peak displacement is attained. Hence, Eq. (1) indirectly expresses the effect of viscous damping in reducing the peak displacement of a linear elastic system. The efficiency of viscous damping is higher when e_ξ approaches 1, and lower when e_ξ approaches 0, i.e. the undamped case.

The influences of earthquake magnitude, fault distance, soil type and fault type on damping efficiency are assessed here under a large set of earthquake ground motions that represent the distribution of such characteristics effectively. Furthermore, the variation of damping efficiency with viscous damping ratio is also investigated.

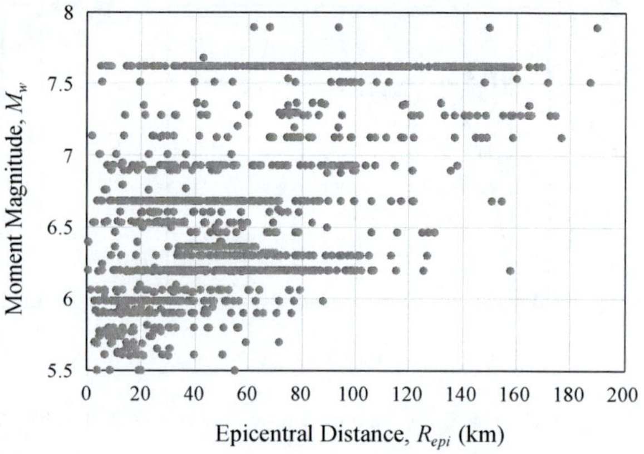

Fig. 2 Magnitude–distance distribution of 1372 ground motions pairs

2 Strong Ground Motions

A large set of free-field strong motion records are selected from the NGA database, which were also employed in a companion study [2, 32, 33]. The data set (Mw ≥ 5.5 and PGA ≥ 0.05g) is composed of 1372 pairs of ground motion records or 2744 horizontal components from 104 earthquakes around the World. Figure 2 shows the scatter diagram of Mw versus Repi for the ground motions in the database. Ground motion sites in the database with V_{S30} > 360 m/s (NEHRP A, B, and C) are designated as stiff soil type, while those with lower V_{S30} values (NEHRP D and E) are specified as soft soil type.

Soil class (S), distance to epicenter (R_{epi}), moment magnitude (Mw), and fault mechanism are selected as the basic parameters in order to characterize source and site properties of ground motions. Based on the employed GM database, four M_w groups (5.5–6.0, 6.0–6.5, 6.5–7.0, > 7.0), two R_{epi} groups (<30 km: Near Fault, > 100 km: Far Fault), three fault types (normal, strike-slip, and reverse), and two soil types (stiff and soft) are considered in the evaluation.

3 Sensitivity of Damping Efficiency to Strong Motions Characteristics

Damping efficiency to several strong motion parameters, namely fault type, magnitude, fault distance and soil type is assessed for 5% damping here. It is assumed that a similar sensitivity holds for other damping ratios as well.

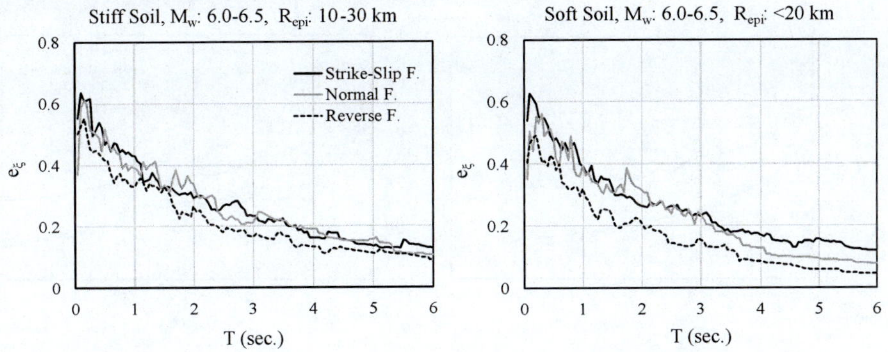

Fig. 3 The effect of fault type on e_ξ for both soil types and two R_{epi} groups

3.1 Fault Type

Damping efficiency spectra is obtained for all three fault types, for the ground motions in the 6.0–6.5 magnitude range, and at near fault distances. The results shown in Fig. 3 do not indicate a consistent sensitivity of damping efficiency to the fault type. The trends are similar for longer fault distances. Accordingly, fault type is eliminated in the following analyses as an influential ground motion source parameter on damping efficiency.

3.2 Magnitude

Figure 4 shows the variations of mean e_ξ for different M_w groups, soft soil type and two R_{epi} groups. It can be observed that damping efficiency is quite sensitive to earthquake magnitude, at both distance ranges. Damping is more efficient at larger magnitudes in dissipating energy, hence in reducing vibration energy. This is due to the fact that large magnitude ground motions impose larger response velocities, accordingly larger damping forces which in turn lead to larger energy dissipation, in comparison to smaller magnitude ground motions.

It can be observed in Fig. 4 that $6 < M_w < 6.5$ and $6.5 < M_w < 7.0$ curves are too close. Same observation holds for the stiff soil group. Accordingly, we have merged the two magnitude groups into one, i.e. $6 < M_w < 7$ in the foregoing evaluations.

3.3 Fault Distance

The sensitivity of damping efficiency to epicentral distance, either near fault or far fault, is shown in Fig. 5. Although all comparisons are not ultimately decisive, the

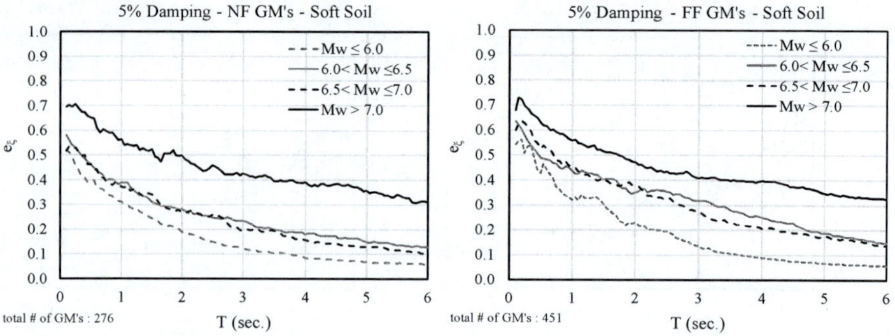

Fig. 4 The effect of M_w on e_ξ for soft soil type and two R_{epi} groups, NF and FF

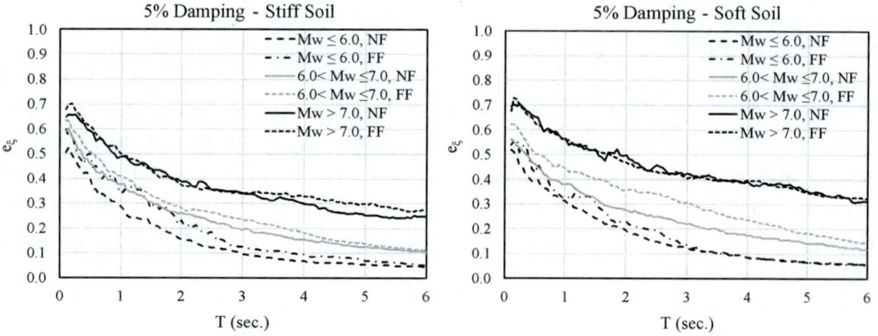

Fig. 5 The effect of R_{epi} (NF and FF) on e_ξ for two soil types and three Mw groups

sensitivity of e_ξ to fault distance is quite weak. Hence, we have decided to ignore epicentral distance as an influential parameter on damping efficiency.

3.4 Soil Type

The effect of soil type (stiff and soft) on damping efficiency is presented in Fig. 6 for two distance ranges and three magnitude groups. Regardless of the distance of ground motions, soil type has some influence on damping efficiency, especially for larger magnitude groups, although it is not very significant. However, since soil type has a strong influence on E_D more than E_I [2], we have decided to retain the soil type as an influential parameter on damping efficiency.

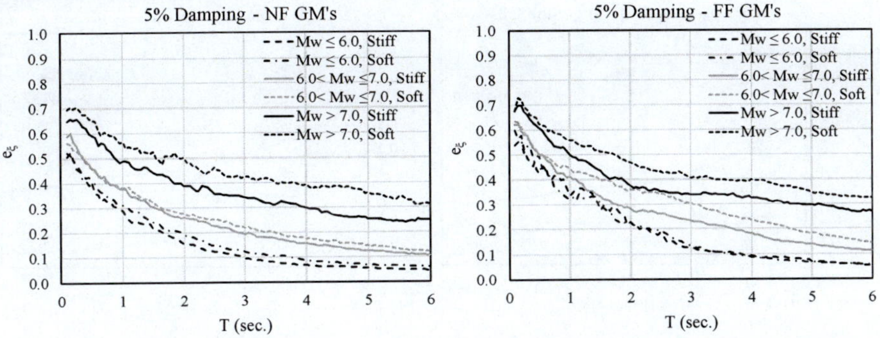

Fig. 6 The effect of soil type (stiff or soft) on e_ξ for distance ranges and three Mw groups

3.5 Concluding Remarks on the Influence of Ground Motion Parameters

The observations summarized above reveal that magnitude is the most influential source parameter and soil type is a significantly influential site parameter on damping efficiency. On the other hand, the influence of fault type and distance to fault on damping efficiency are much less. Therefore, they need not be considered in the evaluation of damping efficiency.

Damping efficiency spectra for the combinations of the magnitude and soil type groups are presented in Fig. 7. The spectral graphs have been calculated for 5, 10 and 20% viscous damping ratios, respectively.

4 Sensitivity of Damping Efficiency to Viscous Damping Ratio

Viscous damping ratio is the sole structural parameter controlling energy dissipation in a SDOF system with a given vibration period T under earthquake ground motions. The sensitivity of damping efficiency to viscous damping ratio is assessed by constructing the damping efficiency spectra for the three magnitude groups and two soil types, for SDOF systems with 5, 10 and 20% viscous damping ratios. The results are presented in Fig. 8. This figure is simply a re-organized form of Fig. 7, where the influence of viscous damping ratio on damping efficiency is shown explicitly.

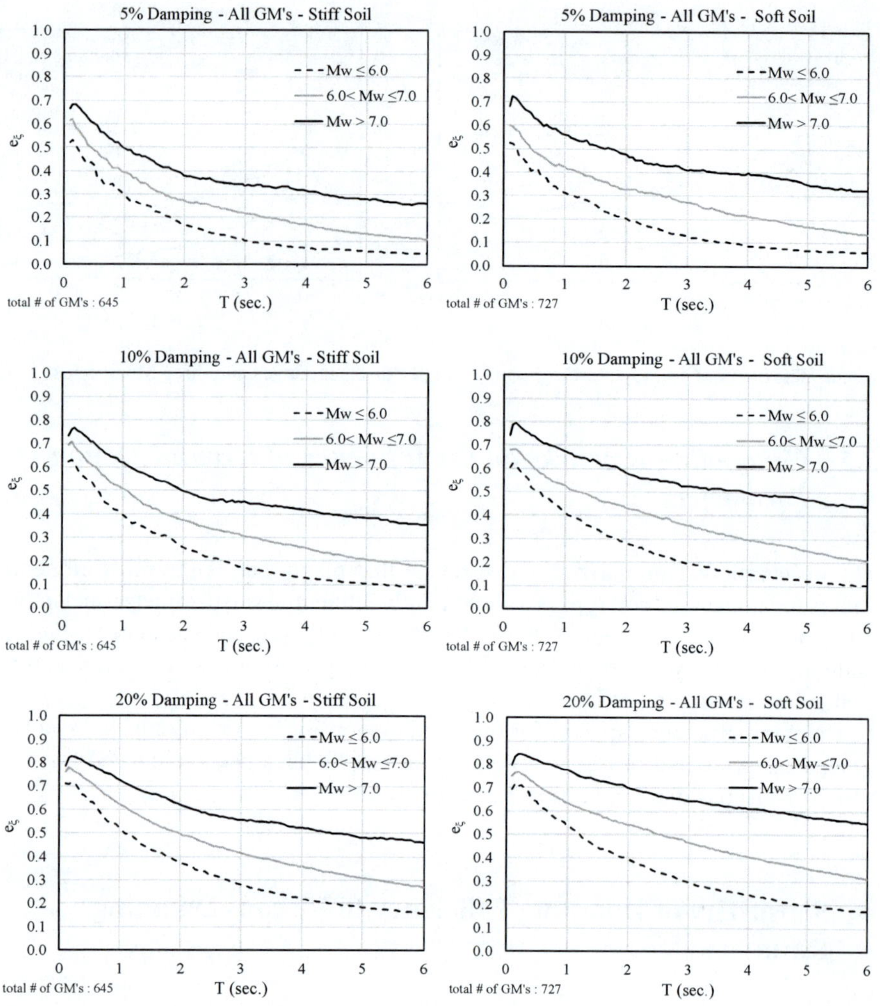

Fig. 7 Damping efficiency spectra for two soil types (stiff or soft), three M_w groups and three damping ratios of 5, 10 and 20%

5 Discussion of Results

Simple observations on Figs. 7 and 8 reveal the influence of magnitude, soil type and viscous damping ratio on the efficiency of viscous damping clearly. We can carry out a comparative assessment for the short-to-medium period ($T < 1$ s) and long period ($T > 3$ s) systems, separately.

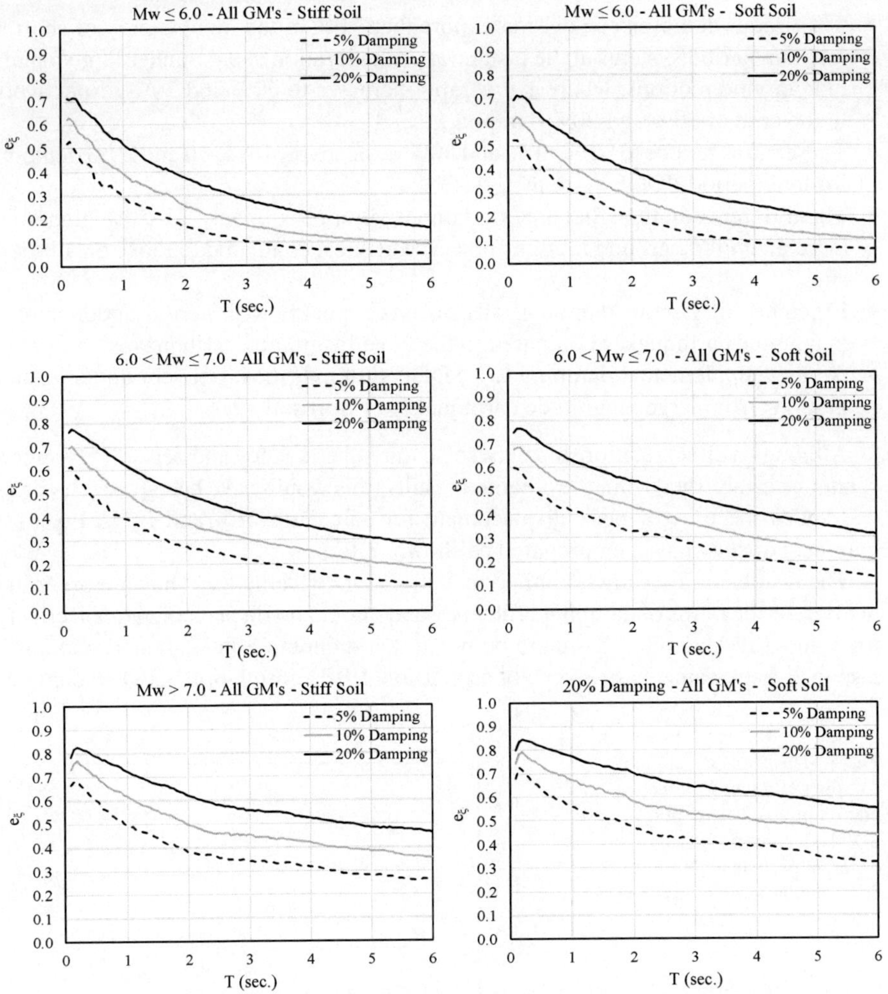

Fig. 8 Damping efficiency spectra for two soil types (stiff or soft), three Mw groups and three damping ratios of 5, 10 and 20%

- Although the efficiency of viscous damping is much higher for shorter period SDOF systems due to higher response velocities, the effect of magnitude and soil type on damping efficiency are less compared to the longer period SDOF systems.
- Along the longer period range, the influence of magnitude primarily, and soil type to a lesser extent, becomes more effective. Nevertheless, supplemental viscous damping in the long period SDOF systems is not efficient under ground motions with Mw < 6.

- 5% viscous damping can dissipate more than 50% of the input energy of short to medium period systems at the instant of peak vibration energy under large magnitude ground motions, whereas this ratio increases to 60% and 70% respectively, in the case of 10% and 20% damping.
- These ratios reduce to 30%, 40% and 50% respectively for 5, 10 and 20% damping for long period SDOF systems.
- The difference in the efficiency of damping is approximately 15–20% higher for large magnitude ground motions on soft soil sites, compared to those on stiff soil sites.
- Efficiency of viscous damping falls off more quickly with period under smaller magnitude earthquakes compared to the large magnitude earthquakes.
- Adding supplemental damping to a SDOF system is most efficient under ground motions from large magnitude earthquakes on soft soil sites.

Although a direct relationship between damping efficiency and peak displacement cannot be established from the obtained results (this would have been possible if t_{max} was set to the time of peak displacement for calculating E_D and E_I in Eq. 1), a graphical display has been prepared as shown in Fig. 9.

The results in Fig. 9 are compiled in Table 1 for convenience. They are presented in terms of the ratios of damping efficiency and maximum displacement respectively, for 5 and 20% damping. It should be noted that damping efficiency and maximum response displacement are inversely proportional. The correlation between damping

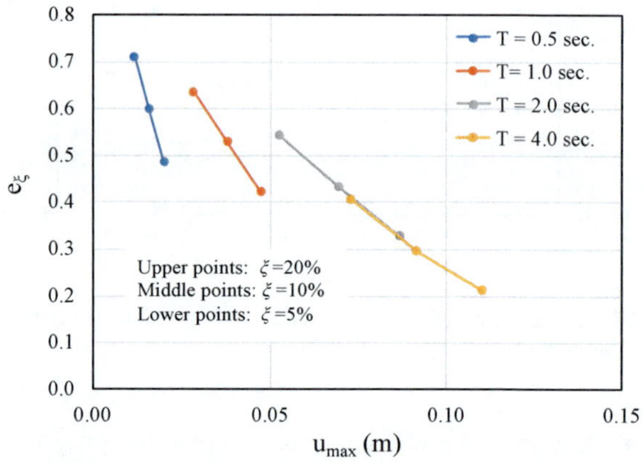

Fig. 9 The effect of damping efficiency on mean maximum displacement of SDOF systems with four different periods and three different damping ratios. 6 < Mw < 7, soft soil

Table 1 Damping efficiency and maximum displacement ratios of SDOF systems

Period T (s)	$e_{0.05}/e_{0.20}$	$u_{\max}$ $(T, 0.20)/u_{\max}$ $(T, 0.05)$
0.5	0.683	0.584
1.0	0.664	0.602
2.0	0.605	0.722
4.0	0.522	0.660

efficiency ratio and maximum displacement ratio is high, but not identical naturally, since the definition of damping efficiency is not directly related to the effect of damping on maximum displacement.

6 Summary and Conclusions

A sensitivity analysis is conducted in this study for determining the strong motion and systems parameters which are influencing the efficiency of viscous damping in SDOF systems in reducing their response to seismic excitations. The main focus of the study is predicting the efficiency of added damping in SDOF systems for performance control. It is revealed that earthquake magnitude, soil type and vibration period are influencing the efficiency of damping most significantly. Damping efficiency is expressed in spectral form for several magnitudes, two soil types and three damping ratios.

References

1. Nurtuğ, A., Sucuoğlu, H.: Prediction of seismic energy dissipation in SDOF systems. Earthq. Eng. Struct. Dyn. **24**, 1215–1223 (1995)
2. Alıcı, F.S., Sucuoğlu, H.: Elastic and inelastic near-fault input energy spectra. Earthq. Spectra **34**, 611–637 (2018)
3. Newmark, N.M., Hall, W.J.: Earthquake Spectra and Design. Earthquake Engineering Research Institute, Berkeley, CA (1982)
4. Wu, J., Hanson, D.: Study of inelastic spectra with high damping. J. Struct. Eng. **115**(6), 1412–1431 (1989)
5. Idriss, I.M.: Procedures for selecting earthquake ground motions at rock sites. A report to the National Institute of Standards and Technology, University of California (1993)
6. Tolis, S.V., Faccioli, E.: Displacement design spectra. J. Earthq. Eng. **3**(1), 107–125 (1999)
7. Bommer, J.J., Elnashai, A.S., Weir, A.G.: Compatible acceleration and displacement spectra for seismic design codes. In: 12th World Conference on Earthquake Engineering, Auckland, New Zealand (2000)
8. Naeim, F., Kircher, C.A.: On the damping adjustment factors for earthquake response spectra. Struct. Des. Tall Build. **10**, 361–369 (2001)
9. Ramirez, O.M., Constantinou, M.C., Kircher, C.A., Whittaker, A.S., Jhonson, M.W., Gomez, J.D.: Development and evaluation of simplified procedures for analysis and design of buildings with passive energy dissipation systems. Technical Report MCEER-00-0010. Multidisciplinary Center for Earthquake Engineering Research, University at Buffalo (SUNY), Buffalo, NY (2000)
10. Ramirez, O.M., Constantinou, M.C., Whittaker, A.S., Kircher, C.A., Chrysostomou, C.Z.: Elastic and inelastic seismic response of buildings with damping systems. Earthq. Spectra **18**(3), 531–547 (2002)
11. Malhotra, P.K.: Smooth spectra of horizontal and vertical ground motions. Bull. Seismol. Soc. Am. **96**(2), 506–518 (2006)
12. Lin, Y.Y.: Statistical study on damping modification factors adopted in Taiwan's seismic isolation design code by using the 21 September 1999 Chi-Chi earthquake, Taiwan. Eng. Struct. **29**, 682–693 (2007)

13. Lin, Y.Y., Chang, K.C.: Effects of site classes on damping reduction factors. J. Struct. Eng. **130**(11), 1667–1675 (2004)
14. Hatzigeorgiou, G.D.: Damping modification factors for SDOF systems subjected to near-fault, far-fault and artificial earthquakes. Earthq. Eng. Struct. Dyn. **39**, 1239–1258 (2010)
15. Castillo, T., Ruiz, S.E.: Reduction factors for seismic design spectra for structures with viscous energy dampers. J. Earthq. Eng. **18**, 232–349 (2014)
16. Palermo, M., Silvestri, S., Trombetti, T.: Stochastic-based damping reduction factors. Soil Dyn. Earthq. Eng. **80**, 168–176 (2016)
17. Stafford, P.J., Mendis, R., Bommer, J.J.: Dependence of damping correction factors for response spectra on duration and numbers of cycles. J. Struct. Eng. **134**(8), 1364–1373 (2008)
18. Zhou, J., Tang, K., Wang, H., Fang, X.: Influence of ground motion duration on damping reduction factor. J. Earthq. Eng. **18**, 816–830 (2014)
19. Hubbard, D.T., Mavroeidis, G.P.: Damping coefficients for near-fault ground motion response spectra. Soil Dyn. Earthq. Eng. **31**, 401–417 (2010)
20. Lin, Y.Y., Chang, K.C.: Study on damping reduction factor for buildings under earthquake ground motions. J. Struct. Eng. **129**(2), 206–214 (2003)
21. Bommer, J.J., Mendis, R.: Scaling of spectral displacement ordinates with damping ratios. Earthq. Eng. Struct. Dyn. **34**, 145–165 (2005)
22. Lin, Y.Y., Miranda, E., Chang, K.C.: Evaluation of damping reduction factors for estimating elastic response of structures with high damping. Earthq. Eng. Struct. Dyn. **34**, 1427–1443 (2005)
23. Cameron, W.I., Green, R.A.: Damping correction factors for horizontal ground-motion response spectra. Bull. Seismol. Soc. Am. **97**(3), 934–960 (2007)
24. Cardone, D., Dolce, M., Rivelli, M.: Evaluation of reduction factors for high-damping design response spectra. Bull. Earthq. Eng. **7**, 273–291 (2009)
25. Hao, A., Zhou, D., Li, Y., Zhang, H.: Effects of moment magnitude, site conditions and closest distance on damping modification factors. Soil Dyn. Earthq. Eng. **31**, 1232–1247 (2011)
26. Rezaeian, S., Bozorgnia, Y., Idriss, I.M., Abrahamson, N., Campbell, K., Silva, W.: Damping scaling of response spectra for shallow crustal earthquakes in active tectonic regions. In: 15th World Conference on Earthquake Engineering, Lisbon, Portugal (2012)
27. Rezaeian, S., Bozorgnia, Y., Idriss, I.M., Abrahamson, N., Campbell, K., Silva, W.: Damping scaling factors for elastic response spectra for shallow crustal earthquakes in active tectonic regions: "average" horizontal component. Earthq. Spectra **30**(2), 939–963 (2014)
28. Akkar, S., Sandıkkaya, M.A., Ay, B.Ö.: Compatible ground-motion prediction equations for damping scaling factors and vertical-to-horizontal spectral amplitude ratios for the broader Europe region. Bull. Earthq. Eng. **12**, 517–547 (2014)
29. Mollaioli, F., Liberatore, L., Lucchini, A.: Displacement damping modification factors for pulse-like and ordinary records. Eng. Struct. **78**, 17–27 (2014)
30. Daneshvar, P., Bouaanani, N., Goda, K., Atkinson, G.M.: Damping reduction factors for crustal, inslab, and interface earthquakes characterizing seismic hazard in Southwestern British Columbia, Canada. Earthq. Spectra **32**(1), 45–74 (2016)
31. Kong, C., Kowalsky, M.J.: Impact of damping scaling factors on direct displacement-based design. Earthq. Spectra **32**(2), 843–859 (2016)
32. Sucuoğlu, H., Alıcı, F.S.: Damping spectra for estimating inelastic deformations from modal response spectrum analysis. Earthq. Eng. Struct. Dyn. **50**, 436–454 (2021)
33. Pacific Earthquake Engineering Center (PEER): Guidelines for performance-based seismic design of tall buildings. PEER Report No. 2017/05. Berkeley, California (2017)

Vector-Valued Intensity Measures to Predict Peak and Hysteretic Energy Demands of 3D R/C Buildings

José I. Torres, Edén Bojórquez, Alfredo Reyes, and Juan Bojórquez

Abstract In this study, several peak and energy vector-valued ground motion intensity measures (*IMs*) are proposed to predict maximum inter-story drift and hysteretic energy demands of 3D reinforced concrete (R/C) buildings subjected to narrow-band motions. The selected vector-valued *IMs* are based on the spectral acceleration, pseudo-velocity, velocity and input energy at first mode of the structure as first component. As the second component, ground motion parameters based on peak, integral and spectral shape proxies such as the well-known N_p are used. The objective of the present study is to provide vector-valued *IMs* whit the ability to predict the maximum inter-story drift and hysteretic energy demands on 3D framed structures. It is observed that vector-valued *IMs* based on N_p provide a high relation whit maximum inter-story drift and hysteretic energy demands of reinforced concrete framed buildings.

Keywords 3D buildings · Hysteretic energy · Intensity measure · Reinforced concrete · Structural dynamics · Vector-valued

1 Introduction

Since the beginning of earthquake engineering and seismology, several studies have been developed to provide a parameter with the ability to characterize the ground motion potential of an earthquake which is known as ground motion intensity

J. I. Torres (✉) · E. Bojórquez · A. Reyes · J. Bojórquez
Facultad de Ingeniería, Universidad Autónoma de Sinaloa, Culiacán 80040, México

E. Bojórquez
e-mail: eden@uas.edu.com.mx

A. Reyes
e-mail: reyes@uas.edu.mx

J. Bojórquez
e-mail: juanbmu@uas.edu.mx

A. Benavent-Climent and F. Mollaioli (eds.), *Energy-Based Seismic Engineering*, Lecture Notes in Civil Engineering 155,
https://doi.org/10.1007/978-3-030-73932-4_18

measure. The most important characteristic of an *IM* is the reduction of the uncertainty in the estimation of the seismic response of structures under earthquakes (efficiency). In others words, the efficiency is defined as the ability to predict the response of structures subjected to earthquakes with low uncertainty. Although, several studies have been developed to propose or to analyze ground motion intensity measures [1–18]. In most of the cases, the proposed *IMs* are based in the prediction of maximum demands such as maximum ductility and inter-story drift. By other hand, nowadays, several studies promote the use of vector-valued or scalar ground motion *IMs* based on spectral shape, because they predict with good accuracy the maximum inter-story drift and maximum ductility of structures subjected to earthquakes [12–15, 17–20]. In particular, vector and scalar ground motion intensity measures based on N_p which are representative of the spectral shape have resulted very well correlated with the nonlinear structural response [18, 20, 21]. However, as it was mentioned before, an appropriated *IMs* should be capable of predicting all types of engineering demand parameters, as example, hysteretic energy demands. Hysteretic energy demands are very important in structures when subjected to long duration narrow-band ground motions [22–28]. In this study, vector-valued ground motion intensity measures have been proposed in order to predict the seismic response of 3D reinforced concrete buildings under narrow-band motions recorded at the soft soil site of Mexico City. The proposed *IMs* are separated in 2 sets, the first one is composed by vector-valued *IMs* with maximum peak ground and integral response, and the second set by vector-valued *IMs* with spectral shape parameters as is the case of the N_p parameter. For the evaluation of the vector-valued *IMs*, the maximum inter-story drift and the hysteretic energy demand are used as engineering demand parameters. It is important to say that hysteretic energy demand is not commonly used in efficiency studies of *IMs* in comparison with the maximum inter-story drift or peak demands. So far, this study differs from others by the use of 3D structures, as well as the use of both performance parameters (maximum inter-story drift and the hysteretic energy demand). Finally, an optimization of the N_p parameter for the selected spectral shape vector-valued *IMs* developed through the values of $Sa(T_1)$, $Sv(T_1)$, $V(T_1)$ and $E_I(T_1)$ is computing in the present study.

2 Methodology

2.1 Vector-Valued Ground Motion Intensity Measures

The prediction of the seismic response is estimated considering 32 different vector-valued ground motion *IMs*, which are divided in 2 sets of 16 *IMs*. The first set is based on traditional peak and integral *IMs*, while the second is based on the proxy of the spectral shape named N_p. For both sets the parameters $Sa(T_1)$, $Sv(T_1)$, $V(T_1)$ and $E_I(T_1)$ are considered: where, *Sa* is the spectral acceleration, *Sv* is the pseudo-velocity, *V* is the velocity, E_I is the input energy and T_1 is the first mode of vibration.

Note that E_I can be defined from the equation of motion of a single degree of freedom system as follows:

$$m\ddot{x}(t) + c\dot{x} + f_s(x, \dot{x}) = -m\ddot{x}_g(t) \tag{1}$$

In Eq. (1), m is the mass of the system; c, the viscous damping coefficient; $f_s(x, \dot{x})$, the non-linear force; $\ddot{x}$, the ground acceleration; and x, the displacement with respect to the base of the system. A dot above x indicates a derivative with respect to time. In case of an elastic linear systems, $f_s(x, \dot{x}) = kx$, where k is the stiffness of the system.

Integrating each member of Eq. (1) with respects to x, yields:

$$\int m\ddot{x}(t)dx + \int c\dot{x}(t)dx + \int f_s(x, \dot{x})dx = -\int m\ddot{x}_g(t)dx \tag{2}$$

Equation (2) can be written as energy balanced equation as follows [29]:

$$E_K + E_D + E_S + E_H = E_I \tag{3}$$

where E_K, E_D, E_S and E_H represent the kinetic (K), viscous damping (D), deformation (S) and dissipated hysteretic (H) energies, respectively; and E_I is the relative input energy.

For the set number 1, the parameters PGA, PGV, t_D and I_D are used as second component of the vector-valued *IMs*; where *PGA* represents peak ground acceleration, *PGV* peak ground velocity, and finally, t_D and I_D, represents the effective duration of the earthquake [30] and the accumulated potential [4], respectively. The effective duration (t_D) is defined as the time interval between 5% and 95% of the Arias Intensity [2], which is obtained from the following equation:

$$I_A = \int_0^T a^2(t)dt \tag{4}$$

where T is the total duration of the earthquake. In addition, the I_D factor is defined as:

$$I_D = \frac{\int_0^T a(t)^2 dt}{PGA \cdot PGV} \tag{5}$$

Table 1 shows a summary of the selected vector-valued *IMs* of the first set. The first column indicates the ground motion *IMs*; the second, third and fourth columns, indicate if the *IMs* are based on peak ground response, duration and/or spectral shape response, respectively.

In the second set of vector-valued *IMs*, the parameter N_p is used as the second component of the vector-valued *IMs* to determine the spectral shape [18]. The general form of the N_p generalized parameter is defined in the following equation:

Table 1 First set of vector-valued *IMs*

Intensity measures	Peak ground response	Duration	Spectral shape
<$Sa(T_1)$, PGA>	*		
<$Sa(T_1)$, PGV>	*		
<$Sa(T_1)$, t_D>	*	*	
<$Sa(T_1)$, I_D>	*	*	
<$Sv(T_1)$, PGA>	*		
<$Sv(T_1)$, PGV>	*		
<$Sv(T_1)$, t_D>	*	*	
<$Sv(T_1)$, I_D>	*	*	
<$V(T_1)$, PGA>	*		
<$V(T_1)$, PGV>	*		
<$V(T_1)$, t_D>	*	*	
<$V(T_1)$, I_D>	*	*	
<$E_I(T_1)$, PGA>	*	*	
<$E_I(T_1)$, PGV>	*	*	
<$E_I(T_1)$, t_D>	*	*	
<$E_I(T_1)$, I_D>	*	*	

$$N_{pg} = \frac{S_{avg}(T_i, \ldots, T_f)}{S(T_j)} \quad (6)$$

From the Eq. (6), $S(T_j)$ represents a spectral parameter taken from any type of spectrum as in the case of acceleration, velocity, displacement, input energy, inelastic parameters and so on, at period T_j. $S_{avg}(T_i, \ldots, T_f)$ is the geometrical mean of a specific spectral parameter between the range of periods T_i and T_f. Note that the periods T_i and T_j could be different. N_{pg} is similar to the traditional definition of N_p [10] but for different types of spectra and a wider range of periods. In such a way that parameters as the traditional N_p or the recently proposed Sa_{Ratio} [31] are particular cases of the generalized spectral shape parameter N_{pg}. If the pseudo-acceleration spectrum is used, and $T_i = T_j = T_1$ (first mode structural vibration period) N_{pg} is equal to the traditional N_p. If the value of N_p is lower than one, we can expect that the average spectrum with negative slope in the range of periods between T_i and T_f, if N_p is equal to one, the average spectrum is to about flat in the same range of periods, and in the case of N_p is larger than one, we can expect an increasing trend of the spectra beyond T_i. In this work, 10 points are used to calculate the average spectrum, a value of $S(T_j)$ equal to T_1, and the values of T_i and T_f, were obtained through an optimization study as will be observed below. In Table 2 a summary of the selected vector-valued *IMs* for the second set is illustrated.

2.2 Structural 3D Reinforced Concrete Frames Models

Two 3D R/C frames having 7 and 10 stories were considered for the studies reported herein. The 3D frames are denoted as F7 and F10 respectively (see Fig. 1). The frames, designed according to the Mexico City Seismic Design Provisions (MCSDP). It was considered that the elements (beams and columns) of the structures have a hysteretic behavior similar to the modified Takeda model to represent the non-linearity of the material. The main characteristics of the 3D R/C models are shown in Table 3. Note that column eight in Table 3 represents the fundamental period of vibration of the structure.

2.3 Earthquake Ground Motion Records

To determine the seismic performance of the two 3D R/C frames, a dynamic analysis is carried out. For this aim, the structures are subjected under the action of seismic records representative of the site in which they are assumed to be displaced. In this work it is assumed that the buildings are located at soft soil of Mexico City and are subject to narrow band seismic ground motions. The following table indicates the narrow band accelerograms obtained from the soft soil of Mexico City used in the present study (Table 4).

2.4 Scaling of Seismic Records

The seismic records or accelerograms were scaled to represent seismic events of different intensities, and thus to determine the responses of the modeled structures in terms of the maximum inter-story drifts and the hysteretic energy demands, which allows to compare the efficiency of the different *IMs* used in this work. For the

Table 2 Second set of vector-valued *IMs*

Intensity measures	Peak ground response	Duration	Spectral shape
$<Sa(T_1), N_{pSa}>$	*		*
$<Sa(T_1), N_{pSv}>$	*		*
$<Sa(T_1), N_{pV}>$	*		*
$<Sa(T_1), N_{pEI}>$	*	*	*
$<Sv(T_1), N_{pSa}>$	*		*
$<Sv(T_1), N_{pSv}>$	*		*
$<Sv(T_1), N_{pV}>$	*		*
$<Sv(T_1), N_{pEI}>$	*	*	*
$<V(T_1), N_{pSa}>$	*		*
$<V(T_1), N_{pSv}>$	*		*
$<V(T_1), N_{pV}>$	*		*
$<V(T_1), N_{pEI}>$	*	*	*
$<E_I(T_1), N_{pSa}>$	*	*	*
$<E_I(T_1), N_{pSv}>$	*	*	*
$<E_I(T_1), N_{pV}>$	*	*	*
$<E_I(T_1), N_{pEI}>$	*	*	*

scaling of the seismic records the combination of the horizontal components of the earthquake was used [see Eq. (7)]. The scaling was done to have specific values in the response spectra in a given period (the period of interest, which is the fundamental period of the structures was used). The parameters to scale the records are $Sa(T_1)$, $Sv(T_1)$, $V(T_1)$ and $E_I(T_1)$.

$$IMs = \sqrt{(IMs_{Comp1})^2 + (IMs_{Comp2})^2} \tag{7}$$

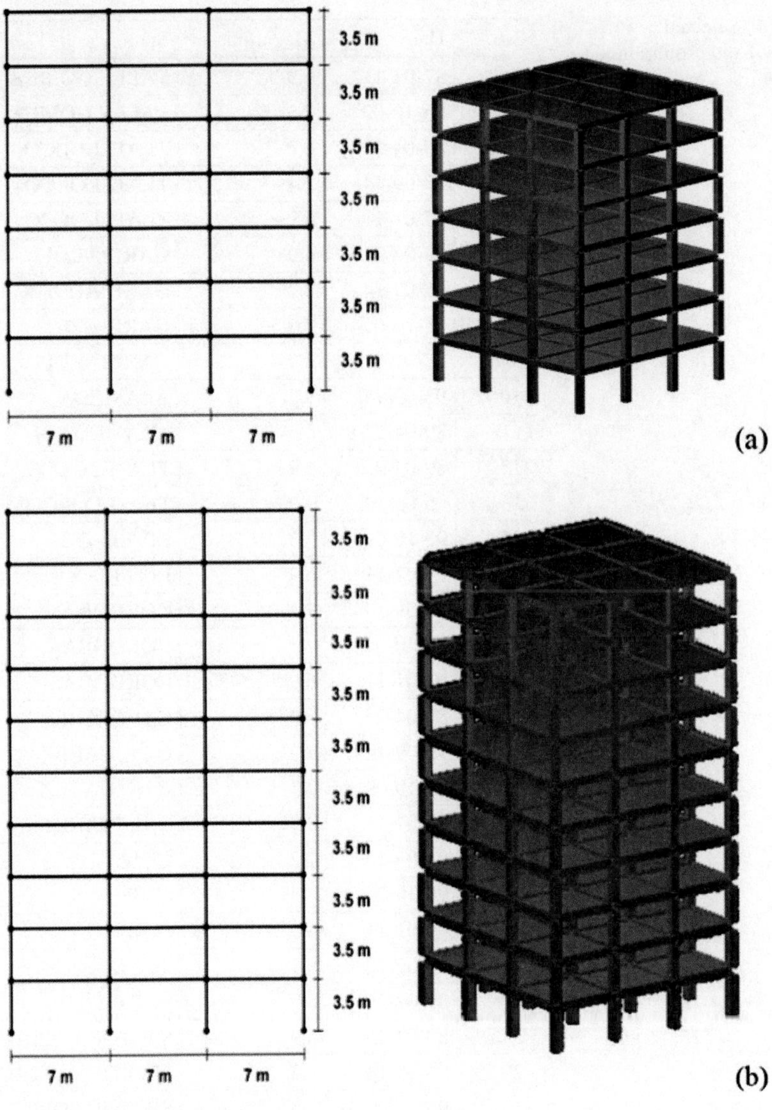

Fig. 1 F4 (**a**) and F10 (**b**) frame models

Table 3 Characteristics of the 3D R/C models

Frame	Number of stories	T_1(s)
F7	7	0.7
F10	10	0.98

Table 4 Selected narrow-band ground motion records

No	Date	Magnitude	Station
1	97-01-11	6.9	VALLE GÓMEZ
2	95-10-09	7.3	VALLE GÓMEZ
3	89-04-25	6.9	TLATELOLCO
4	95-09-14	7.4	TLATELOLCO
5	97-01-11	6.9	TLATELOLCO
6	89-04-25	6.9	GARIBALDI
7	95-09-14	7.2	GARIBALDI
8	95-10-09	7.3	GARIBALDI
9	97-01-11	6.9	GARIBALDI
10	95-09-14	7.2	ALAMEDA
11	89-04-25	6.9	ALAMEDA
12	89-04-25	6.9	TLATELOLCO
13	95-09-14	7.2	TLATELOLCO
14	95-10-09	7.3	LIVERPOOL
15	97-01-11	6.9	LIVERPOOL
16	95-09-14	7.2	CORDOBA
17	95-10-09	7.3	CORDOBA
18	97-01-11	6.9	CORDOBA
19	89-04-25	6.9	C.U. JUAREZ
20	95-09-14	7.2	C.U. JUAREZ
21	95-10-09	7.3	C.U. JUAREZ
22	97-01-11	6.9	C.U. JUAREZ
23	95-09-14	7.2	CUJP
24	95-10-09	7.3	CUJP
25	97-01-11	6.9	CUJP
26	85-09-19	8.1	SCT B-1
27	89-04-25	6.9	SCT B-2
28	89-04-25	6.9	SECTOR POPULAR
29	95-09-14	7.2	SECTOR POPULAR
30	95-10-09	7.3	SECTOR POPULAR
31	97-01-11	6.9	SECTOR POPULAR

In Eq. (7), while *IMs* is the selected intensity measure that represent the combination of both directions of the ground motion (i.e. *Sa, Sv, V* and E_I), IMs_{Comp1} and IMs_{Comp2} are the intensity measure in each horizontal component orthogonal.

2.5 Performance Parameters

The engineering demand parameters selected were the maximum inter-story drift and the normalized hysteretic energy by the mass (E_N), Eq. (8). These parameters are selected due to their importance for seismic design purposes, since they capture both information about maximum demands and the effect on the duration of the earthquake and with the cumulative demands [32]. In fact, currently various damage indexes have been proposed based on hysteretic energy [33–35].

$$E_N = \frac{E_H}{m} \tag{8}$$

3 Numerical Results

3.1 Relation Between Vector-Valued IMs and the Structural Demand of 3D R/C Fames

Baker and Cornell [14] and Bojórquez and Iervolino [18] showed the advantages of using vector-valued ground motion intensity measures instead of scalars. The main advantage is the increasing in the efficiency to predict the structural response. Herein with the aim to obtain the relation between the structural response of R/C frames and the vectors selected; nonlinear incremental dynamic analysis was used to obtain the seismic response of the R/C frames subjected to the 31 ground motion records by using the first parameter of the vector and then the relation between the structural response of the R/C frames and the second parameter of the vector is obtained. Figure 2a shows a general illustrative example of the incremental dynamic

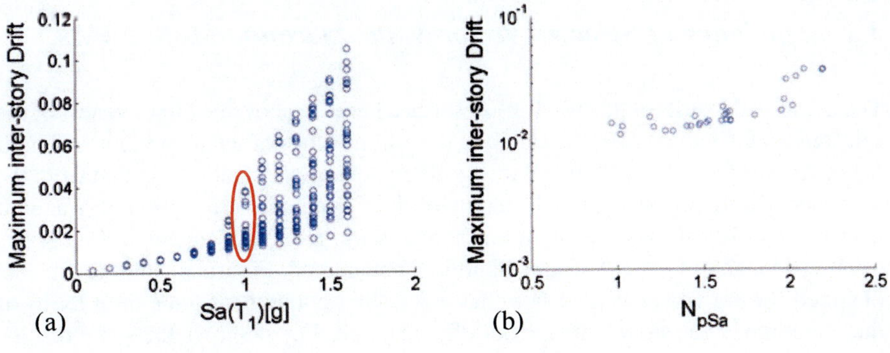

Fig. 2 **a** Illustrative example of incremental dynamic analysis scaling for $Sa(T_1)$; **b** relation between N_{pSa} and Maximum inter-story drift at $Sa(T_1) = 1g$

analysis for Sa(T_1) in terms of the maximum inter-story drift. It is observed a poor relation among Sa(T_1) and maximum inter-story drift, in fact the uncertainty to predict maximum inter-story drift using the spectral acceleration tend to increase with the intensity of the earthquake ground motion. Figure 2b illustrates the relation obtained for <$Sa(T_1)$, N_{pSa}> and the maximum inter-story drift when $Sa(T_1) = 1$ g (see the values in the circle of Fig. 2b). Note the good relation between N_{pSa} and the maximum inter-story drift reflecting the advantage of using the vector-valued ground motion intensity measure. It explains the reduction in the uncertainty associated with the structural response when vector-valued parameters are selected as intensity measures, and this type of intensity measures could be more efficient for nonlinear structural response prediction.

3.2 *Optimization of the* $\mathrm{N_p}$ *Parameter*

As it was mentioned above, a study was carried out to determine the value of T_i and T_f to be used for the N_p spectral shape parameter of the second set of intensity measures. For this reason, pairs of T_i values were created ranging from $0.1 * T_1$ to $1.0 * T_1$, and T_f from $1.1 * T_1$ to $3.0 * T_1$. The value of N_p of all the pairs was calculated, applied to the 16 vector-valued spectral shape *IMs* and were selected the values for which T_i and T_f minimize the standard deviation of the natural logarithm of the seismic response (maximum inter-story drift or hysteretic energy) of the R/C buildings.

A summary of the optimal T_i and T_f values computed for the F7 R/C frame is presented in Table 5 and for the F10 R/C frame in Table 6. It is observed in general, that the value of the initial period T_i is equal to T_1, while for T_f, a value between 2.0 and 3.0 times T_1 is suggested, which is valid for the two selected performance parameters (maximum inter-story drift and normalized hysteretic energy).

3.3 *Efficiency of Selected Vector-Valued Ground Motion* IMs

The numerical results of this work are described here. To show the effectiveness of the selected vector-valued *IMs*, the standard deviation values of the natural logarithm of the maximum inter-story drift and the hysteretic energy demands, obtained from the incremental dynamic analysis at different levels of intensities are presented. Standard deviations were estimated using a linear regression for all cases and scaling levels of *IMs* used, and of the R/C framed building considered. Figure 3 shows the results obtained for the vector-valued *IMs* that use as first component $Sa(T_1)$, in terms of the maximum inter-story drift for both frames; and, Fig. 4 shows the same results, but now, in terms of the hysteretic energy demands. By other hand, Fig. 5 shows the results obtained for vector-valued *IMs* that use EI as first component, in terms of the maximum inter-story drift, for the two 3D R/C frames; and, Fig. 6, shows

Table 5 Optimal T_i and T_f values for the 16 vector-valued spectral shape *IMs* to predict the seismic response of F7 R/C frame

IMs	Inter-story drift		Normalized hysteretic energy	
	T_i/T_1	T_f/T_1	T_i/T_1	T_f/T_1
<$Sa(T_1)$, N_{pSa}>	1	3	1	3
<$Sa(T_1)$, N_{pSv}>	1	3	1	3
<$Sa(T_1)$, N_{pV}>	1	3	1	3
<$Sa(T_1)$, N_{pEI}>	1	3	1	3
<$Sv(T_1)$, N_{pSa}>	1	3	1	3
<$Sv(T_1)$, N_{pSv}>	1	3	1	3
<$Sv(T_1)$, N_{pV}>	1	3	1	3
<$Sv(T_1)$, N_{pEI}>	1	3	1	3
<$V(T_1)$, N_{pSa}>	1	3	1	3
<$V(T_1)$, N_{pSv}>	1	3	1	3
<$V(T_1)$, N_{pV}>	1	3	1	3
<$V(T_1)$, N_{pEI}>	1	3	1	3
<$E_I(T_1)$, N_{pSa}>	0.3	1.1	1	3
<$E_I(T_1)$, N_{pSv}>	0.3	1.1	0.3	1.1
<$E_I(T_1)$, N_{pV}>	1	3	1	3
<$E_I(T_1)$, N_{pEI}>	0.3	1.1	1	3

the same results, but, in terms of the hysteretic energy demands. Note that since the results generated by vector-valued *IMs* that use $Sv(T_1)$ and $V(T_1)$ as first component have low efficiency than $Sa(T_1)$, only the results of the *IMs* that use $Sa(T_1)$ as the first component have been included for the sake of brevity. In addition, with the aim to have a comparison in terms of input energy, also this parameter was plotted when represent the first component of the vector. It can be seen that the vector-valued *IMs* that use spectral shape parameters N_p are more efficient to predict the seismic demands (peak and cumulative) for every frame, especially those based on spectral acceleration or velocity (N_{pSa}, N_{pSv}, N_{pV}). By the other hand, the peak ground response and duration parameters of the first set of *IMs* are not good estimators of the maximum inter-story drift and the normalized hysteretic energy demands, because they produce large uncertainties. Finally, it is observable that the selected vector-valued *IMs* are more efficient in predicting the maximum inter-story drift than the hysteretic energy demands.

Figure 7 shows the comparison between the 16 vector-valued *IMs* of the second set of *IMs*, those that use in their second component an *IMs* based on the spectral shape, for both models F7 and F10 respectively, the results shown in terms of the standard deviation of the natural logarithm of the maximum inter-story drifts and for a median value of 0.03. As seen in Fig. 7, vector-valued *IMs* that use $Sa(T_1)$, or $Sv(T_1)$, have the best efficiency to predict the seismic response, using N_{pSa}, N_{pSv} or N_{pV} as second component.

Table 6 Optimal T_i and T_f values for the 16 vector-valued spectral shape *IMs* to predict the seismic response of F10 R/C frame

IMs	Inter-story drift		Normalized hysteretic energy	
	T_i/T_1	T_f/T_1	T_i/T_1	T_f/T_1
$<Sa(T_1), N_{pSa}>$	1	2.6	1	2.5
$<Sa(T_1), N_{pSv}>$	1	2.5	1	2.4
$<Sa(T_1), N_{pV}>$	1	2.4	1	2.4
$<Sa(T_1), N_{pEI}>$	1	2.2	1	2.3
$<Sv(T_1), N_{pSa}>$	1	2.6	1	2.5
$<Sv(T_1), N_{pSv}>$	1	2.5	1	2.4
$<Sv(T_1), N_{pV}>$	1	2.4	1	2.4
$<Sv(T_1), N_{pEI}>$	1	2.2	1	2.3
$<V(T_1), N_{pSa}>$	1	2.6	1	2.6
$<V(T_1), N_{pSv}>$	1	2.6	1	2.5
$<V(T_1), N_{pV}>$	1	2.6	1	2.6
$<V(T_1), N_{pEI}>$	1	2.7	1	2.6
$<E_I(T_1), N_{pSa}>$	1	2.7	1	2.5
$<E_I(T_1), N_{pSv}>$	1	2.7	1	2.5
$<E_I(T_1), N_{pV}>$	1	2.7	1	2.5
$<E_I(T_1), N_{pEI}>$	0.1	1.2	1	2.7

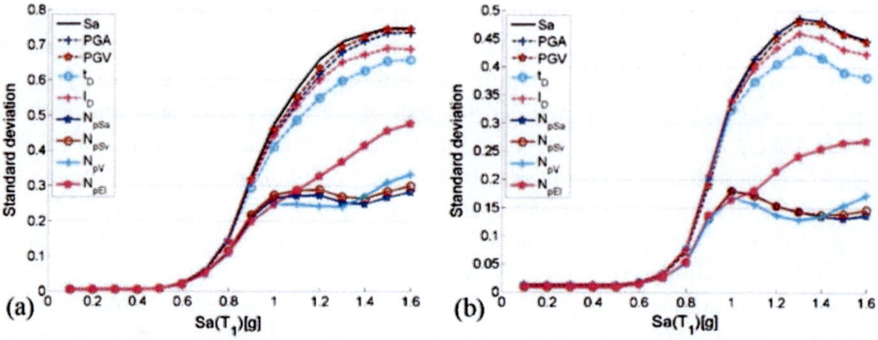

Fig. 3 Efficiency comparison of the standard deviation of the natural logarithm for the maximum inter-story drift of vector-valued *IMs* that use $Sa(T_1)$ as first component for: **a** F7 frame, and **b** F10 frame

4 Conclusions

Several vector-valued ground motion intensity measures have been analyzed with the aim to obtain the best predictor of the structural response in terms of the maximum

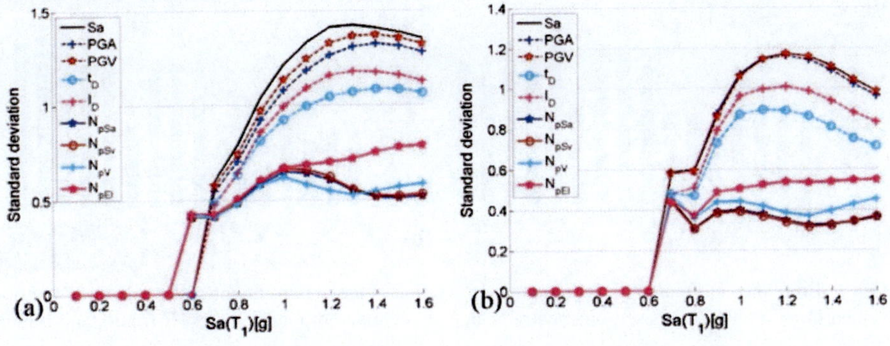

Fig. 4 Efficiency comparison of the standard deviation of the hysteretic energy demands of vector-valued *IMs* that use $Sa(T_1)$ as first component for: **a** F7 frame, and **b** F10 frame

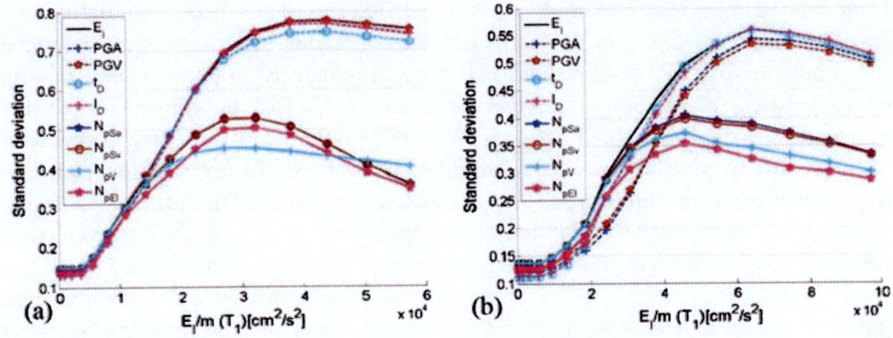

Fig. 5 Efficiency comparison of the standard deviation of the natural logarithm for the maximum inter-story drift of vector-valued *IMs* that use $E_I(T_1)$ as first component for: **a** F7 frame, and **b** F10 frame

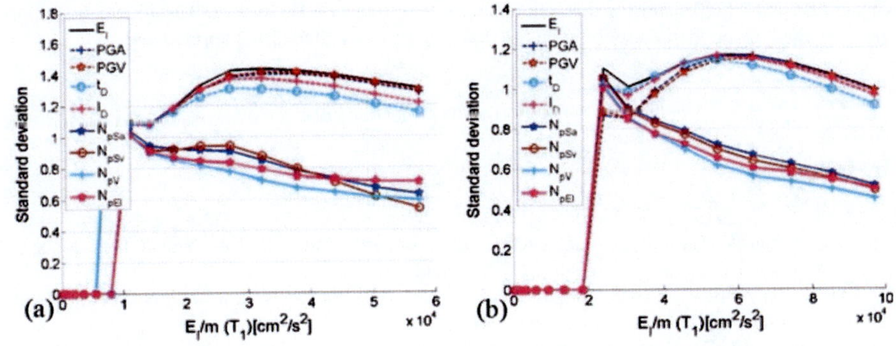

Fig. 6 Efficiency comparison of the standard deviation of the hysteretic energy demands of vector-valued *IMs* that use $E_I(T_1)$ as first component for: **a** F7 frame, and **b** F10 frame

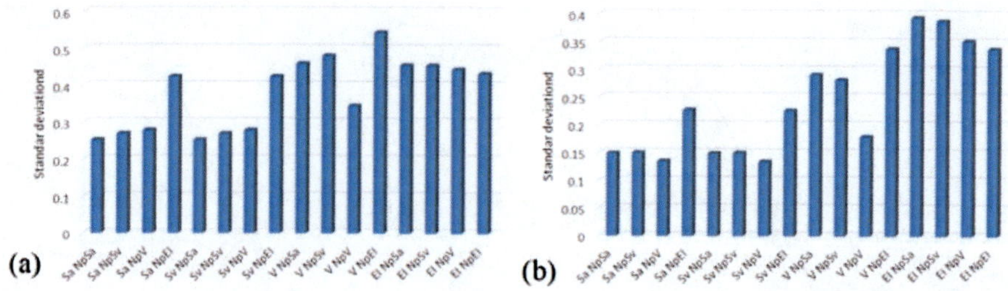

Fig. 7 Efficiency comparison of the standard deviation of the maximum inter-story drifts of vector-valued *IMs* that use *IMs* based on spectral shape as second component for: **a** F7 frame, and **b** F10 frame

inter-story drift and the hysteretic energy demands of 3D R/C frames under narrow-band ground motions. The study considered *IMs* based on peak, cumulative and spectral shape proxies. The numerical study concludes that there is no evidence to support the use of vector-valued *IMs* based exclusively in peak ground motion characteristics for predicting seismic demands of 3D R/C buildings. The results obtained by the vector-valued *IMs* that use $Sa(T_1)$ and $Sv(T_1)$ as the first component of the vector are similar since, they are related, and their efficiency is superior than those obtained using $V(T_1)$ and $E_I(T_1)$ as the first component. The optimization study shows that T_i values equal to T_1, and T_f ranging from 2.0 to 3.0 times T_1, produce an improvement in the prediction of the seismic response. The vectors $<Sa(T_1), N_{psa}>$, $<Sa(T_1), N_{pSv}>$ and $<Sa(T_1), N_{pV}>$ are those that produce the least uncertainty when predicting seismic demands, in such a way that they are very promising to the next generation of advanced vector-valued ground motion intensity measures.

Acknowledgements The authors express their gratitude to the *Consejo Nacional de Ciencia y Tecnología* (CONACYT) in Mexico for funding the research reported in this paper under grant Ciencia Básica 287103 and for the scholarship given to the Ph.D. student. The financial support given by the Universidad Autónoma de Sinaloa under grant PROFAPI is appreciated.

References

1. Housner, G.W.: Limit design of structures to resist earthquakes. In: First World Conference on Earthquake Engineering, Berkeley, California (1956)
2. Arias, A.: A measure of earthquake intensity. In: Seismic Design for Nuclear Power Plants, pp. 438–483. MIT Press, Cambridge, MA (1970)
3. Von-Thun, J.L., Rochin, L.H., Scott, G.A., Wilson, J.A.: Earthquake ground motions for design and analysis of dams. In: Earthquake Engineering and Soil Dynamics II—Recent Advance in Ground-Motion Evaluation, Geotechnical Special Publication 20 ASCE, pp. 463–481 New York (1998)
4. Cosenza, E., Manfredi, G.: A seismic design method including damage effect. In: 11th European Conference on Earthquake Engineering, Paris, France (1998)

5. Mehanny, S.S.: A broad-range power-law form scalar-based seismic intensity measure. Eng. Struct. **31**(7), 1354–1368 (2009)
6. Jamshidiha, H.R., Yakhchalian, M., Mohebi, B.: Advanced scalar intensity measures for collapse capacity prediction of steel moment resisting frames with fluid viscous dampers. Soil Dyn. Earthq. Eng. **109**, 102–118 (2018)
7. Javadi, E., Yakhchalian, M.: Selection of optimal intensity measure for seismic assessment of steel buckling restrained braced frames under near-fault ground motions. J. Rehab. Civ. Eng. **7**(3), 162–181 (2019)
8. Luco, N., Cornell, C.A.: Structure-specific scalar intensity measures for near-source and ordinary earthquake ground motions. Earthq. Spectra **23**(2), 357–392 (2007)
9. Tothong, P., Cornell, C.A.: Structural performance assessment under near-source pulselike ground motions using advanced ground motion intensity measures. Earthq. Eng. Struct. Dyn. **37**(7), 1013–1037 (2008)
10. Baker, J.W., Cornell, C.A.: Vector-valued intensity measures for pulse-like near-fault ground motions. Eng. Struct. **30**(4), 1048–1057 (2008)
11. Yakhchalian, M., Nicknam, A., Amiri, G.G.: Optimal vector-valued intensity measure for seismic collapse assessment of structures. Earthq. Eng. Eng. Vib. **14**(1), 37–54 (2015)
12. Eads, L., Miranda, E., Lignos, D.G.: Average spectral acceleration as an intensity measure for collapse risk assessment. Earthq. Eng. Struct. Dyn. **44**(12), 2057–2073 (2015)
13. Cordova, P.P., Dierlein, G.G., Mehanny, S.S.F., Cornell, C.A.: Development of a two parameter seismic intensity measure and probabilistic assessment procedure. In: The Second U.S.-Japan Workshop on Performance-Based Earthquake Engineering Methodology for Reinforce Concrete Building Structures, Sapporo, Hokkaido, pp. 187–206 (2001)
14. Baker, J.W., Cornell, C.A.: A vector-valued ground motion intensity measure consisting of spectral acceleration and epsilon. Earthq. Eng. Struct. Dyn. **34**, 1193–1217 (2005)
15. Tothong, P., Luco, N.: Probabilistic seismic demand analysis using advanced ground motion intensity measures. Earthq. Eng. Struct. Dyn. **36**, 1837–1860 (2007)
16. Yakut, A., Yilmaz, H.: Correlation of deformation demands with ground motion intensity. J. Struct. Eng. ASCE **134**(12), 1818–1828 (2008)
17. Mehanny, S.S.F.: A broad-range power-law form scalar-based seismic intensity measure. Eng. Struct. **31**, 1354–1368 (2009)
18. Bojórquez, E., Iervolino, I.: Spectral shape proxies and nonlinear structural response. Soil Dyn. Earthq. Eng. **31**(7), 996–1008 (2011)
19. Baker, J.W., Cornell, C.A.: Vector-valued intensity measures incorporating spectral shape for prediction of structural response. J. Earthq. Eng. **12**(4), 534–554 (2008)
20. Buratti, N.: Confronto tra le performance di diverse misure di intensità dello scuotimento sismico. Congreso Nacional de Ingeniería Sísmica Italiano, ANDIS Bari (2011)
21. Buratti, N.: A comparison of the performance of various ground-motion intensity measures. In: The 15th World Conference on Earthquake Engineering (2012)
22. Terán-Gilmore, A.: Consideraciones del uso de la energía plástica en el diseño sísmico. Revista de ingeniería Sísmica SMIS **65**, 81–110 (2001)
23. Bojórquez, E., Ruiz, S.E.: Strength reduction factors for the valley of Mexico taking into account low cycle fatigue effects. In: 13° World Conference on Earthquake Engineering, Vancouver, Canada (2004)
24. Arroyo, D., Ordaz, M.: Hysteretic energy demands for SDOF systems subjected to narrow band earthquake ground motions. Applications to the lake bed zone of Mexico City. J. Earthq. Eng. **11**, 147–165 (2007)
25. Terán-Gilmore, A., Jirsa, J.O.: Energy demands for seismic design against low-cycle fatigue. Earthq. Eng. Struct. Dyn. **36**, 383–404 (2007)
26. Terán-Gilmore, A., Sánchez-Badillo, A., Espinosa Johnson, M.: Performance-based seismic design of reinforced concrete ductile buildings subjected to large energy demands. Earthq. Struct. **1**(1), 69–91 (2010)
27. Bojórquez, E., Ruiz, S.E., Terán-Gilmore, A.: Reliability-based evaluation of steel structures using energy concepts. Eng. Struct. **30**(6), 1745–1759 (2008)

28. Bojórquez, E., Terán-Gilmore, A., Ruiz, S.E., Reyes-Salazar, A.: Evaluation of structural reliability of steel frames: Inter-story drifts versus plastic hysteretic energy. Earthq. Spectra **27**(3), 661–682 (2011)
29. Uang, C.M., Bertero, V.V.: Evaluation of seismic energy in structures. Earthq. Eng. Struct. Dyn. **19**, 77–90 (1990)
30. Trifunac, M.D., Brady, A.G.: A study of the duration of strong earthquake ground motion. Bull. Seismol. Soc. Am. **65**(3), 581–626 (1975)
31. Eads, L., Miranda, E., Lignos, D.: Spectral shape metrics and structural collapse potential. Earthq. Eng. Struct. Dyn. **45**(10), 1643–1659 (2016)
32. Iervolino, I., Manfredi, G., Cosenza, E.: Ground motion duration effects on nonlinear seismic response. Earthq. Eng. Struct. Dyn. **35**, 21–38 (2006)
33. Terán-Gilmore, A., Jirsa, J.O.: A damage model for practical seismic design that accounts for low cycle fatigue. Earthq. Spectra **21**(3), 803–832 (2005)
34. Rodríguez, M.E., Padilla, C.: A damage index for the seismic analysis of reinforced concrete members. J. Earthq. Eng. **13**(3), 364–383 (2008)
35. Bojórquez, E., Reyes-Salazar, A., Terán-Gilmore, A., Ruiz, S.E.: Energy-based damage index for steel structures. J. Steel Compos. Struct. **10**(4), 343–360 (2010)

Evaluation of Earthquake Resistance of Steel Moment Resisting Frames

Satoshi Yamada

Abstract Proper evaluation of the energy dissipation capacity of a structure, along with proper assessment of inputs, is an important basis of energy-based seismic design. Earthquake response analysis is a useful tool for establishing and calibrating a design method. The reliability of the design method depends largely on the reliability of the analysis. The reliability of the analysis result improves according with the accuracy of the hysteresis model applied to the analysis. In this study, first, based on the experimental results of cyclic loading tests conducted with various loading condition and loading protocol, and analytical results of numerical study, realistic hysteresis model of steel structural members subjected to cyclic bending under axial force is modeled. Next, to evaluate the ultimate earthquake resistance of steel moment resisting frames (MRFs), some series of response analyses of steel MRFs adopted to the proposed hysteresis model are conducted. From the analysis results, the effects of various parameters related to the MRFs, such as the deformation capacity of the column determined by local buckling, the elastic stiffness and strength of the column base, on the ultimate earthquake resistance of the steel MRFs are evaluated. Furthermore, seismic performance of MRFs under multiple strong excitations are examined.

Keywords Steel structure · Moment resisting frame · Earthquake resistance · Hysteresis model · Deterioration

1 Introduction

Proper evaluation of the energy dissipation capacity of a structure, along with proper assessment of inputs, is an important basis of energy-based seismic design. Earthquake response analysis is a useful tool for establishing and calibrating a design method. The reliability of the design method depends largely on the reliability of the analysis. The reliability of the analysis result improves according with the accuracy

S. Yamada (✉)
The University of Tokyo, Tokyo 1138656, Japan
e-mail: s_yamada@arch1.t.u-tokyo.ac.jp

A. Benavent-Climent and F. Mollaioli (eds.), *Energy-Based Seismic Engineering*, Lecture Notes in Civil Engineering 155,
https://doi.org/10.1007/978-3-030-73932-4_19

of the hysteresis model applied to the analysis. The most fundamental and reliable method for evaluating the behavior of steel members is the use of modeling based on experimental results. The range of application of the hysteretic model based on experimental results is determined by the range of parameters, for example, the width-to-thickness ratio, axial force and the loading protocol. Even with the combination of different width-to-thickness ratios and axial force only, a large number of experimental results are necessary. In addition, the loading protocol is an important factor. Generally, cyclic loading tests of steel members have been conducted with the standard type loading protocol, for example, an incremental amplitude protocol or a constant amplitude protocol. Therefore, it is necessary to conduct experiments with various loading protocols, and also to conduct analysis that complements the experiments. Response analysis of MRFs taking into consideration the deterioration behavior of members has already been conducted [1–6]. However, the number of the study is limited, because the current hysteresis models of steel members considering the deteriorating behavior are limited [7, 8].

In this study, first, based on the experimental results of cyclic loading tests conducted with various loading condition and loading protocol, and analytical results of numerical study, realistic hysteresis model of steel structural members subjected to cyclic bending under axial force is modeled [8–10]. Next, to evaluate the ultimate earthquake resistance of steel moment resisting frames (MRFs), some series of response analyses of MRFs adopted to the proposed hysteresis model are conducted. From the analysis results, the effects of various parameters related to the MRFs, such as the deformation capacity of the column determined by local buckling, the elastic stiffness and strength of the column base, on the ultimate earthquake resistance of the MRFs are evaluated. Furthermore, seismic performance of MRFs under multiple strong excitations are also examined by response analyses [16, 17].

2 Hysteretic Model of Steel Members Subjected to Random Loading Histories

2.1 Hysteretic Rule of Steel Members Under Random Cyclic Loading

Hysteresis loops of steel members under random cyclic loading can be decomposed into the skeleton curve, the Bauschinger part (in the deteriorating range, it is defined as the strength increasing part), and the unloading part, as shown in Fig. 1 [8–10]. Before the member reaches its maximum strength, the skeleton curve is obtained by connecting parts of the load-deformation relationship of both positive and negative sides sequentially when the member experiences its highest loading for the first time. After the member reaches the maximum strength, the skeleton curve is obtained as the envelope curve of the cumulative load-deformation relationship on both positive and negative sides.

The skeleton curve is empirically known to corresponds with load-deformation relationship under monotonic loading [9, 10]. In the case of square hollow section members generally used columns of steel building structures in Japan, the skeleton curve is obtained through analysis including the deteriorating range governed by local buckling [8, 11]. In the analysis, maximum strength and deteriorating behavior of columns subjected to bending and axial force are predicted based on the stress-strain relationship of stub columns modeled by experimental results. Also, in the case of H-shaped members generally used beams of steel building, the skeleton curve is obtained through analysis up to the maximum strength point determined by ductile fracture [12]. In the analysis, maximum strength point of beam is predicted by flange strain histories of critical section.

For the Bauschinger part before maximum strength, a bi-linear model proposed by Akiyama and Takahashi [10] is applied in both cases of SHS columns and H-shaped beams. For the strength increasing part in the deteriorating range governed by local buckling, a bi-linear model based on experimental results [8] is applied. Elastic stiffness is applied in the unloading part.

2.2 *Calibration of the Hysteresis Model*

Proposed Hysteresis model is examined in comparison with experimental results. Example of the comparison is shown in Fig. 2. Proposed hysteresis model shows good agreement with experimental results up to column lose their restring force.

3 Response Analysis on Steel MRFs Based on the Realistic Behaviors of Members

3.1 *Outline of the Analysis*

Analytical models are low to middle-rise steel MRFs consisted with H shaped section beams and square hollow section columns those are common in Japan. Numbers of stories are 6, 9 and 12 (Fig. 3).

In the analysis, NS component of El-Centro record (1940 Imperial Valley Earthquake), EW component of Taft record (1952 Kern County Earthquake), EW component of Hachinohe record (1968 Tokachi-oki Earthquake), NS component of JMA-Kobe record (1995 Kobe earthquake) and NS component of JMA-Sendai record (2011 Tohoku earthquake) are used as input wave. In calculation, amplification factor is multiplied to the acceleration data to change intensity of excitation. Thus, the analytical results are continuously obtained, according to the intensity of excitation. In this study, the ultimate state of the MRF is defined as the state reached just before any one of the structural elements loses its restoring force.

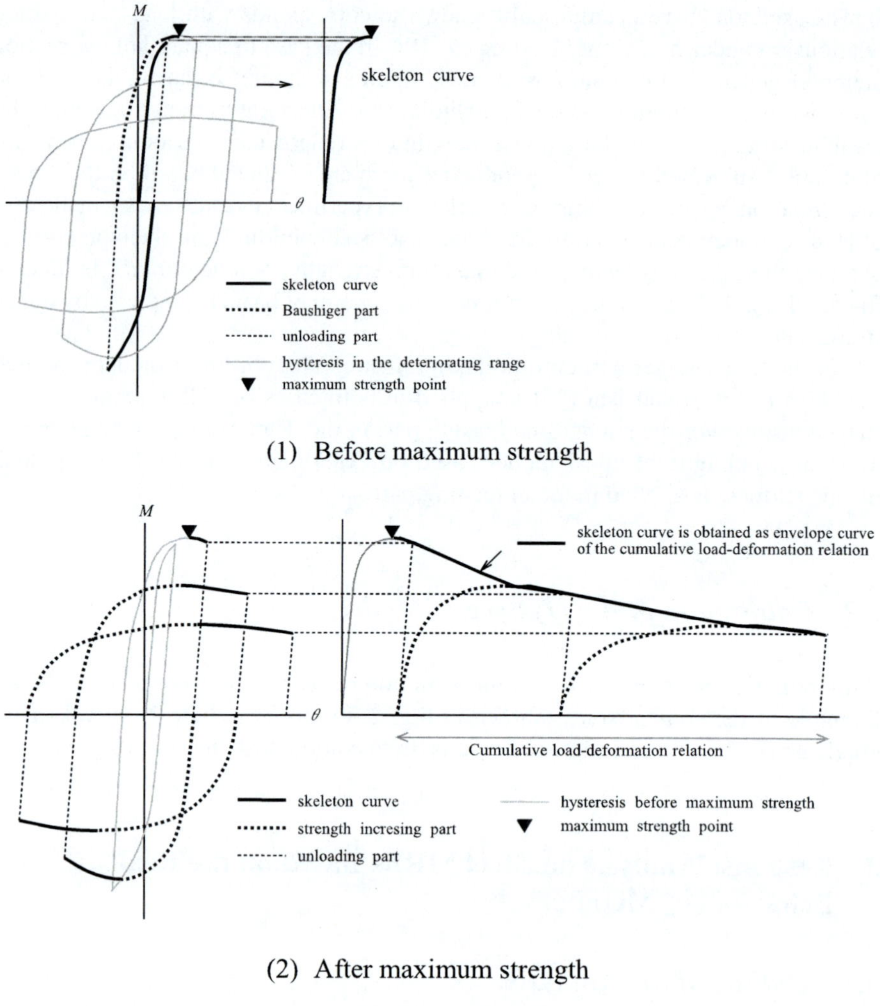

Fig. 1 Decomposition of the hysteresis of steel members [8]

3.2 Ultimate Earthquake Resistance of Steel MRFs Determined by Deterioration and Losing Restoring of Members Due to Local Buckling [1, 2]

To evaluate the ultimate earthquake resistance of multi-story steel frames determined by deterioration and losing restoring force of members due to local buckling, inelastic response analysis based on the realistic behavior of steel members are carried out. Analytical models are designed by a plastic design method based on the deformation capacities of members proposed by Akiyama [13, 14]. Parameters of analysis models are ductility of beams and columns determined by local buckling (width-to-thickness

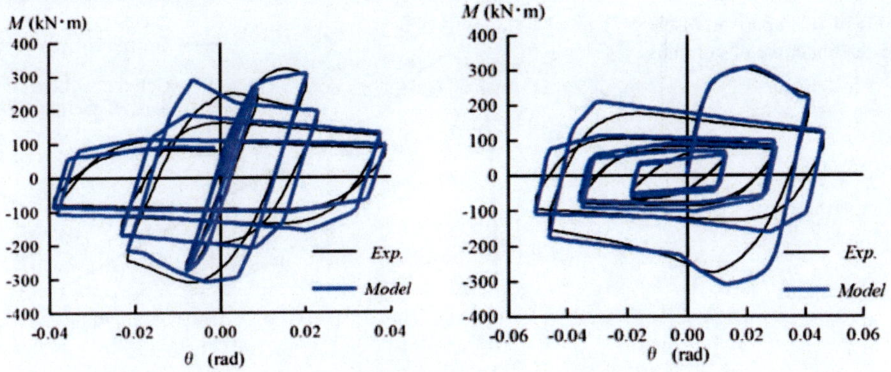

Fig. 2 Comparison of hysteresis model and experimental results (SHS columns subjected to incremental and decremental loading amplitude) [8]

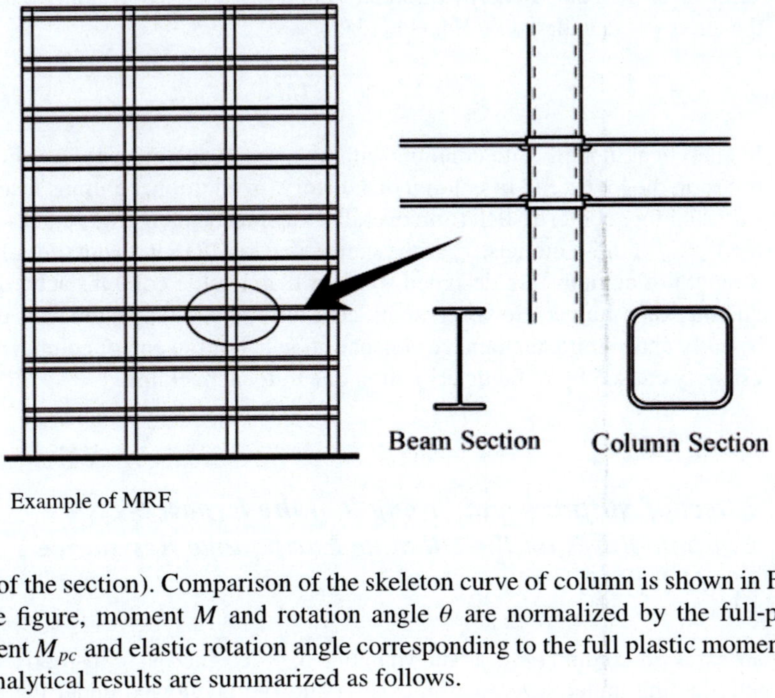

Fig. 3 Example of MRF

ratio of the section). Comparison of the skeleton curve of column is shown in Fig. 4. In the figure, moment M and rotation angle θ are normalized by the full-plastic moment M_{pc} and elastic rotation angle corresponding to the full plastic moment θ_{pc}.

Analytical results are summarized as follows.

(1) By making the strength distribution of structural members corresponding to the deformation capacities assumed yield mechanism of weak columns, weak beams and weak panel zone, overall sway mechanism of the MRF can be formed.

(2) In analytical models with ductile columns, damage disperses over beams and panels. Dissipated energy of the models until ultimate state significantly exceeds design demands. Figure 5 (1) shows ultimate earthquake resistance of

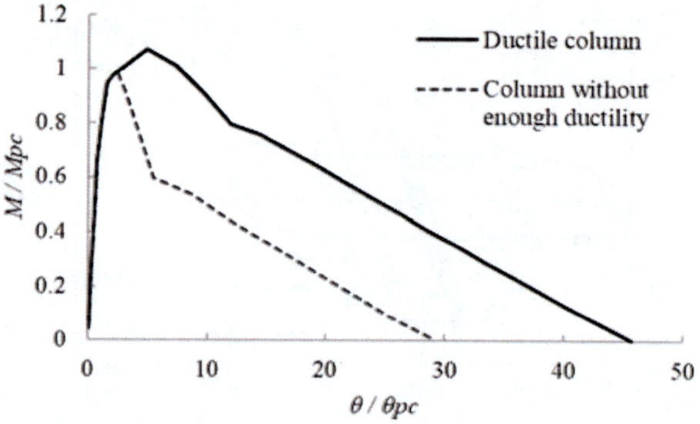

Fig. 4 Comparison of the skeleton curve of columns

the model as equivalent velocity of dissipated energy V_u with the relationship of maximum story drift angle R_{max}. In the figure, V_u is divided by equivalent velocity of demanded energy dissipation in design V_d. Equivalent velocity of the energy is calculated by Eq. (1) [13].

$$V = \sqrt{2E/M} \tag{1}$$

(3) In analytical models with columns without enough ductility, damage concentrates on the lower end of column of 1st story. In addition, columns lose those restoring force before MRF form overall Sway mechanism. Figure 5 (2) shows the V_u/V_d of those models. V_d is the same value as MRFs with ductile columns. Strength of columns are designed stronger than ductile columns according to ductility of columns. However, significant difference is appeared. It is caused by early and significant damage concentration at bottom end of columns of the 1st story caused by rapid deterioration due to local buckling.

3.3 *Effect of Stiffness and Strength of the Exposed-Type Column-Bases on the Ultimate Earthquake Resistance of MRF [3–5]*

In Japan, exposed-column base, as shown in Fig. 6, is designed as semi-rigid connection, and yielding under severe earthquake is allowed is allowed under the condition that using of anchor bolts with guaranteed deformation capacity and additional demand horizontal strength on the 1st story [15]. To evaluate the effect of elastic stiffness and slip behavior of exposed column base on the ultimate earthquake resistance, response analyses of MRFs are conducted. Parameter of the analyses are elastic stiffness of column base and strength of column base. Hysteresis model of exposed-type column base is shown in Fig. 6.

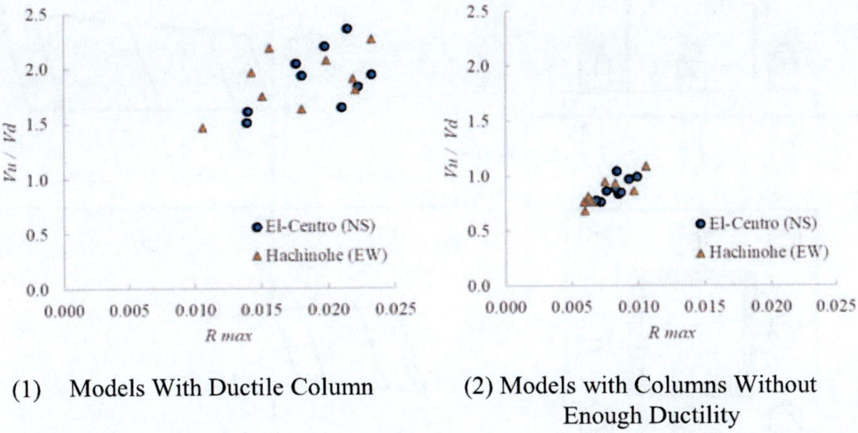

Fig. 5 Ultimate earthquake resistance of MRF [2] (modified)

Analytical results are summarized as follows.

(1) As the elastic stiffness of the column base decreases, the damage concentration on the bottom end of the column of the first story is reduced, and the ultimate earthquake resistance of the MRF is improved.
(2) The intensity of the damage concentration at the structural element of the 1st story in the MRFs with weak exposed-type column bases were much less than that of the MRFs with pin type column bases and the MRFs with fixed column bases.
(3) Ultimate earthquake resistance of the MRFs with weak exposed type column bases are similar to that of the MRFs with pin type column bases.
(4) Ultimate earthquake resistance of the MRFs with weak exposed type column bases are better than that of the MRFs with fixed column bases.
(5) Contribution of the exposed column base for the energy dissipation is very small comparing to the other structural elements.

3.4 *Seismic Performance of MRFs Subjected to Multiple Strong Ground Motions [16, 17]*

In current seismic design, MRFs are designed not to collapse under single severe earthquake. However, its performance under multiple severe earthquakes is not clarified. To evaluate the seismic performance of steel MRFs under multiple strong ground motions, response analyses of MRFs subjected to multiple excitations are conducted. From analytical results, following conclusions are obtained.

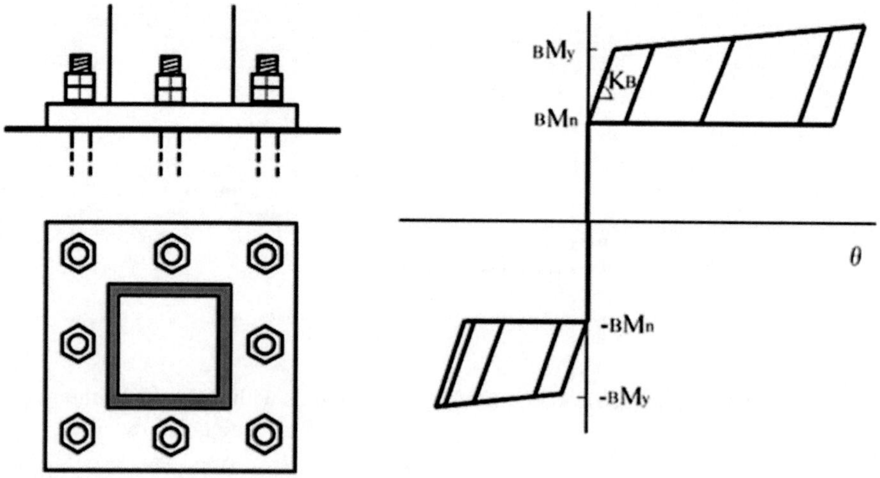

Fig. 6 Exposed column base with hysteresis model

(1) In case of ductile MRFs, if deterioration caused by local buckling are not occurred, stable behavior with a linear increment in cumulative damage D is found in most cases as shown in Fig. 7, especially when the ground motion intensity is equal to the design level.

(2) In case of ductile MRFs, when the intensity is greater than the design level, both ductile fracture and weak story collapse are observed in several cases, even deterioration caused by local buckling are not occurred.

(3) Although MRFs were designed to form an overall sway mechanism by making the columns 1.5 times stronger than the beams, considering local buckling in columns, in many cases weak story collapse occurred due to the deterioration of columns under multiple strong excitations.

4 Conclusion

To improve the accuracy of evaluation of earthquake resistance of steel MRFs, realistic hysteresis model of steel member is proposed. The proposed model can be applied random loading history with enough accuracy including deteriorating behavior of SHS columns governed by local buckling. Adopting the model, some series of response analyses are conducted to evaluate the effect of structural behaviors of members such as local buckling of SHS columns and elasto-plastic behaviors of exposed type column bases. Furthermore, seismic behavior of MRFs under multiple strong excitations are examined.

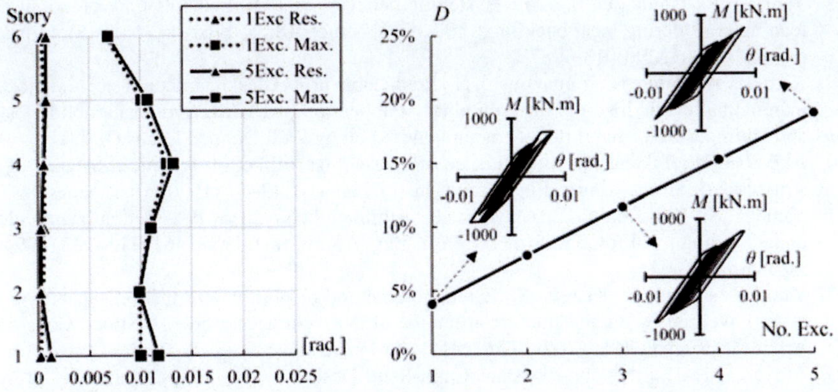

(1) Maximum and Residual Story Drift angle (2) Cumulative Damage of Beam

Fig. 7 Examples of the response of the ductile SMRFs excited by multiple ground accelerations [16]

Acknowledgements This article is a compilation of previous studies. The author thanks to Prof. Hiroshi Akiyama, Prof. Takanori Ishida, Prof. Shoichi Kishiki, Prof. Yu Jiao and other co-workers.

References

1. Yamada, S., Akiyama, H.: Inelastic response analysis of multi-story steel frames based on the realistic behaviors of members governed by local buckling. J. Struct. Constr. Eng. Archit. Inst. Japan 463, 125–133 (1994). (in Japanese)
2. Yamada, S., Akiyama, H., Mizutani, H.: Ultimate earthquake resistance of multi-story steel frames accompanied by the plastic deformation of panel zone. J. Struct. Constr. Eng. Archit. Inst. Japan 481, 137–144 (1996). (in Japanese)
3. Yamada, S., Akiyama, H.: Influence of the rigidity of column bases on the ultimate earthquake resistance of multi-story steel moment frames. J. Struct. Constr. Eng. Archit. Inst. Japan 498, 113–118 (1997). (in Japanese)
4. Yamada, S., Akiyama, H., Sadamoto, M.: Influence of the elasto-plastic behavior of column bases with slip-type hysteresis characteristics on the ultimate earthquake resistance of multi-story steel moment frames. J. Struct. Constr. Eng. Archit. Inst. Japan 502, 141–147 (1997). (in Japanese)
5. Ishida, T., Yamada. S., Kubota, K.: Evaluation of Ds value of steel moment resisting frames with -type column bases. J. Struct. Constr. Eng. Archit. Inst. Japan 720, 357–367 (2016). (in Japanese)
6. Bai, Y., Shi, Y., Deng, K.: Collapse analysis of high-rise steel moment frames incorporating deterioration effects of column axial force—bending moment interaction. Eng. Struct. **127**, 402–415 (2016)
7. Lignos, D., Krawinkler, H.: Deterioration modeling of steel components in support of collapse prediction of steel moment frames under earthquake loading. J. Struct. Eng. ASCE **137**(11), 1291–1302 (2011)

8. Yamada, S., Ishida, T., Jiao, Y.: Hysteretic behavior of RHS columns under random cyclic loading considering local buckling. Int. J. Steel Struct. **18**(5), 1761–1771 (2018). https://doi.org/10.1007/s13296-018-0087-x
9. Kato, B., Akiyama, H., Yamanouchi, H.: Predictable properties of structural steels subjected to incremental cyclic loading. In: Preliminary Publication of IABSE Symposium on Resistance and Ultimate Deformability of Structures Acted on by Well Defined Loads (1973)
10. Akiyama, H., Takahashi, M.: Influence of Bauschinger effect on seismic resistance of steel structures. J. Struct. Constr. Eng. Archit. Inst. Japan 418, 49–57 (1990). (in Japanese)
11. Yamada, S., Kuwamura, H., Akiyama, H.: Ultimate behavior of box-section beam-columns under varying axial force. J. Struct. Constr. Eng. Archit. Inst. Japan 461, 115–122 (1994). (in Japanese)
12. Yamada, S., Jiao, Y., Kishiki, S.: Evaluation method of plastic deformation capacity of steel beam governed by ductile fracture at the toe of the weld access hole. J. Struct. Constr. Eng. Archit. Inst. Japan 711, 1767–1777 (2015). (in Japanese)
13. Akiyama, H.: Earthquake-Resistant Limit-State Design for Buildings. University of Tokyo Press, Japan (1985)
14. Architectural Institute of Japan.: Ultimate Strength and Deformation Capacity of Buildings in Seismic Design (1990). (in Japanese)
15. Building Standard Law Japan, Building Center Japan
16. Tenderan, R., Ishida, T., Jiao, Y., Yamada, S.: Seismic performance of ductile steel moment-resisting frames subjected to multiple strong ground motions. Earthq. Spectra **35**(1), 289–310 (2019). https://doi.org/10.1193/111217EQS235M
17. Tenderan, R., Ishida, T., Jiao, Y., Yamada, S.: Structural behavior of steel moment-resisting frames subjected to multiple earthquakes considering column strength deterioration. In: 16th East Asia-Pacific Conference on Structural Engineering and Construction (2019)

Energy-Based Design Theory for Self-Centering Structures

Ge Song, Ying Zhou, and T. Y. Yang

Abstract Although conventional structures can be designed to avoid collapse under seismic actions, unrecoverable nonlinear deformations would still occur to dissipate earthquake energy. It enlarges structural damages and residual deformations, which leads to repair costs increased and downtime prolonged. Self-centering structures are introduced for the resilience demand in seismic engineering. Inelastic behaviors can be limited within specific areas, which would prevent key components from unrecoverable damages. This paper presents an energy-based design theory (EBDT) for self-centering structures. A damage model considering both residual deformation and hysteretic energy E_{H} is proposed. Based on the design energy spectrum and the damage model, E_{H} is introduced as a design parameter and accounted in the design procedure. EBDT enables designers to select multiple performance objectives for different seismic hazards. The design procedure is elaborated with an example. The results show that EBDT can provide a reliable design procedure for self-centering structures.

Keywords Structural damages · Energy-based design theory · Self-centering structures · Hysteretic energy

G. Song · Y. Zhou (✉)
State Key Laboratory of Disaster Reduction in Civil Engineering, Tongji University, Shanghai 200092, China
e-mail: yingzhou@tongji.edu.cn

T. Y. Yang
International Joint Research Laboratory of Earthquake Engineering, Tongji University, Shanghai 200092, China

Department of Civil Engineering, University of British Columbia, Vancouver V6T1Z4, Canada

A. Benavent-Climent and F. Mollaioli (eds.), *Energy-Based Seismic Engineering*, Lecture Notes in Civil Engineering 155,
https://doi.org/10.1007/978-3-030-73932-4_20

1 Introduction

To satisfy with the enhanced resilience demand, self-centering systems are introduced in recent decades. They exhibit superior aseismic capacities, which can limit damages developed in structures and unrecoverable residual deformations after earthquakes [1–5]. The direct displacement-based design (DDBD) is commonly adopted to design self-centering systems [6]. DDBD select the maximum deformation as the index to ensure satisfactory strength requirements and deformation limits under seismic actions. However, for self-centering systems, the cumulative damages caused by cyclic loadings are proved to play a significant role in behaviors and post-earthquake performance [7]. By introducing the hysteretic energy demand E_{H} as a key design index, the energy part that impacts structural damages greatly can be considered. Therefore, an energy-based design procedure can provide a more efficient framework for self-centering structures.

Housner firstly addressed the energy concept in the First World Conference on Earthquake Engineering [8]. Many studies have investigated and proposed energy-based design theories (EBDT) for conventional systems [9–15], yet EBDT specific to self-centering systems is still limited. This paper proposes an EBDT for self-centering structures. E_{H} is considered in design procedures by introducing a damage model considering both residual deformation and hysteretic energy. EBDT enables designers to select multiple performance objectives for different seismic hazards.

2 Energy-Based Design Theory

Since the hysteretic energy E_{H} is the main parameter that needs to be considered in EBDTs, the premise of EBDTs is to quantify E_{H} demand in structures accurately. To do this, the structural energy-dissipation mechanism shall be controllable, which is the critical step to implement the design method into component level. Due to the variety of structural types and discrete features in ground motions, conventional structures would exhibit uncontrollable energy-dissipation mechanism under seismic actions. It leads to challenges in quantifying E_{H} for structures and key components. Therefore, the controllability of energy-dissipation mechanism is the first issue that needs to be investigated in EBDTs. In addition, E_{H} shall be transformed as a design index to consider the energy influences in EBDTs.

2.1 *Energy-Dissipation Mechanism Design for Self-Centering Structures*

The structural energy-dissipation mechanism design is the premise of EBDTs and the critical step to implement the design method into component level. Compared

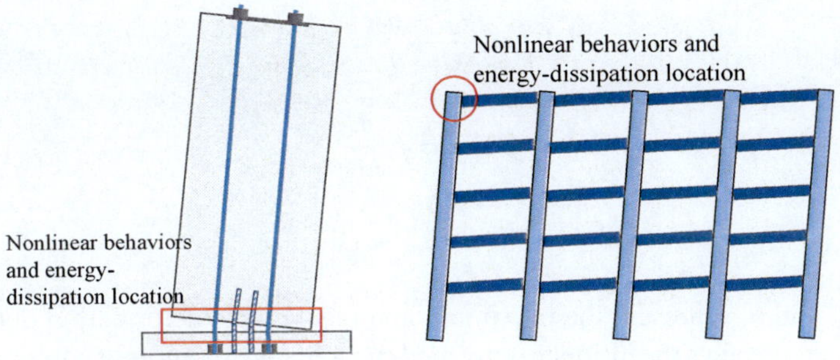

(a) Self-centering shear wall (b) Self-centering frame

Fig. 1 Deformed structures

with traditional systems, self-centering structures enable opening-closing behaviors occurred in joints. Meanwhile, prestressing is introduced in structures to provide resilient capacity, while energy-dissipation devices are installed in joints where large deformations occurred to dissipate input seismic energy. Figure 1 shows a typical deformed self-centering shear wall and frame. It has been proved that critical components, including columns, beams and wall panels, can be prevented from damage under seismic actions, while all nonlinearity would be limited in wall-base joints or beam-column joints. The opening-closing behaviors and the nonlinearity developed at joint sections are the only locations where dissipates input energy. It leads to the yielding mechanism and damage development of self-centering systems stable and controllable. Therefore, self-centering structures can satisfy with the design premise of EBDTs automatically.

2.2 *Transformation of E_H in EBDTs*

The performance-based seismic design (PBSD) has become the most commonly adopted design theory in earthquake engineering. It enables designers to select desired performance levels corresponding to different seismic intensities to achieve multiple structural requirements under earthquakes. In addition, it provides design objectives for the force- and displacement-based design theories. To enhance the resilience demand, the damage development and controllability shall be considered in design procedures as well.

Damage models and damage index (DI) are normally adopted to quantify and evaluate structure damages, where DI is a dimensionless parameter representing the damages developed in structures [16]. To account for damages in EBDTs and transform E_{H} as a design parameter, a damage model is utilized in the proposed design theory [7], as shown in Eq. (1).

$$\mathrm{DI}_{\theta_c} = \frac{\theta_m - \theta_r}{\theta_u - \theta_r} = \frac{\theta_c}{\theta_u - \theta_r} \tag{1a}$$

$$\mathrm{DI}_{E_H} = \beta \frac{E_H}{M_y \theta_u} \tag{1b}$$

$$\mathrm{DI} = \mathrm{DI}_{\theta_c} + \mathrm{DI}_{E_H} \tag{1c}$$

where θ_m and θ_r represent the maximum and recoverable deformation under earthquakes; θ_u represents the ultimate deformed capacity of structures or components; M_y stands for the design moment of structures or components; β is a non-negative factor, differs with different systems; $\theta_c = \theta_m - \theta_r$ represents the residual deformation after earthquakes.

Equation (1) can consider both the deformation- and hysteretic energy-induced damages for structures. Meanwhile, the residual deformation, which is a key index for evaluation of post-earthquake resilient capacity for self-centering systems, is included in the damage model as well. It has been proved that the damage model can be adopted to quantify and evaluate damage development for self-centering structures [7].

Equation (2) can be converted from Eq. (1) as:

$$M_y = \frac{\beta E_H}{(\mathrm{DI} - \frac{\theta_c}{\theta_u - \theta_r})\theta_u} \tag{2}$$

It shows that E_H is transformed as a design parameter, while both the DI and deformation indexes are considered in determining M_y. According to Eq. (2), designers can select desired DI, θ_m, θ_r, and θ_c to achieve multiple performance objectives corresponding to certain seismic intensities. This indicates that the method combines the proposed EBDT with the performance-based design theory closely. It should be noted that the non-negative parameter β is related to energy-dissipation capacity of structures or components, which needs to be investigated specifically.

2.3 *Quantification of E_H*

E_H is the key parameter that directly related to nonlinear behaviors and structural damages. The premise of EBDTs is to identify and quantify E_H for overall systems and components accurately. Previous studies have proven that the hysteretic energy demand and input energy not only related to structural features, such as the hysteretic model, damping ratio, stiffness *etc.*, but also closely related to earthquake features. Therefore, E_H and E_I should be determined according to different structural systems and earthquake types. Practical design energy spectra are commonly used to obtain E_H and E_I. Many methodologies have been proposed to construct energy spectra [17, 18], yet their feasibility for self-centering systems is still debatable. Zhou et al.

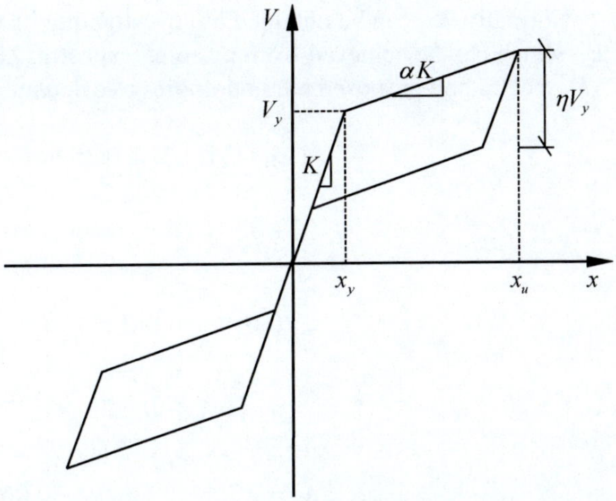

Fig. 2 Flag-shaped hysteretic model

[19] investigated influences of structural features and earthquake characteristics on E_{I} and E_{H}, and proposed methodologies to construct design input energy spectra and hysteretic energy spectra for self-centering systems. The proposed spectra are constructed based on the typical flag-shaped hysteretic model, as shown in Fig. 2. The ground motions applied to perform nonlinear time history analyses are selected based on site types specified in the Chinese code [20].

Figure 3 presents the basic design input energy spectrum proposed by Zhou et al. [19]. The spectrum is divided into three portions as the function of structural period. It considers both the structural features, including the damping ration, hysteretic model and ductility, and ground motion types specified in the Chinese code [20].

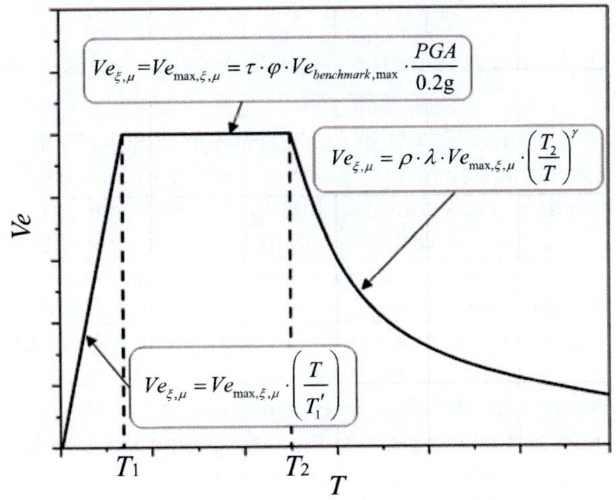

Fig. 3 Basic design energy spectrum

Normally, E_H can be obtained by investigating its relations with the input energy E_I, which can be achieved from hysteretic spectra. Zhou et al. [19] investigated E_H / E_I spectra and proposed methodologies to calculate it, as listed in Eq. (3).

$$(E_H/E_I)_{\xi,\mu,\eta} = 0.35I_1 \cdot I_2 \cdot I_3 \tag{3a}$$

$$I_1 = 5.75 \cdot \xi^{0.28} \cdot \eta^{1.06+0.17\cdot\ln(\xi-0.0087)} \tag{3b}$$

$$I_2 = (0.014 + 0.08\eta) \cdot \xi^{-0.56-0.52\cdot 0.23^{\eta}} \tag{3c}$$

$$I_3 = A + B \cdot T + C \cdot T^2 \tag{3d}$$

$$A = 0.35 + 0.52 \cdot \mu - 0.06 \cdot \mu^2 \tag{3e}$$

$$B = -0.11 \cdot \ln(\mu - 1.19) \tag{3f}$$

$$C = 0.02 - 0.04 \cdot 0.80^{\mu} \tag{3g}$$

As E_H of the overall system being obtained, the hysteretic energy demand of the energy-dissipation locations shown in Fig. 1 needs to be quantified. It is critical for implementing the design method into component level. Previous studies have focused on the energy distribution in traditional structures [21–24], yet its applicability in self-centering structures is still debatable. Since there are various self-centering systems, it is a case-specific issue that needs to be investigated based on the target structure.

2.4 Performance Objectives for Self-Centering Systems

As shown in Eq. (2), designers can select desired DI, θ_m, θ_r, and θ_c to achieve multiple performance objectives corresponding to target seismic intensities. DI is a dimensionless index implying the damage states of structures or components. Park-Ang [16] proposed damage intervals for traditional concrete components, as listed in Table 1. Designers can choose various DI based on desired damage states under certain seismic intensities to control damages developed in structures. Based on Eq. (2), bigger DI leads to smaller M_y, which would result in more damages occurred in structures. In addition, Park-Ang select DI between the moderate and severe damage intervals as the repairable damage limit [16]. It should be noted that, to enhance the post-earthquake repairability, it is significant to select conservative DI to ensure the performance capacity of structures.

Although Park-Ang and other studies have suggested damage intervals for different structures and components, their applicability in self-centering systems is

Table 1 Damage intervals

Damage state	Park-Ang	Zhou
No damage	[0, 0.1)	[0, 0.1)
Minor damage	[0.1, 0.25)	[0.1, 0.3)
Moderate damage	[0.25, 0.4)	[0.3, 0.5)
Severe damage	[0.4, 0.8)	[0.5, 0.9)
Collapse	[0.8, +∞)	[0.9, +∞)

still debatable. Like the distribution of E_H in structures, it is a case-specific issue that needs to be investigated for different systems. Zhou et al. [7] evaluated the damage development for self-centering concrete frames and suggested damage intervals for hybrid joints, as shown in Table 1. The extension of damage limits indicates the superior capacity enhancement and damage controllability of self-centering structures. DI = 0.5 can be adopted as the post-earthquake repairable limit according to Park-Ang.

As for the selection of θ_m, θ_r, and θ_c, it should be consistent with performance levels and capacity requirements specified in different codes. Zhou et al. [7] suggested performance levels for self-centering systems according to the new four-level seismic fortification targets specified in the Chinese code [20], as shown in Table 2.

In addition, Zhou et al. [7] suggested seismic performance indexes for self-centering shear wall structures and frames respectively according to the performance levels listed in Table 2, as shown in Table 3. It should be noted that performance indexes for other self-centering systems should be determined specifically. On the other hand, designers can choose various performance levels and indexes according to different capacity and application requirements. Figure 4 illustrates the main steps of EBDT proposed in this paper.

Table 2 Seismic performance levels

Seismic fortification targets	Level 1	Level 2	Level 3	Level 4
Performance	Fully operation	Fully operation	Repairability	Life safety

Table 3 Seismic performance indexes for self-centering shear wall structures

Seismic fortification targets	Level 1	Level 2	Level 3	Level 4
Self-centering wall structures				
θ_m (%)	–	0.1	–	2
θ_c (%)	0.2	0.2	0.5	–
Self-centering frames				
Deformation index	1st	2nd	3rd	4th
θ_m (%)	0.18	–	2	3.5
θ_c (%)	0.2	0.2	0.5	–

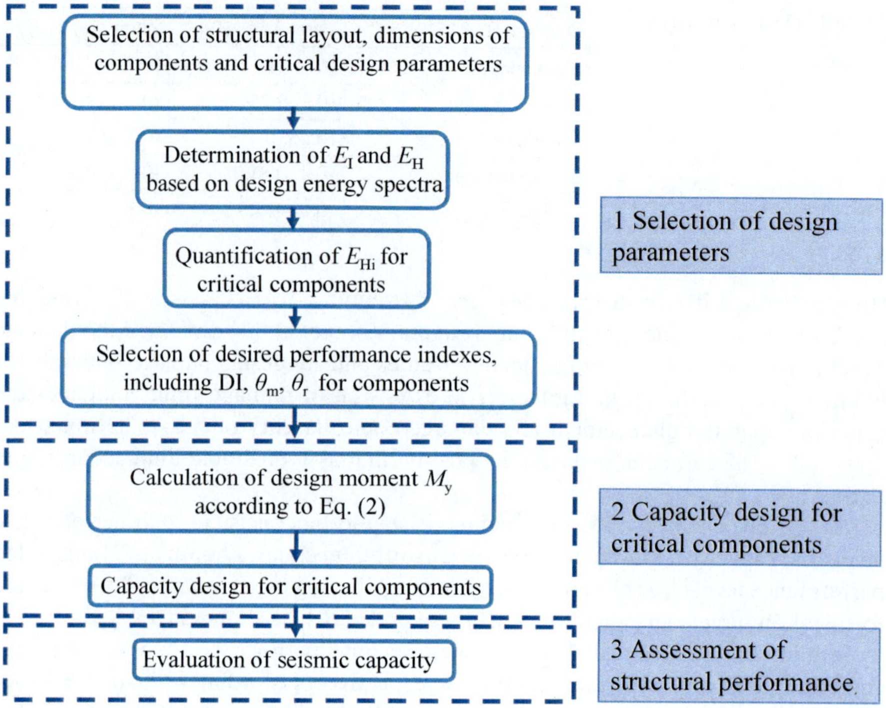

Fig. 4 Flow chart for EBDT

3 Design Example

The design example elaborated in the PRESSS Design Handbook is selected as the prototype structure [25]. Figure 5 illustrates the elevation view of the target self-centering concrete frame. The story height is 3.8 m, while each span is 7.5 m.

Fig. 5 Elevation view of the prototype building

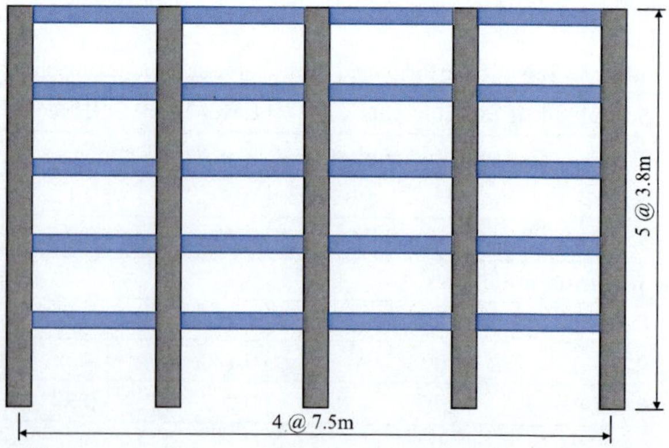

According to the Chinese code [20], the seismic fortification intensity of the structure site is assumed to be 7. The seismic site is Group 2 and Site Class II. Dimensions of beams and columns are 400 mm × 650 mm and 700 mm × 700 mm, respectively. Detailed material information can be derived from the reference [25].

3.1 Selection of Design Parameters

Figure 6 shows the design energy spectra constructed based on the Chinese code according to the site information of the prototype building.

It can be obtained from Fig. 6 that E_I of the system equals to 4255 KN m, while the hysteretic energy demand of the system E_H equals to 2919 KN m. It should be noted that, as the critical substructures in self-centering frames, hybrid joints need to be designed specifically for the PT tendons and energy-dissipation devices. The typical hybrid joint, specified in ACI T1.2-03 [26] is adopted in the prototype building, which has equal top and bottom energy-dissipation mild steels in beam sections and unboned PT tendons installed in the centroid of beams.

To determine M_y for hybrid joints with Eq. (2), E_H of each hybrid joint shall be quantified. Song et al. investigated the energy distribution and development in self-centering concrete frames, and quantify E_H for each hybrid joint in the structure [27]. Table 4 lists the ratio of energy dissipated by hybrid joints at the ith floor $E_{Hi,joint}$ to that by all hybrid joints $E_{H,\,joint}$. As concluded in the study, $E_{Hi,\,joint}$ would be equally

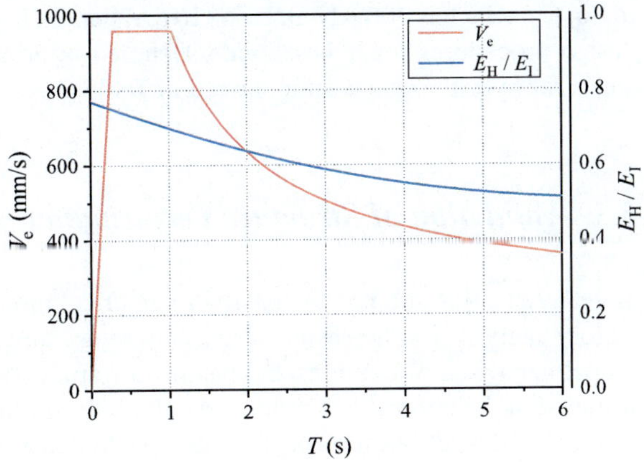

Fig. 6 Design energy spectra

Table 4 EH distribution in the frame

Hi/H	0	0.2	0.4	0.6	0.8	1.0
$E_{Hi,\,joint}$	0	0.212	0.261	0.236	0.197	0.168

Table 5 Design input parameters

Parameter	Value	Parameter	Value
E_I (KN m)	4255	E_H (KN m)	2919
$E_{H,joint}$ (KN m)	2199	$\int dE$ (KN m)	72
DI	0.5	θ_c (%)	0.4
θ_r (%)	0.834	θ_u (%)	3.4

Table 6 Detailed information of reinforcement for hybrid joints

Model	Hybrid joint					Column	
	PT number	Initial PT force (KN)	Initial PT stress ($/f_{ptu}$)	Mild steel type	Mild steel number (each)	Mild steel type	Mild steel number (each)
EBSD	7	960.31	0.55	D22	3	D25	3

dissipated by joints at the same story, implying that 1/8 of $E_{Hi,joint}$ for the joints at the ith floor. Therefore, E_H for each joint equals to 72 KN m.

Table 5 lists design parameters needed in Eq. (2) for EBDT of the prototype frame.

3.2 *Capacity Design of Hybrid Joints*

As the design parameters are selected, M_y of each joint can be calculated according to Eq. (2), which is 406 KN m. NZS3101 Appendix B [28] specifies detailed capacity design procedures for hybrid joints. Detailed information of PT tendons and mild steels for hybrid joints is summarized in Table 6.

3.3 *Validation of Structure Performance*

OpenSees [29] is adopted to construct numerical model and perform nonlinear time history analyses for the designed self-centering frame. Fiber elements are utilized to model behaviors of columns and beams in frames. A zero-length spring element is installed in the interface of beam and column to simulate behaviors of hybrid joints. 22 pairs of far-field earthquake records are adopted to perform nonlinear time history analyses.

Table 7 presents the maximum of DI, θ_c, and θ_m from hybrid joints under each seismic action.

It can be found from the table that mean values of θ_c and θ_m obtained from 44 earthquakes show good agreement with the selected design indexes in EBDT. Although DI = 0.306, which is less than the selected performance indexes, it indicates conservative design results of EBDT. In the design procedure, DI = 0.5 was adopted

Table 7 Analytical results

GM	DI	θ_c	θ_m
Target	0.500	0.004	0.012
Mean	0.306	0.003	0.012

Table 8 Comparison of energy distribution

Hi/H		0	0.2	0.4	0.6	0.8	1.0
$E_{\mathrm{Hi,joint}}$	EBDT	0	0.212	0.261	0.236	0.197	0.168
	Mean	0	0.195	0.242	0.196	0.188	0.175

as the damage index to satisfy with the requirement of post-earthquake repairability. As listed in Table 1, DI of moderate damage state for self-centering systems is defined as [0.3, 0.5]. DI = 0.306 belongs to the moderate damage state, which satisfies with the resilient demand of design objectives.

Table 8 compares mean values of distribution of E_H along structural height to the design parameters listed in Table 4. The EBDT designed structures present good agreement of distribution of E_H compared with the design objectives.

4 Conclusion

This paper presents an energy-based design theory (EBDT) for self-centering structures. The hysteretic energy demand is adopted in capacity design of structures by introducing a damage model considering both residual deformation and hysteretic energy E_H. Performance levels and objectives are proposed based on the Chinese code. EBDT enables designers to conduct multi-level performance design for desired seismic hazards and structural demands. The design procedure is elaborated with a five-story self-centering frame. *OpenSees* was adopted to perform nonlinear time history analyses for validation of structural performance. The results show that EBDT can provide a reliable design procedure for self-centering structures.

Acknowledgements The authors are grateful for the financial support received from the National Natural Science Foundation of China (Grant No. 52025083, Grant No. 51778486), Shanghai Committee of Science and Technology (Grant No. 19DZ1201200) and China Scholarship Council during a visiting study in University of British Columbia (No.201906260206).

References

1. Priestley, M.J.N., Tao, J.: Seismic response of precast prestressed concrete frames with partially debonded tendons. PCI J. **38**(1), 58–69 (1993)
2. Cheok, G., Stone, W.: Performance of 1/3-Scale Model Precast Concrete Beam–Column Connections Subjected to Cyclic Inelastic Loads—Report no. 4. NISTIR 5436. National Institute of Standards and Technology, NIST, Gaithersburg (1994)
3. El-Sheikh, M., Sause, R., Pessiki, S., Lu, L.W.: Seismic behavior and design of unbonded post-tensioned precast concrete frames. PCI J. **44**(3), 54–71 (1999)
4. El-Sheikh, M., Pessiki, S., Sause, R., Lu, L.W.: Moment rotation behavior of unbonded post-tensioned precast concrete beam–column connections. ACI Struct. J. **1**, 122–132 (2000)
5. Korkmaz, H.H., Tankut, T.: Performance of a precast concrete beam-to-beam connection subject to reversed cyclic loading. Eng. Struct. **27**(9), 1392–1407 (2005)
6. Priestley, M.J.N.: Direct displacement-based design of precast/prestressed concrete buildings. PCI J. **47**(6), 66–78 (2002)
7. Zhou, Y., Song, G., Huang, W.: Performance-based damage evaluation of hybrid joints. B Earthq. Eng. (2020). https://doi.org/10.1007/s10518-020-00838-8
8. Housner, G.W.: Limit design of structures to resist earthquakes. In: Proceedings of the First World Conference on Earthquake Engineering (1959)
9. Uang, C.M., Bertero, V.V.: Evaluation of seismic energy in structures. Earthq. Eng. Struct. Dyn. **19**(2), 77–90 (1990)
10. Akiyama, H.: Earthquake Resistant Limit State Design for Building. University of Tokyo Press, Tokyo (1985)
11. Fajfar, P., Gaspersic, P.: The N2 method for the seismic damage analyses of RC buildings. Earthq. Eng. Struct. Dyn. **25**(1), 31–46 (1996)
12. Akbas, B., Shen, J., Hao, H.: Energy approach in performance-based seismic design of steel moment resisting frames for basic safety objective. Struct. Des. Tall. Spec. Build. **10**(3), 193–217 (2001)
13. Chou, C.C., Uang, C.M.: A procedure for evaluating seismic energy demand of framed structures. Earthq. Eng. Struct. Dyn. **32**(2), 229–244 (2003)
14. Yang, T.Y., Tung, D., Li, Y.J.: Equivalent energy-based design procedure for earthquake resilient fused structures. Earthq. Spectra **34**(2), 1–21 (2018)
15. Yang, T.Y., Atkinson, J., Tobber, L., Tung, D.P., Neville, B.: Seismic design of outrigger systems using equivalent energy design procedure. Struct. Des. Tall. Spec. Build. **29**(10), e1743 (2020)
16. Park, Y.J., Ang, H.S.: Mechanistic seismic damage model for reinforced concrete. J. Struct. Eng. **111**(4), 722–739 (1985)
17. Zhou, Y., Song, G., Huang, S.M., Wu, H.: Input energy spectra for self-centering SDOF systems. Soil Dyn. Earthq. Eng. **121**, 293–305 (2019)
18. Zhou, Y., Song, G., Tan, P.: Hysteretic energy demand for self-centering SDOF systems. Soil Dyn. Earthq. Eng. **125**, 105703 (2019)
19. Zhou, Y., Song, G.: Design energy spectra for self-centering SDOF systems. Earthq. Eng. Eng. Dyn. **40**(2), 1–2 (2020). (in Chinese)
20. Code for Seismic Design of Buildings [GB 50011-2010 (2016 Edition)]. Ministry of Housing and Urban-rural Construction of the People's Republic of China (2016)
21. Bojorquez, E., Ruiz, S.E., Teran, A.: Reliability-based evaluation of steel structures using energy concepts. Eng. Struct. **30**(6), 1745–1759 (2008). https://doi.org/10.1016/j.engstruct.2007.11.014
22. Lopez-Barraza, A., Ruiz, E., Reyes-Salazar, A., Bojorquez, E.: Demands and distribution of hysteretic energy in moment resistant self-centering steel frames. Steel Compos. Struct. **20**(5), 1155–1171 (2016)
23. Du, B., He, Z., Huang, G.H.: An estimate on distribution of hysteretic energy demand in seismic precast concrete frame structures. J. Earthq. Eng. https://doi.org/10.1080/13632469.2019.1605950

24. Tu, B.B., Zhao, D.: Distribution of accumulated irrecoverable hysteretic energy in MDOF structures. Multidiscip. Model. Mater. Struct. **14**(2), 202–215 (2018). https://doi.org/10.1108/MMMS-05-2017-0028
25. Pampanin, S., Marriott, D., Palermo, A.: PRESSS Design Handbook. New Zealand Concrete Society Industry, Auckland (2010)
26. ACI Innovation Task Group 1. Special Hybrid Moment Frames Composed of Discretely Jointed Precast and Post-Tensioned Concrete Members: ACI T1.2-03. American Concrete Institute, Farmington Hill (2003)
27. Song, G., Yang, T.Y., Zhou, Y.: Distribution of hysteretic energy demands in self-centering concrete frames with hybrid joints. Soil Dyn. Earthq. Eng. (under review)
28. NZS3101: Appendix B: Special Provisions for the Seismic Design of Ductile Jointed Precast Concrete Structural Systems. Standards New Zealand, Wellington (2006)
29. PEER: Open System of Earthquake Engineering Simulation (OpenSees). University of California, Berkeley, CA. Pacific Earthquake Engineering Research Center (2000)